Arun K. Bhunia (Ed.)

Sensors for Food Safety and Quality

MDPI

This book is a reprint of the Special Issue that appeared in the online, open access journal, *Sensors* (ISSN 1424-8220) from 2015–2016 (available at: http://www.mdpi.com/journal/sensors/special_issues/food_sensors).

Guest Editor
Arun K. Bhunia
Purdue University
USA

Editorial Office
MDPI AG
Klybeckstrasse 64
Basel, Switzerland

Publisher
Shu-Kun Lin

Managing Editor
Lin Li

1. Edition 2016

MDPI • Basel • Beijing • Wuhan • Barcelona

ISBN 978-3-03842-199-3 (Hbk)
ISBN 978-3-03842-200-6 (PDF)

Table of Contents

List of Contributors

Badr Abdullah Faculty of Engineering and Technology, Liverpool John Moores University, Henry Cotton Building, 15-21 Webster Street, Liverpool L3 2ET, UK.

Ahmed Al-Shamma'a Faculty of Engineering and Technology, Liverpool John Moores University, Henry Cotton Building, 15-21 Webster Street, Liverpool L3 2ET, UK.

Ole Alvseike ANIMALIA, Norwegian Meat and Poultry Research Centre, Lørenveien 38, Postboks 396 Økern, Oslo 0513, Norway.

Silvana Andreescu Department of Chemistry and Biomolecular Science, Clarkson University, Potsdam, NY 13699-5810, USA.

Xingjian Bai Molecular Food Microbiology Laboratory, Department of Food Science, Purdue University, West Lafayette, IN 47907, USA.

Manuela Baietto Dipartimento di Scienze Agrarie e Ambientali-Produzione, Territorio, Agroenergia (DISAA), Università degli Studi di Milano, via Celoria 2- 20133 Milano, Italy.

Yidan Bao College of Biosystems Engineering and Food Science, Zhejiang University, Hangzhou 310058, China.

Arun K. Bhunia Molecular Food Microbiology Laboratory, Department of Food Science, Purdue University, West Lafayette, IN 47907, USA; Department of Comparative Pathobiology, Purdue University, West Lafayette, IN 47907, USA.

Stefania Gudrun Bjarnadottir ANIMALIA, Norwegian Meat and Poultry Research Centre, Lørenveien 38, Postboks 396 Økern, Oslo 0513, Norway.

Jeffrey D. Brewster Molecular Characterization of Foodborne Pathogens Research Unit, United States Department of Agriculture-Northeast Area, Agricultural Research Service, Eastern Regional Research Center, Wyndmoor, PA 19038, USA.

Gonca Bülbül Department of Chemistry and Biomolecular Science, Clarkson University, Potsdam, NY 13699-5810, USA.

Shengsheng Chen School of Electrical Engineering, Henan University of Technology, Zhengzhou 450007, China.

Rosa M. Garcia Departmentof Engineering, University of Almería, 04120 Almería, Spain; BITAL (Research Center on Agricultural and Food Biotechnology), University of Almeria, 04120 Almeria, Spain.

Jose A. Gazquez Departmentof Engineering, University of Almería, 04120 Almería, Spain; BITAL (Research Center on Agricultural and Food Biotechnology), University of Almeria, 04120 Almeria, Spain.

Andrew G. Gehring Molecular Characterization of Foodborne Pathogens Research Unit, United States Department of Agriculture-Northeast Area, Agricultural Research Service, Eastern Regional Research Center, Wyndmoor, PA 19038, USA.

Francesco Genova Center for Integrated Research—CIR, Unit of Food Science and Human Nutrition, Università Campus Bio-Medico, Via Alvaro del Portillo 21– Rome 00128, Italy.

Simone Grasso Center for Integrated Research—CIR, Unit of Electronics for Sensor Systems, Università Campus Bio-Medico, Via Alvaro del Portillo 21– Rome 00128, Italy.

Akhtar Hayat Interdisciplinary Research Centre in Biomedical Materials, COMSAT Institute of Information Technology (CIIT), Defence Road, Off Raiwind Road, Lahore 54000, Pakistan.

Yiping He Molecular Characterization of Foodborne Pathogens Research Unit, United States Department of Agriculture-Northeast Area, Agricultural Research Service, Eastern Regional Research Center, Wyndmoor, PA 19038, USA.

Yong He College of Biosystems Engineering and Food Science, Zhejiang University, Hangzhou 310058, China.

Yanbo Hui School of Electrical Engineering, Henan University of Technology, Zhengzhou 450007, China.

Hidekazu Ikezaki Intelligent Sensor Technology, Inc., 5-1-1 Onna, Atsugi-shi, Kanagawa 243-0032, Japan.

Peter L. Irwin Molecular Characterization of Foodborne Pathogens Research Unit, United States Department of Agriculture-Northeast Area, Agricultural Research Service, Eastern Regional Research Center, Wyndmoor, PA 19038, USA.

Shin-Ichiro Isobe Department of Applied Chemistry and Biochemistry, Faculty of Engineering, Kyushu Sangyo University, 2-3-1 Matsukadai, Kasi-i, Higashi-ku, Fukuoka 813-8503, Japan.

Huali Jin School of Food Science & Engineering, Henan University of Technology, Zhengzhou 450007, China.

Dalia El Khaled Departmentof Engineering, University of Almería, 04120 Almería, Spain; BITAL (Research Center on Agricultural and Food Biotechnology), University of Almeria, 04120 Almeria, Spain.

Kwang-Pyo Kim Molecular Food Microbiology Laboratory, Department of Food Science, Purdue University, West Lafayette, IN 47907, USA; Department of Food Science and Technology, College of Agriculture and Life Sciences, Chonbuk National University, Jeonbuk 561756, Korea.

Ryoiti Kiyama Biomedical Research Institute, National Institute of Advanced Industrial Science and Technology (AIST), 1-1-1 Higashi, Tsukuba, Ibaraki 305-8566, Japan.

Olga Korostynska Faculty of Engineering and Technology, Liverpool John Moores University, Henry Cotton Building, 15-21 Webster Street, Liverpool L3 2ET, UK.

Lena Leprun Molecular Food Microbiology Laboratory, Department of Food Science, Purdue University, West Lafayette, IN 47907, USA; Present address CROUS de Dijon, Dijon Cedex 21012, France.

Kathrine Lunde ANIMALIA, Norwegian Meat and Poultry Research Centre, Lørenveien 38, Postboks 396 Økern, Oslo 0513, Norway.

Francisco Manzano-Agugliaro Departmentof Engineering, University of Almería, 04120 Almería, Spain; BITAL (Research Center on Agricultural and Food Biotechnology), University of Almeria, 04120 Almeria, Spain.

Alex Mason Faculty of Engineering and Technology, Liverpool John Moores University, Henry Cotton Building, 15-21 Webster Street, Liverpool L3 2ET, UK.

Tanya M. Monro The University of South Australia, Adelaide 5000, Australia.

Magomed Muradov Faculty of Engineering and Technology, Liverpool John Moores University, Henry Cotton Building, 15-21 Webster Street, Liverpool L3 2ET, UK.

Yoshinobu Naito Intelligent Sensor Technology, Inc., 5-1-1 Onna, Atsugi-shi, Kanagawa 243-0032, Japan.

Kentaro Nishi Department of Applied Chemistry and Biochemistry, Faculty of Engineering, Kyushu Sangyo University, 2-3-1 Matsukadai, Kasi-i, Higashi-ku, Fukuoka 813-8503, Japan.

Qunfeng Niu School of Electrical Engineering, Henan University of Technology, Zhengzhou 450007, China.

Junpei Noda Graduate School of Information Science and Electrical Engineering, Kyushu University, Nishi-ku, Fukuoka 819-0395, Japan.

Nuria Novas Departmentof Engineering, University of Almería, 04120 Almería, Spain; BITAL (Research Center on Agricultural and Food Biotechnology), University of Almeria, 04120 Almeria, Spain.

George C. Paoli Molecular Characterization of Foodborne Pathogens Research Unit, United States Department of Agriculture-Northeast Area, Agricultural Research Service, Eastern Regional Research Center, Wyndmoor, PA 19038, USA.

Francesca Romana Parente Department of Industrial and Information Engineering and Economics, University of L'Aquila, Via Gronchi 18–L'Aquila 67100, Italy.

Jiyu Peng College of Biosystems Engineering and Food Science, Zhejiang University, Hangzhou 310058, China.

Giorgio Pennazza Center for Integrated Research—CIR, Unit of Electronics for Sensor Systems, Università Campus Bio-Medico, Via Alvaro del Portillo 21– Rome 00128, Italy.

Marco Santonico Center for Integrated Research—CIR, Unit of Electronics for Sensor Systems, Università Campus Bio-Medico, Via Alvaro del Portillo 21– Rome 00128, Italy.

Tawana Simons Molecular Characterization of Foodborne Pathogens Research Unit, United States Department of Agriculture-Northeast Area, Agricultural Research Service, Eastern Regional Research Center, Wyndmoor, PA 19038, USA.

Atul K. Singh Molecular Food Microbiology Laboratory, Department of Food Science, Purdue University, West Lafayette, IN 47907, USA.

George K. Skouroumounis School of Agriculture, Food and Wine, the University of Adelaide, Waite Campus, PMB 1, Glen Osmond 5064, Australia.

Yusuke Tahara Graduate School of Information Science and Electrical Engineering, Kyushu University, Nishi-ku, Fukuoka 819-0395, Japan.

Dennis Taylor School of Agriculture, Food and Wine, the University of Adelaide, Waite Campus, PMB 1, Glen Osmond 5064, Australia.

Kiyoshi Toko Research and Development Center for Taste and Odor Sensing, Kyushu University, Nishi-ku, Fukuoka 819-0395, Japan; Graduate School of Information Science and Electrical Engineering, Kyushu University, Nishi-ku, Fukuoka 819-0395, Japan.

Shu-I Tu Molecular Characterization of Foodborne Pathogens Research Unit, United States Department of Agriculture-Northeast Area, Agricultural Research Service, Eastern Regional Research Center, Wyndmoor, PA 19038, USA.

Joseph Uknalis Molecular Characterization of Foodborne Pathogens Research Unit, United States Department of Agriculture-Northeast Area, Agricultural Research Service, Eastern Regional Research Center, Wyndmoor, PA 19038, USA.

Hailong Wang College of Biosystems Engineering and Food Science, Zhejiang University, Hangzhou 310058, China.

Li Wang School of Electrical Engineering, Henan University of Technology, Zhengzhou 450007, China.

Alphus D. Wilson USDA Forest Service, Center for Bottomland Hardwoods Research, Southern Hardwoods Laboratory, 432 Stoneville Road, Stoneville, MS 38776-0227, USA.

Chuanqi Xie College of Biosystems Engineering and Food Science, Zhejiang University, Hangzhou 310058, China.

Rui Yatabe Research and Development Center for Taste and Odor Sensing, Kyushu University, Nishi-ku, Fukuoka 819-0395, Japan.

Wenqi Zhang Institute for Photonics and Advanced Sensing and School of Physical Sciences, the University of Adelaide, Adelaide 5005, Australia.

Yun Zhu Biomedical Research Institute, National Institute of Advanced Industrial Science and Technology (AIST), 1-1-1 Higashi, Tsukuba, Ibaraki 305-8566, Japan; Scinet Company, 4-21-12 Takanawa, Minato-ku, Tokyo 108-0074, Japan.

Alessandro Zompanti Center for Integrated Research—CIR, Unit of Electronics for Sensor Systems, Università Campus Bio-Medico, Via Alvaro del Portillo 21– Rome 00128, Italy.

About the Guest Editor

Arun K. Bhunia received his bachelor degree in veterinary medicine (BVSc) in India, his PhD from the University of Wyoming (USA), and his postdoctoral training at the University of Arkansas (USA). Currently, he is Professor of Food Microbiology at Purdue University (USA) and the Chair of the Microbiology Training Group for the Purdue University Interdisciplinary Life Sciences Program (PULSe). His expertise is in the area of microbial food safety, foodborne pathogen detection, microbial pathogenesis, and probiotic vaccines. He has coauthored 154 research publications, two text books (Fundamental Food Microbiology; Foodborne Microbial Pathogens), edited three books, and delivered over 114 talks in national and international venues. He teaches both graduate and under graduate level courses on foodborne pathogens and mechanism of pathogenesis, microbial techniques, and food safety and public health. He is serving as editor or as an editorial board member for several journals, and is a member of the USDA National Advisory Committee on Microbiological Criteria for Foods (NACMCF; 2013-Present). Prof. Bhunia has received the Purdue Agriculture Research Award (2003), Purdue Faculty Scholar Award (2005), Purdue Team Award (2006), IFT R&D Award (2009), Outstanding Graduate Educator Award from the Department of Food Science (2010, 2011) and the College of Agriculture (2013), and has been granted the High-End Foreign Experts Recruitment Program (China) fellowship (2014–2016) and the Fulbright Specialist Roster (2016).

Editorial

This Special Issue is dedicated to publishing articles that describe the novel sensors or high-throughput screening technologies that are involved in detecting multiple pathogens, spoilage microorganisms, microbial communities, indicator microorganisms, microbial or non-microbial toxins, and non-microbial parameters (water activity, pH, metabolic by-products) relevant to improving the safety, quality, and security of foods.

The progressive thinking of the editorial team of "Sensors" to publish specialized articles highlighting the recent developments in the field of food safety and food quality is commendable! In this Special Issue of "Sensors for Food Safety and Quality", a total of twelve high quality articles were published, five of which are "review" articles, and seven are "research" articles. All articles were peer-reviewed to accept only the highest quality papers. Articles were of broad interest addressing both food safety and food quality issues. In the area of food safety, detection and high throughput screening platforms for pathogens and toxins, heavy metals, pesticides, allergens, etc. were reported. For food quality assessment, the articles focused primarily on analyzing sensory attributes, freshness, ripeness, taste, flavors, water loss, and food adulteration. A variety of sensing platforms were also introduced such as nanoparticle-based sensors, fluorescence-based sensors, dielectric sensing, infrared spectroscopy, hyperspectral imaging, light scattering sensors, microwave cavity sensors, electronic tongue, and electronic nose. Studies also emphasized wireless monitoring systems to obtain a real-time data. Specific information on each published article is summarized below:

Review Articles: Bulbul *et al.* provided a comprehensive review on the portable nanoparticle (NP)-based sensors for food safety assessment that has been developed over the past decade. The authors discussed several approaches that include the use of nanoparticles and nanostructures to enhance sensitivity and selectivity, design new detection schemes, improve sample preparation and increase portability. They used selected examples of NP-based detection schemes with colorimetric and electrochemical detection for chemical and biological contaminants including pesticides, heavy metals, bacterial pathogens and natural toxins.

Nishi *et al.* presented an in-depth review on fluorescence-based bioassays for detection and evaluation of varieties of food components that may cause allergy, food poisoning and toxicity, and foodborne infection. They summarized a list of fluorescent dyes that are used in different bioanalytical platforms including antibody/protein microarrays, bead suspension arrays, capillary sensor arrays, DNA microarrays, polymerase chain reaction (PCR)-based arrays, glycan/lectin arrays, immunoassays, enzyme-linked immunosorbent assay (ELISA)-based arrays, microfluidic chips and tissue arrays.

Three additional review articles in this special edition, focused on the development of sensing technologies for assessment of fruits and vegetable quality. In one study, El Khaled *et al.* reviewed the application of dielectric spectroscopy to assess fruit and vegetable quality. Dielectric properties of fresh fruits or vegetables differ from that of the damaged or spoiled produce. Rapid assessment of dielectric properties using sensing platforms would be an asset for evaluating horticulture crops. In the review article by Wang et al., the authors summarized application of visible and near infrared (Vis/NIR) spectroscopy, multispectral imaging and hyperspectral imaging techniques for food quality attributes' measurement. In addition, they also reported the discrimination ability of these techniques for various fruit species, i.e., apple, orange, kiwifruit, peach, grape, strawberry, grape, banana, mango and others. Baietto and Wilson summarized the application of an electronic-nose (e-nose) for fruit identification, ripeness and quality grading. Fruits produce a wide range of volatile organic compounds that provide characteristically distinct aromas and contribute to unique flavor characteristics, which dictate consumer acceptance. These authors emphasized potential application of the e-nose in discriminating complex mixtures of fruit volatiles to replace more conventional expensive analytical methods and as a specialized gas-sensing instrument for fruit identification, cultivar discrimination, ripeness assessment and fruit grading for assuring fruit quality in commercial markets.

Research Articles: Among the seven research articles, five articles reported sensor development and application in improving food quality while two articles addressed foodborne pathogen and toxin detection. Wang *et al.* reported the development of an electronic tongue (e-tongue) to assess peanut meal taste after enzymatic hydrolysis. Peanut meal is primarily composed of proteins, which have complex tastes after enzymatic hydrolysis. Conventional sensory analysis is used to assess taste, which is labor intensive and expensive. Here, the authors show the feasibility of using an e-tongue for such tests and show strong correlation with the sensory analysis.

Yatabe *et al.* took a different approach in analyzing taste, primarily bitterness. They developed a taste sensor consisting of a lipid/polymer membrane where the change in the surface structure of the membrane was evaluated by measuring the contact angle and surface zeta potential, using Fourier transform infrared spectroscopy (FTIR), X-ray photon spectroscopy (XPS) and gas cluster ion beam time-of-flight secondary ion mass spectrometry.

Santonico *et al.* designed a multi-sensor platform to detect adulteration in olive oil. Olive oil is prone to adulteration and fraud since authentic extra-virgin olive oil can fetch premier price. In this study, the authors used a novel artificial sensorial system, based on gas and liquid analysis coupled with an innovative electronic interface to validate authenticity/adulteration in olive oil.

Zhang *et al.* reported a wireless system to monitor ullage and temperature in wine barrels. This low cost sensor can be imbedded within the bung of the barrel and can provide early warning to the wine maker to take corrective actions.

Mason *et al.* employed a microwave spectroscopy to measure the drip loss in pork loin. Water loss during cutting, hanging, handling and processing of meat is critical in determining both product quality and value. Even a 1% water loss may cost a large meat company as much as €50,000/day. Current methods takes 1–3 days, while a novel microwave cavity sensor can estimate drip loss within minutes.

Two research articles on foodborne pathogen and toxin detection were published in this issue. Kim *et al.* developed a novel PCR assay coupled with a laser light scattering sensor to detect Listeria species in food. While Gehring *et al.* developed an antibody-based microarray method for detection of E. coli O157:H7 and Shiga-toxin on a microtiter plate in an array format enabling multiple sample testing at a time for the presence of pathogens or other toxins very quickly.

Finally, it is with great anticipation that I offer this collection of articles, addressing specific niche areas in food safety and food quality, as a great resource to our readers in advancing the field of sensors.

Professor Arun K. Bhunia
Guest Editor
March 2, 2016

Electronic-Nose Applications for Fruit Identification, Ripeness and Quality Grading

Manuela Baietto and Alphus D. Wilson

Abstract: Fruits produce a wide range of volatile organic compounds that impart their characteristically distinct aromas and contribute to unique flavor characteristics. Fruit aroma and flavor characteristics are of key importance in determining consumer acceptance in commercial fruit markets based on individual preference. Fruit producers, suppliers and retailers traditionally utilize and rely on human testers or panels to evaluate fruit quality and aroma characters for assessing fruit salability in fresh markets. We explore the current and potential utilization of electronic-nose devices (with specialized sensor arrays), instruments that are very effective in discriminating complex mixtures of fruit volatiles, as new effective tools for more efficient fruit aroma analyses to replace conventional expensive methods used in fruit aroma assessments. We review the chemical nature of fruit volatiles during all stages of the agro-fruit production process, describe some of the more important applications that electronic nose (e-nose) technologies have provided for fruit aroma characterizations, and summarize recent research providing e-nose data on the effectiveness of these specialized gas-sensing instruments for fruit identifications, cultivar discriminations, ripeness assessments and fruit grading for assuring fruit quality in commercial markets.

Reprinted from *Sensors*. Cite as: Baietto, M.; Wilson, A.D. Electronic-Nose Applications for Fruit Identification, Ripeness and Quality Grading. *Sensors* **2015**, *15*, 899–931.

1. Introduction

Fruit quality is judged by consumers primarily from their perception of the acceptability of fruits based on characteristics including visual appeal (lack of blemishes, color, size, and texture), ripeness, aroma and flavor. The quality of fruits (as measured by aroma, flavor, color, and textural characteristics) constantly changes during fruit development from pre-harvest through post-harvest stages as fruits grow and ripen, and during maintenance in storage [1]. Personal consumer preferences for different types of fruits are reflected in their particular choices of fruit varieties or cultivars selected for purchase. Fruit varieties vary widely in aroma characteristics due to differences in the composition of aromatic volatiles present in fruit aromas which are ultimately determined by plant genetics [2,3].

1

Previously, professional human graders and panelists have been used to judge fruit quality based on visual and aroma characteristics for selecting and evaluating fruits for ripeness at harvest and salability in commercial fruit markets [4]. The advent of electronic-nose (e-nose) devices has offered new alternative tools for grading fruits and other perishable foods using more consistent qualitative and quantitative measures of aroma characteristics that avoid the highly variable subjective opinions of human graders [5,6]. These instruments provide new means for characterizing fruit aromas for numerous applications ranging from the development of new fruit varieties by geneticists or fruit breeders to the timing of fruit harvests, transportation, storage operations (handling), and final selection by commercial dealers and retailers in fresh produce markets.

Fruit aroma is often the most valued characteristic determining fruit quality and consumer choice because aroma is usually the best indicator of fruit flavor. Electronic-noses are ideal digital, electronic devices for identifying, characterizing and grading fruit aromas from different fruits and fruit varieties because these instruments are capable of rapidly and consistently evaluating complex volatile gaseous mixtures without having to identify all of the chemical constituents present in the bouquet of fruit aromas [5,7]. E-noses contain a sensor array that evaluates all of the chemical constituents present in an aroma mixture (as a whole sample) and coverts the electronic output signals (via a transducer) from all of the sensors in the array and collectively assembles them to form a distinct digital pattern, sometimes referred to as an Electronic Aroma Signature Pattern (EASP) that is highly unique and specific to the particular gas mixture being analyzed [8,9]. In this way, the instrument output generates an aroma signature or smell-print that can be used to identify the particular type and variety of fruit being analyzed.

Fruits produce and release a wide variety of Volatile Organic Compounds (VOCs) that make up their characteristic aromas with esters, terpenoids, lactones and derivatives of amino acids, fatty-acids and phenolic compounds being the dominant classes of organic volatiles represented in fruit aromas [3]. Even though different fruits share some aromatic characteristics, each fruit has a distinctive aroma that depends upon the specific combination of VOCs present in the aroma mixture [10]. Whereas some specific volatiles are common to different fruit types, other fruit volatiles are specific to only one or only a few related fruits. Production and emission of volatiles from fruits is markedly influenced by numerous factors that interact in complex ways to determine fruit volatile composition. Multiple biochemical pathways are responsible for determining the final composition of volatile compounds released from different fruit types.

The purpose of this review is to summarize some of main chemical characteristics of fruit volatile gaseous mixtures which are conducive to characterization and analysis by electronic-nose technologies, to describe the diverse potential applications of e-nose technologies in the agro-fruit production sector of the agricultural production industry, and to provide examples of research that have demonstrated many ways in which e-nose devices have been utilized to distinguish between the fruit volatiles of different plant species and varieties for the purpose of analyzing and grading fruit quality and aroma characteristics.

2. Chemical Characteristics of Fruit Volatiles

Fruit aromas consist of a complex mixture of VOCs whose composition is specific to plant species and fruit variety [2,3]. Although different fruits often share many aromatic characteristics, each fruit has a distinctive aroma that depends upon the combination of volatiles, the concentration and the perception threshold of individual volatile compounds [10]. The most important aroma compounds include amino acid-derived compounds, lipid-derived compounds, phenolic derivatives, and mono- and sesquiterpenes [3]. Although fruit aromas are generally complex mixtures of a wide range of compounds, volatile esters often represent the major components of aroma volatiles present in rosaceous fruits such as apple (*Malaus domestica* Borkh.) and peach (*Prunus persica* L.) [11,12].

Fruit volatiles are mainly composed of VOCs in relatively few chemical classes, including primarily aliphatic esters, alcohols, aldehydes, ketones, lactones, terpenoids (monoterpenes, sesquitepenes) and apocarotenoids [13]. However, there are many thousands of volatile compounds represented by these chemical classes that make up the complex aroma mixtures of the numerous fruit types cultivated for agronomic markets of the world. Some of the predominant VOC principal components, comprising the distinctive aroma mixtures of selective fruit types which are representatives of the most common chemical classes found in fruit volatiles, are presented in Table 1. Volatile compounds in the aliphatic ester chemical class are the most abundant types of organic compounds found in many fruit volatiles; and esters are most responsible for the sweet smell of flowers and fruits of most angiosperms or seed plants [14]. In aromatic melon varieties for example, volatile esters predominate in fruit aromas that also contain aldehydes, short-chain alcohols, sesquiterpenes, norisoprenes, and aromatic sulfur-containing compounds [15]. Non-aromatic fruit varieties often have much lower levels of total volatiles and lack volatile esters [16]. Esters are a particularly important component of strawberry fruit aroma, accounting for 90% of the total number of volatiles in ripe strawberry fruit [17]. Esters also are the key volatiles responsible for the flavor characteristics of citrus [18].

Fruit volatiles, in addition to chemical class categorizations, may be classified as primary compounds (present in intact fruit tissue) or secondary compounds (produced as a result of tissue disruption of fruit tissue) [13]. Consequently, the condition of the fruit tissue being analyzed, either intact or disrupted, will influence the characteristics (chemical composition) and output patterns of the resulting aroma profiles. Some aroma compounds may be released only from cell disruption due to physical damage or injury to the fruit. Other fruit volatiles are more bound internally in the fruit, perhaps due to lower volatility or lack of direct exposure to the air as a result of fruit tissue compartmentalization.

Table 1. Chemical classes of VOCs that are principal components of distinctive fruit aromas.

Fruit Type	Chemical Class	Example Compounds	Chemical Structure	Reference
Apple	Aliphatic esters	ethyl butanoate		[18–20]
Apricots	Aliphatic alcohols	1-hexanol		[21,22]
Banana	Aliphatic esters	isoamyl acetate		[23]
Caraway	Terpenoids	carvone		[24,25]
Cantaloupe	Thiobutyrates	S-methyl thiobutanoate		[26]
Kiwifruit	Aliphatic aldehydes	hexanal		[27]
Peach	Lactones	γ-decalactone		[28]
Peach	Aliphatic esters	hexyl acetate		[29]
Pineapple	Organic acids	hexanoic acid		[30]
Raspberry	Aliphatic ketones	raspberry ketone		[31,32]
Strawberry	Furanone lactones	furaneol		[17]
Tomato	Apocarotenoids	β-ionone		[33]

Numerous factors affect fruit volatile chemical composition during all phases of the agronomic production process, including plant genetics, harvest time, fruit maturity, and agronomic environmental conditions, as well as postharvest handling, transportation and storage. Plant genetics, plant hormones, and environmental factors strongly influence the biosynthesis pathways responsible for the release of volatile aroma compounds from fruits under various conditions over time. The availability of primary precursor substrates for biosynthetic pathways producing aroma compounds is highly regulated both in amount and composition during fruit development [34]. All of these factors have varying effects on the aroma volatiles released at different stages of fruit development and after fruit harvest when the detached fruit is no longer influenced by the biochemical processes of the plant.

Fatty acids are the major primary precursor substrates of many aroma volatiles in most fruit types [2]. Aliphatic alcohols, aldehydes, ketones, organic acids, esters and lactones, ranging from C_1 to C_{20}, are all derived from fatty acid precursors through three key biosynthetic processes: α-oxidation, β-oxidation and the lipoxygenase pathway [35]. Volatiles derived from fatty acid precursors are important character-impact aroma compounds responsible for fresh fruit flavors at high concentrations.

The terpenoids comprise the largest class of plant secondary metabolites (about 20,000 identified), derived from the universal C_5 precursor isopentenyl diphosphate (IPP) and its allylic isomer dimethylallyl diphosphate (DMAPP) from two independent pathways, the mevalonic acid (MVA) and methylerythritol phosphate (MEP) biochemical pathways, with many volatile products represented [36,37]. Terpenes are classified into monoterpenes, diterpenes and sesquiterpenes, depending on the number of repeating units of a 5-carbon molecule (isoprene), the structural unit of all terpenoids, present in the molecule. Hemiterpenes (C_5), monoterpenes (C_{10}), sesquiterpenes (C_{15}), homoterpenes (C_{11} and C_{16}), and some diterpenes (C_{20}) are quite volatile VOCs because they have a high vapor pressure, allowing their rapid release into the atmosphere [13].

The complex gaseous mixtures of VOCs released from various fruit types (plant species), detectable by e-nose instruments, depend on the extent to which different metabolic pathways predominate in the generation of fruit volatiles as determined by genetic and environmental factors. Certain types of VOCs are more frequently associated with specific fruits as a result of unique combinations of metabolic pathways that control primary and secondary metabolite production in fruit tissues. The major chemical classes and representative VOCs found in fruit volatile mixtures, most associated as principal components derived from specific fruit types, are presented in Table 2. These associations do not preclude the occurrence of VOCs from many other chemical classes in volatile mixtures from each fruit type, but occur in different relative proportions (molar ratios) as lesser components in VOC mixtures derived from various fruit types.

Table 2. Principal volatile compounds comprising the distinctive aromas of different fruit cultivars.

Fruit Type	Cultivars/Varieties	Principal Volatile Compounds in Aroma [†]	References
Apple	Cox orange	Acetaldehyde, ethyl butanoate, ethyl methyl propanoate, 2-methyl butanol	[18]
	Elstar	Ethyl butanoate, ethyl 2-methyl butanoate	[18]
	Fuji	Ethyl 2-methyl butanoate, 2-methyl butyl acetate, hexyl acetate	[18]
	Pink Lady	Butyl acetate, hexyl acetate, 2-methylbutyl acetate, hexyl butanoate, hexyl 2-methyl butanoate, hexyl hexanoate	[20]
Banana	Cavendish	(E)-2-hexenal, acetoin	[38]
	Frayssinette	2, 3-Butanediol, solerol	[39]
	Plantain	(E)-2-hexenal, hexanal	[23,38,40]
Blackberry	Black Diamond	Furaneol, 2-heptanol, β-ionone, linalool	[41,42]
	Marion	Furaneol, hexanal, β-ionone, linalool	[41,42]
Blueberries	Primadonna, Jewel	Many aliphatic esters, C_6-aldehydes	[43]
	Snowchaser, FL02-40	Primarily terpenoids, less aliphatic esters	[43]
Grape	Cabernet Sauvignon	Benzene derivatives, monoterpenes, and sesquiterpenes, (also primarily alcohols)	[44–46]
	Muscat	Citral, citronellol, diendiol I, diendiol II, geraniol, linalool, rose oxide, nerol	[44–46]
	Riesling	Geraniol, α-muurolene, (also primarily esters and aldehydes)	[44–46]
Mango (Columbian)	Haden Irwin, Manila	δ-3-Carene	[47]
	Hilacha, Vallenato	α-Pinene	[47]
	Van Dyke	α-Phellandrene	[47]
	Yulima	Terpinolene	[47]
Pineapple	Cayenne	Ethyl 2-methylbutanoate, ethyl hexanoate, 2, 5-dimethyl-4-hydroxy-3(2H)-furanone (DMHF), decanal, ethyl 3-(methylthio) propionate, ethyl butanoate, (E)-3-ethyl hexenoate	[48]
	Tainong No. 4	Furaneol, 3-(methylthio) propanoic acid methyl ester, 3-(methylthio) propanoic acid ethyl ester, δ-octalactone	[49]
	Tainong No. 6	Ethyl-2-methylbutyrate, methyl-2-methylbutyrate, 3-(methylthio) propanoic acid ethyl ester, ethyl hexanoate, decanal	[50]

[†] Principal chemicals (VOCs) found in complex fruit volatile mixtures are listed in alphabetical order, not in order of relative abundance by quantity within mixtures analyzed from individual fruit cultivars or varieties.

The total number of aromatic compounds that contribute to fruit aromas vary considerably in different fruit types, but the complex mixture of VOCs found in individual fruit aromas usually is an extensive list of volatiles from different classes of organic compounds. Fruit aromas from fresh apples (*Malus domestica* Borkr) have been reported to contain at least 300 volatile compounds [51]. The total number and concentration of VOCs emitted by ripening apples are cultivar specific [52].

The epidermal tissue (peel) of apples produces a greater quantity of volatile compounds than internal fleshy (pericarp) tissues [53]. This higher capacity for aroma production by peel tissue is due to either the abundance of fatty acid substrates or higher metabolic activity in the peel [54,55]. Peach (*Prunus persica* (L.) Batsch) fruit aromas consist of about 100 volatile compounds, among them, C6 aldehydes and alcohols provide the green-note aroma, while lactones and esters are responsible for fruity aromas [56–58]. Esters, including hexyl acetate and (Z)-3-hexenyl acetate, are key odorants influencing the flavor characteristics of peach fruit [59]. Changes in these volatiles occur during fruit development and postharvest ripening [60]. Aldehydes tend to decline, while esters increase in the fruit during development. Postharvest treatments, low temperature and controlled atmosphere, can influence changes in peach aroma quality [12,61]. More than 300 VOCs have been identified in pear fruit (*Pyrus pyrifolia* Nak.) [62]. Methyl and hexyl esters of decadienoate are the main character-imparting compounds of European pear [62,63].

Apricot (*Prunus armeniaca* L.) fruit aromas have more than 200 different volatile compounds [64]. The most abundant volatile compounds by concentration were aldehydes, primarily hexanal and (E)-2-hexenal, that decreased in concentration during ripening [21,22].

Banana (*Musa* spp.) fruit aroma has about 250 VOCs [40], although the characteristic banana fruity top notes are from volatile esters, such as isoamyl acetate and isobutyl acetate that tend to increase in concentration during ripening [23,40]. Volatile compounds in citrus fruits accumulate in oil glands of flavedo and in the oil bodies of the juice sacs from which 100 VOCs have been identified, but varietal differences in the volatile profiles are primarily quantitative and only a few compounds are variety-specific [65–68].

Approximately 42 volatiles has been associated with the fruit aromas of southern highbush blueberry (*Vaccinium* species) cultivars [43]. Certain varieties contain a large amount of esters and C_6 aldehydes, whereas others produce more terpenoids and less esters.

Melon (*Cucumis* and *Citrullus* species) fruit aromas have more than 240 VOCs identified in different varieties [69]. Melon fruits release numerous compounds, particularly C_9 aliphatic compounds that are the major determinants of fruit quality as perceived by consumers. These compounds are strongly dependent on variety and particular physiological characteristics of the fruit. For example, climacteric melons (cantaloupes) have greater aroma intensity and shorter shelf life than less climacteric melons (honeydew melons) [70]. Volatiles derived from amino acids are major contributors to the aromas of both aromatic and non-aromatic melon varieties [71,72].

Grape (*Vitis vinifera* L.) fruit aromas contain many VOCs, including monoterpenes, C_{13} norisoprenoids, alcohols, esters and carbonyl compounds [73]. Grape varieties may be divided into aromatic and nonaromatic categories. Terpenoids are major volatiles in both red and white grapes [44]. At veraison, terpene production (in both Riesling and Cabernet Sauvignon varieties) generally is low, but Riesling grapes produced some terpenes (geraniol and α-muurolene) post-veraison.

Kiwi (*Actinidia* species) fruit aromas consist of more than 80 compounds with the major volatile components being methyl and ethyl butanoate, (Z)- and (E)-2-hexenal, hexanal, (Z)- and (E)-3-hexenol, and methyl benzoate [27]. Some important variations in fruit aroma volatiles produced by different kiwifruit varieties have been found to be due to the presence or absence of diverse sulfur-containing VOCs.

Mango (*Mangifera indica* L.) fruit aromas contain more than 270 VOCs in different mango varieties [74]. Monoterpenes are the most important compounds contributing to mango flavor [75]. Generally, terpenes are the major class of compounds in New World and Colombian mangoes whereas alcohols, ketones, and esters are mainly responsible for the characteristic aroma of Old World mangoes [76,77].

At least 280 VOCs have been found in pineapple (*Ananas comosus* L. Merr.) fruit aromas [30]. Esters and hydrocarbons were found to be the major constituents of fruit aromas, whereas octenoic acid, methyl ester, hexanoic acid, octanoic acid and ethyl ester were minor aromatic components. The relative content of different volatiles in pineapple aroma varied significantly during fruit development [49].

Raspberry (*Rubus idaeus* x *ursinus*) fruit aromas are composed of at least 200 volatile compounds that vary in concentrations for different cultivars. Many alcohols, aldehydes and ketones (including raspberry ketone, α-ionone, β-ionone, linalool, (Z)-3-hexenol, geraniol, nerol, α-terpineol, furaneol, hexanal, β-ocimene, 1-octanol, β-pinene, β-damascenone, ethyl 2-methylpropanoate, (E)-2-hexenal, heptanal, and benzaldehyde have been identified in raspberry aroma [13]. Among them, α-ionone, β-ionone, geraniol, nerol, linalool, and raspberry ketone probably contribute most to red raspberry aroma [31].

The complex fruit aroma of strawberry (*Fragaria* x ananassa Duch.) contains approximately 350 volatile compounds [3,78]. The furanone compound (furaneol), 2,5-dimethyl-4-hydroxy-3(2H)-furanone, and its methyl derivative (mesifurane) are considered the dominating compounds that contribute the typical caramel-like, sweet, floral and fruity aroma [17]. Aldehydes and alcohols (such as hexanal, *trans*-2-hexenal and *cis*-3-hexen-1-ol) contribute the unripe notes to green strawberry aroma in which the concentrations of these components are cultivar and ripeness dependent [17].

Notice from comparisons of key volatiles (principal components) that distinguish between different fruit types that certain variations in fruit volatiles from specific chemical classes often are most useful for fruit aroma discriminations. For example, apple and pineapple varieties may be distinguished primarily by differences in aroma ester composition, *i.e.*, variations in ester volatiles present in fruit aromas, whereas grapes and mango varieties are discriminated mostly by terpene volatiles that are detected in the aroma. By contrast, banana varieties are distinguished mainly by aldehydes and aliphatic alcohol volatiles, but ketones, furaneols, and alcohols are more important for distinguishing between blackberry and raspberry varieties. Discriminations between blueberry varieties are determined predominantly by the presence of esters, aldehydes, and terpenoids in the fruit aromas.

Despite the fact that a very large number of VOCs have been detected in various fresh fruit types, only a small fraction of these compounds have been identified as contributing significantly to the impact components (top and middle notes) of fruit aroma based on quantitative abundance and human olfactory thresholds [79]. The human olfactory threshold (for a particular aroma) is defined as the lowest concentration of aromatic compounds present in an aroma in which human subjects (usually 50 percent of a human panel) can smell the presence of the aroma [80].

Differences in the aroma characteristics of different fruit varieties are attributed to variations in their chemical profiles, based on the types of VOCs present and the relative concentrations of individual volatiles found in the aroma mixture. The principal VOCs, found in the aromas of specific fruit cultivars, may be used to distinguish between different fruit varieties. These differences in aroma composition and relative abundance of fruit volatiles in different fruit varieties are the means by which e-nose devices are capable of recognizing differences in fruit aromas and discriminating between fruit cultivars based on their distinct aroma signature patterns resulting from variations in e-nose response (sensor-array outputs) to different fruit aromas. Thus, e-nose discriminations of fruit aroma are determined both by the volatiles present and molar ratios of individual components found in each aroma (gaseous mixture)s.

3. Electronic-Nose Applications for Fruit Aroma Characterizations

Initial interest in the use of electronic noses as a non-destructive method to study the characteristics of fruits was shown by Benady *et al.* [81] who developed a sensing machine with a single semiconductor gas sensor in 1995, located within a small cup, placed on the surface of fruits of three different muskmelon cultivars. The instrument could discriminate between ripe and unripe fruits with an accuracy of 90.2%, and sort fruits into three ripeness categories (unripe, half-ripe and fully ripe) with an accuracy of 83%. The same research group worked on blueberries (the same year) to determine the variability in e-nose response among blueberry cultivars and to assess ripening stage and fruit quality (Simon *et al.*) [82].

Since these pioneering works, researchers have focused on developing and testing non-destructive sensorial techniques for the evaluation of the many and diverse characteristics and qualities of various fruits, particularly fruit maturity stages, shelf life and genotypic effects on aroma or bouquet characteristics. A comprehensive list of e-nose applications for characterizations and chemical discrimination of fruit volatiles for many fruit types is given in Table 3.

Table 3. Applications of electronic-nose devices for fruit aroma characterizations.

Common Name	Scientific Name	Family	E-Nose Type [†]	Discrimination Type	Reference
Apple	*Malus domestica* Borkr	Rosaceae	FOX 4000 [1]	Post-harvest treatments	[83]
			Smart Nose [2]	Shelf life	[84]
			Prototype MOS [3]	Maturity stage at harvest	[85]
			Prototype MOS	Shelf life	[86]
			Prototype MOS	Shelf life	[87]
			Prototype MOS	Shelf life	[88]
			Prototype QMBs [4]	Shelf life	[89]
			Prototype MOS	Prediction of storage time	[90]
			Cyranose 320 [5]	Aroma profile during deteriorative shelf life	[91]
			Unspecified	Cultivar effect	[92]
			Cyranose 320	Maturity stage at harvest	[93]
			FOX 4000	Cultivar effect	[94]
			Libra Nose [6]	Maturity stage at harvest	[95]
			Libra Nose	Shelf life	[96]
			Unspecified	Quality assessment	[97]
			FOX 4000	Maturity stage at harvest	[98]
Apricot	*Prunus armeniaca* L.	Rosaceae	EOS835 [7]	Ripening stage after harvest	[99]
			PEN2 [8]	Cultivar effect	[100]
			FOX 4000	Cultivar effect	[101]
Banana	*Musa x paradisiaca* L.	Musaceae	Prototype MOS	Ripening stage after harvest	[102]
Bell pepper	*Capsicum annuum* L.	Solanaceae	Unspecified	Quality assessment	[103]
Bergamot	*Citrus bergamia* Risso and Poiteau	Rutaceae	ISE Nose 2000 [9]	Cultivar effect; geographic effect; adulteration	[104]
Blackberry	*Rubus glaucus* Benth	Rosaceae	PEN3 [6]	Maturity stage at harvest	[105]
			Unspecified	Maturity stage at harvest	[106]
Bilberry	*Vaccinium meridionale* Swartz	Ericaceae	PEN3	Maturity stage at harvest	[105]
Blueberry	*Vaccinium* spp.	Ericaceae	Prototype MOS	Ripening stage after harvest; quality control	[82]
Cucumber	*Cucumis sativus* L.	Cucurbitaceae	FOX 4000	Genotypic effect	[107]
Date	*Phoenix dactylifera* L.	Arecaceae	FOX 4000	Cultivar effect	[108]
Durian	*Durio* spp.	Malvaceae	Unspecified	Maturity stage at harvest	[109]
Grape	*Vitis vinifera* L.	Vitaceae	Cyranose 320	Maturity stage at harvest	[110]
			enQbe [10]	Dehydration time	[111]
			enQbe	Dehydration time	[112]
			enQbe	Dehydration time	[113]
			enQbe	Post-harvest treatments	[114]
			zNose [11]		
			Cyranose 320	Canopy side effect	[115]
Hazelnut	*Corylus avellana* L.	Betulaceae	E-Nose 4000 [12]	Cultivar effect	[116]
			Moses II [13]	Cultivar effect	[117]
Loquat	*Eriobotrya japonica* (Thunb.) Lindl.	Rosaceae	Unspecified	Cultivar effect	[118]
Mandarin	*Citrus reticulate* Blanco	Rutaceae	PEN2	Maturity stage at harvest	[119]
			PEN2	Post-harvest treatments	[120]
			PEN2	Maturity stage at harvest	[121]
Mango	*Mangifera indica* L.	Anacardiaceae	Unspecified	Maturity stage at harvest	[122]
			FOX 4000	Cultivar effect, post-harvest treatments	[123]
			FOX 4000	Cultivar effect, maturity stage at harvest, shelf life	[124]
			Cyranose 320	Maturity stage at harvest	[125]
Muskmelon	*Cucumis melo* L.	Cucurbitaceae	Not specified	Maturity stage at harvest	[81]
Onion, spring onion garlic, shallot, leek	*Allium* spp.	Liliaceae	Aromascan CP [14]	Species effect	[126]
Onion	*Allium cepa* L.	Liliaceae	Aromascan	Fertilization, soil type effect	[127]

10

Table 3. *Cont.*

Common Name	Scientific Name	Family	E-Nose Type [†]	Discrimination Type	Reference
			ISENose 2000 [15]	Ecotype effect	[128]
Orange	*Citrus x sinensis (L.) Osbeck*	Rutaceae	Libra Nose	Shelf life	[129]
Pear	*Pyrus communis* L.	Rosaceae	Prototype MOS	Maturity stage at harvest	[85]
			Prototype MOS	Shelf life	[87]
			Aromascan	Maturity stage at harvest	[130]
			Prototype MOS	Maturity stage at harvest; quality assessment	[131]
			Prototype MOS	Maturity stage at harvest; quality assessment	[132]
Peach	*Prunus persica* (L.) Batsch	Rosaceae	PEN2	Shelf life and cultivar effect	[133]
			Prototype MOS	Maturity stage at harvest	[85]
			Libra Nose	Sensorial assessment	[134]
			Libra Nose	Cultivar effect, quality assessment	[135]
			Prototype MOS	Shelf life, quality assessment	[87]
			EOS 835	Cultivar effect, shelf life	[136]
			EOS 835	Cultivar effect	[137]
			PEN3	Maturity stage at harvest	[138]
			FOX 4000	Cultivar effect	[139]
			FOX 4000	Prediction of harvest time, quality assessment	[140]
			Prototype MOS	Prediction of harvest time, quality assessment	[141]
			Prototype MOS	Prediction of harvest time, quality assessment	[142]
			FOX 4000	Shelf life	[61]
Pepper	*Piper nigrum* L.	Piperaceae	Alpha Gemini	Genotype effect	[143]
Persimmon	*Diospirus Kaki* L.f.	Ebenaceae	Prototype MOS	Cultivar effect	[144]
		Ebenace	PEN3	Maturity stage at harvest and shelf life	[145]
Pineapple	*Ananas comosus* (L.) Merr.	Bromeliaceae	PEN2	Shelf life	[146]
Sapodilla	*Manilkara zapota* (L.)P. Royen	Sapotaceae	Prototype MOS	Maturity stage at harvest	[147]
Snake fruit	*Salacca edulis* Reinw.	Arecaceae	FOX 4000	Maturity stage at harvest	[148]
Sour citrus	*Citrus nagato-yuzukichi* Tanaka	Rutaceae	Unspecified	Species effect	[149]
Soursoup	*Annona muricata* L.	Annonaceae	PEN3	Maturity stage at harvest	[150]
Strawberry	*Fragaria x ananassa* Duch.	Rosaceae	Unspecified	Fertilizations effect	[151]
			PEN2	Processing approaches effect	[152]
			PEN2	Processing approaches effect	[153]
Tomato	*Solanum lycopersicon* L.	Solanaceae	Unspecified	Cultivar effect, shelf life	[154]
			Libra Nose	Cultivar effect, shelf life	[155]
			Unspecified	Mechanical deterioration effect	[156]
			PEN2	Maturity stage at harvest	[157]
			PEN2	Post-harvest treatments effect	[158]
Cherry tomato	*Lycopersicon esculentum* var. *cerasiforme*	Solanaceae	PEN2	Shelf life of processed fruits	[159]
Tomato	*Solanum lycopersicon* L.	Solanaceae	e-Nose 4000 [12]	Harvesting and postharvest handling treatments effect	[160]
			e-Nose 4000	Post-harvest treatments	[161]
			enQbe	Cultivation techniques effect	[162]
			Unspecified	Maturity stage at harvest and shelf life	[163]
Ussurian pear	*Pyrus ussuriensis* Maxim.	Rosaceae	Unspecified	Maturity stage at harvest	[164]

[†] Electronic nose (e-nose) names, types and manufacturers: 1 = Alpha MOS, Toulouse, France; 2 = SmartNose BV, Amersfoort, The Netherlands; 3 = Self-made prototype equipped with an array of commercial MOS gas sensors; 4 = Quartz microbalances (QMB) gas sensors; 5 = Cyrano Sciences Inc., Pasadena, CA, USA; 6 = Technobiochip, Marciana, Italy; 7 = Sacmi Imola s.c.a.r.l., Imola, Italy; 8 = Airsense Analytics GmbH, Schwerin, Germany; 9 = ISE, Pisa, Italia; 10 = University of Rome 'Tor Vergata', Rome, Italy; 11 = Electronic Sensor Technology, Newbury Park, CA, USA; 12 = EEV Inc., Amsford, NJ, USA; 13 = Lennartz Electronic GmbH, Tübingen, Germany; 14 = Osmetech Inc., Wobum, MA, USA; 15 = Labservice Analytica, Bologna, Italy. Unspecified e-nose types were not determinable from descriptions given in the methods section.

3.1. Apples and Pears

The fruit species receiving the most interest among e-nose researchers is apple (*Malus domestica* Borkr). Bai *et al.* [83] examined changes in the aroma profile on freshly-cut Gala apple slices treated with ethanol vapor, heat and 1-methylcyclopropene to prolong visual shelf-life. The FOX 4000 e-nose system (Alpha MOS, Toulouse, France) was utilized to assess aroma quality and the results demonstrated that pretreatments with ethanol and heat are effective in prolonging visual shelf life, but at the expense of aroma quality.

11

One of the main needs of fruits producers is to determine optimal picking date for different fruit types in order to assure the presence of preferred traits expected by consumers for maximum salability. Harvesting fruits at the optimal physiological condition results in fruits with the highest quality characteristics (aroma, firmness, color, flavor), and an extended shelf life. Fruit picked too early will not ripen sufficiently after storage and may suffer from physiological disorders, whereas those picked too late will be mealy and too soft following storage. Traditional techniques for assessing apple quality are destructive, requiring that only random samples are tested. For these reasons, electronic noses have been preferred for assessing the maturity stage of apples (at harvest) since 1999 to determine the optimal picking date. Hines *et al.* [88] first tested a prototype e-nose equipped with four commercial tin-oxide gas sensors on "Golden Delicious" apples to accurately classify fruit ripeness. The same year, Young *et al.* [98] worked on Gala apples, discriminating the ripening stage at harvest in four different classes. Saevels *et al.* [95] used the Technobiochip (Elba Island, Italy) Lybra Nose for predicting the optimal harvest date of apples as well as the cultivar effect on aroma of "Jonagold" and "Braeburn" varieties. The e-nose used in this study contained a sensor array based on Quartz Crystal Microbalance (QCM) sensors coated with metalloporphyrins and related compounds. Data were collected for two years and the results yielded a good predictive model developed for each cultivar based on one year of data, but a similar model based on two years of data was less effective at predicting optimal harvest dates.

Another commercial electronic nose (Cyranose 320, Cyrano Sciences Inc., Pasadena, CA, USA) was employed by Pathange *et al.* [93] to assess apple maturity for the apple cultivar "Gala". This instrument could positively classify the fruits into three groups according to their maturity stage (immature, mature and over mature) with an accuracy of 83%. Brezmes *et al.* [85–87] published several works on the assessment of apple ripeness. They initially worked on "Pink Lady" apples using a prototype e-nose equipped with 21 commercial tin oxide or Metal Oxide Semiconducting (MOS) sensors. The accuracy of e-nose classification of fruit maturity stage depended on the statistical classification technique used on e-nose data. PCA analysis did not show any clustering behavior that could be attributed to ripening, whereas a neural network classification algorithm provided good results. Brezmes *et al.* [85] could not determine the correct maturity stage of some apple cultivars, but had very good results on peaches and pears. More recently, Baumgartner *et al.* [84] showed very good results in discriminating the ripening stage of "Golden Delicious" apples, using a lesser-known Swiss e-nose (SMart Nose, SMart Nose SA, Marin-Epanier, Switzerland).

Of all commercial fruit species, apples rank highest with the largest number of experimental and commercial cultivars (including ancient, traditional and modern varieties) available in world markets. Apples have the greatest diversity of flesh and pericarp colors, flesh firmness, shelf life, and especially flavors and aromas. The cultivar effect, or variation in aroma characteristics or bouquet of apples, was studied by Marrazzo *et al.* [92] using a Cyrano Science prototype e-nose equipped with an array of 31 chemical sensors (Pasadena, CA, USA) to discriminate between "McIntosh", "Delicious" and "Gala" apple varieties. This electronic-nose prototype discriminated between fruit-aroma classes (varieties) based on data from the first day after harvest when the cultivar effect on aroma was more intense for freshly-harvested fruits. A more difficult test was recently performed by Pruteanu *et al.* [94] who successfully employed a FOX 4000 e-nose to discriminate between seven different varieties of Romanian apples.

One of the most important quality features of any fruit species or variety is the duration or longevity that optimal fruit characteristics can be maintained, referred to as shelf life, prior to decline to an unsalable state. The ultimate goal is to develop fruit cultivars with a very low after-harvest perishability coefficient, indicating a low cull-rate following commercial display to consumers. The merchantability of fresh products is largely influenced by shelf life. Traditionally, shelf life is measured and assessed through an evaluation of the chemical and physical properties (factors) that most determine ripening, maturation and post-harvest deterioration. Some of the key parameters measured included color, soluble solids content (SSC), percent of sugar (brix), and titratable acidity (TA). All of these parameters are commonly used by researchers and industry fruit graders, but none of these parameters are utilized by or matched with criteria used in fruit selection from the consumer perspective. Thus, measuring and evaluating post-harvest perishability through one single parameter, changes in volatile compounds directly associated with aroma, is market-oriented. Herrmann *et al.* [89] probably was the first researcher to become interested in monitoring aroma of apples during shelf life. It was well known at that time that the ratio of aldehydes within apple headspace volatiles could be used as indicators of ripeness for many apple varieties. Consequently, a QCM prototype electronic nose containing sensors coated with aldehyde-sensitive materials was used to monitor the increase in *trans*-2-hexenal concentration as an indicator of post-harvest development over time. Saevels *et al.* [96] and Baumgartner *et al.* [84] monitored the post-harvest shelf life as well, obtaining very interesting good results, as did Guohua *et al.* [90] with "Fuji" apples, employing an electronic nose prototype equipped with an array of eight tin oxide MOS sensors.

Li *et al.* [91] examined the aroma bouquet of deteriorating apples during the shelf life period. Physical damage to apples dramatically reduced economic value due to changes in color, shape, flavor and aroma, as well as increased susceptibility to attack by various post-harvest pathogens. Results of this work demonstrate that differences in numbers of physical cuts to the fruits had effects on volatile compound emissions. Apples subjected to two and three cuts generated aroma bouquets significantly different from uncut fruits. A similar study was published by Di Natale *et al.* [129] who used the Technobiochip Lybranose QCM e-nose to successfully discriminate between no cuts, one and two cut fruits. Xiaobo *et al.* [97] utilized a more comprehensive approach by combining three different sensors (a near-infrared spectrophotometer, a machine vision system and an electronic nose) to classify "Fuji" apples according to several quality parameters.

Other researchers have focused on pears (*Pyrus* species) to assess the maturity stage at harvest and shelf-life. Oshita *et al.* [130] successfully classified "La France" pears into three groups according to three storage treatments applied after harvest to prolong the shelf-life by using an Aromascan conducting polymer (CP) e-nose. Zhang *et al.* [131,132], published similar papers on the prediction of acidity, soluble solids content and firmness of "Xuequing" pears by employing a prototype e-nose equipped with eight different MOS sensors and an artificial neural network to analyze their response. Finally, Li *et al.* [164] worked on a Chinese species of *Pyrus* (*Pyrus ussuriensis* Maxim.) to characterize its volatile organic compounds (VOCs) at different ripening stages by traditional methods compared to an electronic nose.

3.2. Peach and Apricot

Three research groups have worked on apricot (*Prunus armeniaca* L.), obtaining good results by the use of three different commercial electronic noses [99–101]. The cultivar effect on aroma bouquet was tested using a PEN2 electronic nose, a portable (AlphaMos, Schweirin, Germany) and light-weight sensing machine [100]. It consists of a sensor array composed of 10 different doped semi-conductive MOS sensors positioned in a small chamber. The signal of the electronic nose was statistically analyzed by a trained artificial neural network (ANN) which is a data-processing tool that mimics the structure of the biological neural system, exhibiting brain characteristics of learning. In association with gas chromatography-mass spectroscopy (GC-MS), a FOX 4000 e-nose was used in the second study to characterize and discriminate between eight different apricot cultivars with promising results. Non-destructive cultivar assessment is very important in particular for apricot fruits to classify unknown samples and to prevent adulterations.

Defilippi *et al.* [99] worked on post-harvest quality of apricots, assessed by changes in VOCs detected with an EOS 835 electronic nose (Sacmi scarl, Imola, Italy). In order to determine differences in aroma profile, apricots were harvested at two maturity stages and stored at 0 °C and 20 °C (shelf-life simulation) for 15 and 30 days, then analyzed by a trained panel test and electronic nose. This e-nose could not classify maturity level of cold samples, but could only be classified after simulated shelf-life and panel test.

14

Peaches, *Prunus persica* (L.) Batsch and *P. persica* var. *nucipersica*, have been thoroughly studied by researchers interested in finding new ways of characterizing fruits in a non-destructive way. The first report come from Brezmes *et al.* [85]. The initial approach was aimed at discriminating between cultivars in an attempt to assess the feasibility of a fast, non-destructive, and cheaper way to classify unknown samples and decrease food frauds [133,135–137,139,140].

Peaches, nectarines and most varieties of *P. persica* are climacteric and particularly perishable at harvest and during storage, requires that these fruits be maintained at 0 °C for only a few days because shelf-life of this species is particularly short. These are only some of the reasons why producers utilize many alternative methods to evaluate the maturity stage of the fruit directly on the tree, traditionally based on personal experience. The same thing is asked by industry and retailers who need non-destructive, fast and systematic methods to evaluate the shelf life of fruits and other perishable products. Most of the previously mentioned researchers, including Brezmes *et al.* [87] and Zhang *et al.* [142], have tried to test the feasibility of using an electronic nose to assess shelf life of fruits. Benedetti *et al.* [133] employed a PEN2 e-nose to successfully classify samples of four different cultivars of peach according to their ripening stage. Performing Principal Components Analysis (PCA) on sensor data, peaches showed a linear data distribution for PC1 (from right to left), with increasing days of shelf-life. They concluded that no more than three sensors had a high influence on the sensor-output pattern for the fruit aroma, but only one sensor was relevant in the discrimination of peaches on the basis of shelf-life. They interpreted this single-sensor response to indicate that the sensor signal was directly linked to ethylene production, responsible for ripeness of peaches. A similar response was published by Rizzolo *et al.* [138] by using a similar but more advanced electronic nose (PEN3) manufactured by the same Company. In this case, only three sensors seemed to be associated with fruit ripeness and a linear correlation between PCs and quality indices indicated PC1 was related to ethylene production as well. More recently, Guohua and colleagues developed a model for the prediction of peach freshness based on a home-made electronic nose [165].

Monitoring the sensorial qualities of stored (refrigerated) fruits, in particular the loss of flavor and aroma of peaches throughout the period between harvest and the arrival at retail stores and during transportation to far-away markets, is one of the main problems facing fresh-fruit exporting companies and producers. Electronic noses were successfully employed by Infante *et al.* [136] and Zhang *et al.* [143] to evaluate the development (or loss) of aroma during transportation to markets. In particular, Infante *et al.* [137] could discriminate between aroma qualities of four cultivars, showing that "Tardibelle" peaches have the highest quality attributes even after 42 days of cold storage following harvest.

Zhang *et al.* [141,142,166] attempted to establish a quality index model to describe the effects of different picking dates of peaches by the use of a self-made prototype e-nose equipped with eight commercial tin semiconductor sensors and a commercial acquisition card. Sensor responses were validated by traditional peach quality parameters such as firmness, sugar content and pH at three different picking times. With the aim of predicting fruit quality based on these parameters, Principal Component Regression (PCR) and Partial Least Squares (PLS) regression where applied. The results showed that the two methods allow the determination of firmness, sugar content and pH by use of electronic nose. A similar research was performed by Su *et al.* [140] who tried to assess harvest season and quality of 39 cultivars of peaches and nectarines. By using a FOX 4000 e-nose and manipulating data by Discriminant Function Analysis (DFA), they successfully linked sensor responses to total soluble solids (TSS) concentration and titratable acidity (TA) in all harvest seasons, but only for those samples with a low or high concentration of TSS and a low or high TSS/TA ratio. Infante *et al.* [137] employed the Italian EOS 835 (Sacmi, Imola) e-nose to predict the quality of four peach cultivars by applying a multiple linear regression (MLR) to sensors-response data. They were able to describe the quality attributes "acidity", "sweetness" and the more general "acceptability" by the use of the e-nose, concluding that the instrument could discriminate between peach varieties through descriptors that mainly determine acceptability by the peach consumer. A similar paper was published by Di Natale *et al.* [134] who applied both electronic nose and panel sensor analysis to determine through an advanced data analysis, some proper sensorial indicators for the classification of peach fruits according to consumer palatability.

3.3. Citrus

Fruits belonging to the complex genus Citrus are commonly called citruses. These fruits are well known since ancient times for their nutraceutical properties, providing medical and health benefits,and the unmistakable pungent notes of their aroma. Hernandez Gomez *et al.* [119,121] worked on mandarin (*C. reticulata* Blanco) by using a PEN2 e-nose to associate sensor responses to harvest date (five different dates = aroma classes). No more than three sensors were employed in the Linear Discriminant Analysis (LDA), which gave clearer results than PCA. The later work by Hernandez Gomez *et al.* [120] was focused on evaluating the change in aroma bouquet emitted by mandarins during different storage treatments (plastic bag, paper box and refrigerator) and time (shelf-life), as well as other quality parameters. In this case, no significant predictive results were shown, while quality indices such as firmness were predicted by e-nose sensor responses.

Di Natale *et al.* [129] conducted research on oranges using a QCM prototype electronic nose. They distinguished between the different storage days (duration) for oranges, whereas Russo *et al.* [104] recently were able to discriminate between genuine bergamot (*C. bergamia* Risso et Poiteau) essential oils from other non-genuine types in order to defend the uniqueness of this very economically-important Southern Italian product. Akakabe *et al.* [149] worked on some Chinese species and varieties of citruses (including *C. nagato-yusukichi* Tanaka, *C. sudachi* Hort. ex Shirai, *C. junos* Siebold. ex Tanaka, *C. sphaerocarpa* Tanaka) to evaluate the capability of an electronic nose to discriminate between species and varieties by their aroma profiles.

3.4. Grape

One of the most important and widely-studied applications of electronic gas-sensing machines (electronic noses and tongues) in the food industry concerns analysis of the aroma characteristics of wine, the alcoholic product of fermented grapes [167–173]. However, some researchers have assessed the possibility of employing these gas-sensing machines directly on grapes in the field, or immediately after harvest, to provide important data on fruit quality and physiochemical parameters that are vital to the production success of quality wines and to limit food frauds or adulterations by dilutions with cheaper products. In this regard, Zoecklein *et al.* [114] studied the effects of ethanol treatments on volatile aromatic compounds emitted by two *V. vinifera* L. varieties ("Cabernet Franc" and "Merlot") at the onset of ripening (veraison). Both a conducting polymer (CP) e-nose and a surface acoustic wave-based (SAW) e-nose were capable of successfully discriminating between treated and untreated fruits.

A very interesting paper was published by Devarajan *et al.* [115] who discovered that the canopy side (north versus south and east versus west) has an effect on the aroma bouquet emitted by fruits (and wine) of Cabernet franc. Data were processed by using two different electronic noses (a Cyranose 320 and a zNose 730) on the basis of two growing seasons. Both sensing systems provided effective discrimination of canopy sides for grapes VOCs using canonical discriminant analysis.

The maturity level at harvest over two seasons was evaluated by Athamneh *et al.* [110] using a portable Cyranose 320 e-nose on "Cabernet Sauvignon" grape samples picked at three different maturity stages. The instrument proved capable of discriminating between different stages of maturity and fruits from different vine canopy sides.

Post-harvest dehydration is one of the most important steps in the wine-making process. Although the volatile fraction of a wine can be formed by hundreds of chemically-different compounds, the aroma compounds formed during drying have significant effects on wine quality. Currently, winemakers judge optimum drying times in terms of sugar concentration (brix) or water loss rather than based on more precise continuous monitoring of aroma profiles by e-noses until optimum drying conditions are met. Thus, some researchers have tried to determine the optimum drying times for wines using an electronic nose to assess the quality of aroma bouquets derived from dehydrated grapes [111–113]. In all cases, very good results were obtained by the use of a QCM prototype electronic nose developed by the University of Rome Tor Vergata.

3.5. Strawberry and Other Berries

To our knowledge, only three papers have been published on different applications of electronic noses for the evaluation of aroma characteristics in strawberry (*Fragaria x ananassa* Duch.) fruits. Agulheiro-Santos [151] reported on the capability of an electronic nose to assess the influence of different nitrogen fertilizations on the aroma quality of "Camarosa" strawberry. This represents one of the few examples of scientific research into electronic-nose technologies being applied directly to evaluate and modify growth protocols and agronomic techniques for the improvement of fruit aromatic and flavor qualities.

Strawberry fruit maturity was evaluated using a MOS electronic nose containing 18-metal oxide gas sensors [152]. The instrument discriminated between five stages of fruit maturity (from white to overripe) and three picking dates for two varieties. A PEN2 e-nose was employed by Qiu *et al.* [153] to characterize the aroma of five strawberry varieties, derived from freshly-squeezed juice produced according to a squeezing-processing technique.

Blackberry (*Rubus glaucus* Benth), bilberry (*Vaccinium meridionale* Swartz) and blueberry (*Vaccinium* spp.), comprise a group of fruits called "berries" with peculiarities of shape and size (always round and small fruits), with colors-ranging from yellow to black-blue-and flavor (sweet or sour). This definition of berry is not linked to the botanical meaning of the term "berry". Berries are among the most perishable fruits before and after harvesting. Thus, they have to be very carefully harvested and processed at the proper time and in specific ways to maintain quality and shelf life. Quality control in this field is particularly important. Simon *et al.* [82] tested a very early e-nose prototype, equipped with only two commercial sensors, to assess the maturity level of harvested fruits based on aromatic profile, and to detect damaged fruits in a closed container. The nutraceutical role of blackberries and bilberries was assessed by Bernal *et al.* [105]. In addition, changes in volatile components during different stages of fruit maturity were evaluated by employing a PEN3 electronic nose [106].

3.6. Mango and Other Tropical Fruits

Among so-called "tropical fruits", mango (*Mangifera indica* L.) is the most studied species in the field of gas-sensing e-nose machines. The first report by Lebrun *et al.* [123] attempted to assess the optimal harvest date. Whole and homogenated mango fruits were sampled and the aroma from each fruit was analyzed using a FOX 4000 e-nose. Although a deep study on the optimal dilution of the sample was carried out, this e-nose was not effective in determining the harvest date. Similar mediocre results were obtained by Kitthawee *et al.* [122], involving a study of hard green mangoes. The electronic nose could correctly classify only 68% of fruits according to their ripening stage, too low to be introduced into the retail market as a non-destructive method to assess ripening stage in mangoes. Afterwards, several researchers pursued the same goal some years later. Lebrun *et al.* [124] worked on three different mango cultivars ('Cogshall', 'Kent' and 'Keitt') harvested at different fruit maturities. This e-nose could separate fruits from different picking dates as well as fruits from different varieties. Zakaria *et al.* [125] published similar good results employing a Cyranose 320 e-nose.

A pioneering work by Llobet *et al.* [102], utilizing a self-made electronic prototype equipped with a tin-oxide MOS commercial sensor array, was able to discriminate between different stages of fruit maturity in banana (*Musa x paradisiaca L.*) fruits, and also predict the maturity stage of unknown samples, applying a neural-net classifier.

The Italian research group of Torri *et al.* [146] worked on pineapple (*Ananas comosus* L. Merr.) fruits. One experiment was aimed at monitoring the freshness of minimally-processed slices of pineapple during storage by use of a PEN2 electronic nose. Results indicated that this e-nose was successfully employed in the field to determine that pineapple fruit is particularly perishable and that even minimally-processed pineapple fruits lose their aroma characteristics very quickly.

Other researchers have evaluated the maturity stage for harvesting tropical fruits. Supriyadi *et al.* [148] employed a FOX 4000 e-nose to discriminate between ripe and unripe snake fruits (*Salacca edulis* Reinw.). Pokhum *et al.* [109] identified the ripeness stage of durian fruits (*Durio* spp.) by use of a MOS e-nose. Márquez Cardozo *et al.* [150] worked on the Columbian exotic and highly perishable soursop fruit (*Annona muricata* L.) a member of the custard apple tree family (Anonaceae). In this case, a PEN3 e-nose easily classified fruit samples as unripe, half ripe, ripe or overripe. Nugroho *et al.* [147] recently utilized an array of four commercial tin-semiconductor sensors, their prototype electronic nose, to discriminate between different maturity-classes sapodilla (*Manilkara zapota* (L.) P. Royen) fruits.

3.7. Other Miscellaneous Fruits

A paper by Lebrun *et al.* [108] hitherto has been the only research study to report on the application of an electronic nose for the rapid and non-destructive discrimination of date (*Phoenix dactylifera* L.) varieties. The research group of Alasalvar worked successfully on five raw and eighteen roasted Turkish hazelnuts (*Corylus avellana* L.) in an attempt at characterizing differences in aroma bouquet according to variety [116,117].

Garcia-Breijo *et al.* [144] and Li *et al.* [145] investigated persimmon (*Diospyros kaki* L.) fruits to discriminate between two different cultivars using a semiconductor commercial e-nose sensor array to assess fruit ripening stage and storage life, applying PCA and LDA statistical methods to a PEN3 e-nose sensor output data. They also attempted to determine the fewest number of sensors that could explain all the variance. Guarrasi *et al.* [118,174] worked on loquats [*Eriobotrya japonica* (Thunb.) Lindl.], a plant belonging to the Rosaceae family native to Japan and China and also wide spread in the Mediterranean Regions of Italy. The fruits were characterized chemically (SSC, TA, pH), morphologically, electronically using an e-nose, and olfactorily (via a trained human panel test). Although cross-data from traditional instrumental techniques and e-nose could identify aroma features, neither the panel test nor the electronic nose could discriminate between the four different sample cultivars.

Very recently, a paper by Ghouhua *et al.* [175] reported about a quality forecasting method using a home-made electronic nose based on an eight MOS sensors array. Samples of Winter jujube (*Ziziphus jujuba* Mill.) were analyzed each day, for 8 days, for physical and chemical indexes as well as via EN. PCA results indicated that jujubes under different time had an approximate trend, but the samples could not be qualitative or quantitatively discriminated from each other.

3.8. Vegetable Fruits

From a botanic point of view, tomatoes and other vegetables are actually fruits because they are derived from the ripened ovary of a flower [176]. Nevertheless, the general public views these fruits as vegetables especially from a culinary point of view. Quality attributes like the perfect maturity stage at harvest, long shelf-life and attractive visual appearance are critical factors that must be taken into account when evaluating agricultural protocols. Within all vegetable-type fruits, tomato (*Solanum lycopersicon* L.) is the species receiving the most attention in scientific research efforts in the field of electronic sensing since 1997. Maul *et al.* [160,161] published some early results on the evaluation of an electronic nose to identify and discriminate tomatoes, exposed to different harvesting and postharvest handling treatments, in order to measure quality diversities (in flavor, aroma and other quality parameters) between physiological (portable) maturity and market maturity, highly influenced by long-distance handling (during transportation) and even more by marketing systems.

Sinesio *et al.* [162] employed a prototype electronic nose and a panel test to discriminate between fruit samples based on different qualities of tomatoes. Fruits were harvested from two different Italian farms (one of them conducting traditional agriculture, the other organic farming) and classified as "very good", "good", "fair" and "poor", according to visual selection for the presence of injuries and physical damage. A QCM electronic nose was used, containing eight sensors coated with different metalloporphyrines. Their data showed that through the use of a neural-net statistical algorithm, the e-nose could discriminate between classes better than a trained human panel test.

Electronic noses can actually discriminate between different levels of mechanical damage in fruits like tomatoes which apparently release different VOCs according to different degrees of damage [156]. Bruising fruits usually lead to enhanced ripening at a rate proportion to the amount of damage.

Shelf life of tomatoes was assessed using electronic noses by Berna *et al.* [154,155] and by Hernandez Gomez *et al.* [158] in 2008. In both cases utilizing three different e-noses (a Technobiochip Lybranose, a University of Rome "Tor Vergata" enQbe, and a AlphaMos PEN2), a clear separation of different aroma classes was not achieved, whereas very good results were obtained by Hernandez Gomez *et al.* [157] in discriminating between different fruit maturity stages at harvest. Wang and Zhou [163] also obtained good results by crossing PCA sensor data with firmness data.

Hong and Wang [159] recently found that it was relatively simple to identify high quality fruits visually at harvest, but it is very difficult to assess the freshness of squeezed cherry tomatoes (in 100% tomato juices) unless traditional and time-consuming instruments are employed. Unfortunately, their electronic nose system did not give very convincing results.

Abbey *et al.* [126,127] and Russo *et al.* [128] worked on bulbs of *Allium* ssp. which are not named "fruits" either from a botanical or culinary point of view. We report these studies here because these species are somehow used as "fleshy" vegetables in kitchens. The initial work on *Allium* species was aimed at assessing the ability of an electronic nose to discriminate between different species (*A. sativum* L., *A. ampeloprasum* var. *porrum*, *A. cepa* var. *aggregatum*, and *A. cepa* L.) on the basis of their aroma profile. Based on pyruvic acid and thiosulphinates content (which characterize these species and varieties), the e-nose could successfully discriminate between them. In a secondary experiment, an electronic nose was employed to evaluate the effects of some agronomic factors, such as fertilizations with nitrogen (N) and sulfur (S) as well as the soil effects on plant aroma characteristics in the field and greenhouse. This is an interesting application of an e-nose instrument devoted to classifying different aroma bouquets in agricultural products whose final quality is so influenced by aroma profiles, such as for white pepper (*Piper nigrum* L.). Mamatha and Prakash [177] could easily discriminate between three cultivars of pepper, whereas a recent work of Liu *et al.* [143] was devoted to examining flavor quality of five new pepper genotypes. In all cases, the α-Gemini (Alpha MOS SA, Toulouse, France) e-nose, equipped with an array of 6 MOS sensors that can be chosen and customized by users, was useful in identifying known and unknown samples. Another example of this application in which an e-nose was used to identify unknown genotypes of fruit vegetables was published by Zawirska-Wojtasiak *et al.* [107]. Theyattempted to use the electronic nose to differentiate between transgenic lines of cucumber (*Cucumis sativus* L.) compared to controls.

The protection of specific geographical labels (as sources of fruits) from frauds and adulterations is one of the main concerns of producers, industry and final users as well as those who want to be extremely sure about the quality and the exact origin of fruits used in their businesses. Italy has the highest number of European protected denominations concerning fresh products [103]. An electronic nose was used to identify the geographical origins of two local varieties of bell pepper (*Capsicum annuum* L.) and this e-nose was applied in association with other traditional methods to characterize the product according to morphometric, qualitative, spectroscopic and aromatic data.

4. Conclusions

Electronic-nose devices have been utilized in a wide range of diverse applications in the agriculture and forestry industries to improve the effectiveness, efficiency and safety of processes involved in the production of quality food and fiber plant-based products [5]. As summarized in this review, e-nose instruments also offer many new potential applications for the fruit-production industry to facilitate many tasks involving fruit aroma evaluations during all stages of the agro-fruit production process from early cultivation activities, field-applied pest control applications, and timing of fruit harvests to many post-harvest stages including fruit transportation, storage, and finally the maintenance of fruit shelf life during display in commercial markets.

The potential for future developments and new aroma-based applications of electronic-nose devices in fruit production processes include e-nose detection of pesticide residues on harvested fruit surfaces to facilitate enforcement of human health regulations of the Environmental Protection Agency (EPA) [178–180], post-harvest fruit disease detection and management [5,8], and monitoring gases released from fruits in storage to control fruit ripening (maintain fruit shelf life) and fruit quality.

Acknowledgments: The authors would like to thank Gonzalo Pajares Martinsanz (University Complutense of Madrid, Spain), and Pablo Gonzalez-de-Santos (Centre for Automation and Robotics, Arganda del Rey, Madrid, Spain) for the invitation to write this review article on the applications of electronic-nose technologies for fruit identification, discrimination, ripeness and quality grading for the new Sensors special issue "Agriculture and Forestry: Sensors, Technologies and Procedures". We also appreciate the assistance of Charisse Oberle who helped compile and check the references and provided useful edits for the manuscript.

Conflicts of Interest: The authors declare no conflict of interest.

References

1. Ogundiwin, E.A.; Peace, C.P.; Gradziel, T.M.; Parfitt, D.E.; Bliss, F.A.I.; Crisosto, C.H.A. Fruit quality gene map of *Prunus*. *BMC Genomics* **2009**, *10*.

2. Sanz, C.; Olias, J.M.; Perez, A.G. Aroma biochemistry of fruits and vegetables. In *Phytochemistry of Fruit and Vegetables*; Oxford University Press Inc: New York, NY, USA, 1997; pp. 125–155.

3. Schwab, W.; Davidovich-Rikanati, R.; Lewinsohn, E. Biosynthesis of plant-derived flavor compounds. *Plant J.* **2008**, *54*, 712–732.

4. Eng, T.G.; Tat, M.M. Quality control in fruit processing. *JAOCS* **1985**, *62*, 287–291.

5. Wilson, A.D. Diverse applications of electronic-nose technologies in agriculture and forestry. *Sensors* **2013**, *13*, 2295–2348.

6. Wilson, A.D.; Oberle, C.S.; Oberle, D.F. Detection of off-flavor in catfish using a conducting polymer electronic-nose technology. *Sensors* **2013**, *13*, 15968–15984.

7. Wilson, A.D.; Baietto, M. Applications and advances in electronic-nose technologies. *Sensors* **2009**, *9*, 5099–5148.

8. Wilson, A.D.; Lester, D.G.; Oberle, C.S. Development of conductive polymer analysis for the rapid detection and identification of phytopathogenic microbes. *Phytopathology* **2004**, *94*, 419–431.

9. Wilson, A.D.; Lester, D.G.; Oberle, C.S. Application of conductive polymer analysis for wood and woody plant identifications. *For. Ecol. Managem.* **2005**, *209*, 207–224.

10. Tucker, G.A. Introduction. In *Biochemistry of Fruit Ripening*; Seymour, G.B., Taylo, R.J.E., Tucker, G.A., Eds.; Chapman & Hall: London, UK, 1993; pp. 1–51.

11. Ortiz, A.; Graell, J.; Lara, I. Volatile ester-synthesising capacity throughout on-tree maturation of "Golden Reinders" apples. *Sci. Hortic.* **2011**, *131*, 6–14.

12. Ortiz, A.; Graell, J.; López, M.L.; Echeverría, G.; Lara, I. Volatile ester-synthesising capacity in "Tardibelle" peach fruit in response to controlled atmosphere and 1-MCP treatment. *Food Chem.* **2010**, *123*, 698–704.

13. El Hadi, M.A.M.; Zhang, F.-J.; Wu, F.-F.; Zhou, C.-H.; Tao, J. Advances in fruit aroma volatile research. *Molecules* **2013**, *18*, 8200–8229.

14. Dudareva, N.; Pichersky, E.; Gershenzon, J. Biochemistry of plant volatiles. *Plant Physiol.* **2004**, *135*, 1893–1902.

15. Portnoy, V.; Benyamini, Y.; Bar, E. The molecular and biochemical basis for varietal variation in sesquiterpene content in melon (*Cucumis melo* L.) rinds. *Plant Mol. Biol.* **2008**, *66*, 647–661.

16. Burger, Y.; Sáar, U.; Paris, H.S.; Lewinsohn, E.; Katzir, N.; Tadmor, Y.; Schaffer, A.A. Genetic variability for valuable fruit quality traits in *Cucumis melo*. *Israel J. Plant Sci.* **2006**, *54*, 233–242.

17. Jetti, R.R.; Yang, E.; Kurnianta, A.; Finn, C.; Qian, M.C. Quantification of selected aroma-active compounds in strawberries by headspace solid-phase microextraction gas chromatography and correlation with sensory descriptive analysis. *J. Food Sci.* **2007**, *72*, S487–S496.

18. Berger, R.G. *Flavours and Fragrances-Chemistry, Bioprocessing and Sustainability*; Springer-Verlag: Berlin, Germany, 2007.

19. Holland, D.; Larkov, O.; Bar-Yaákov, I.; Bar, E.; Zax, A.; Brandeis, E. Developmental and varietal differences in volatile ester formation and acetyl-CoA: Alcohol acetyl transferase activities in apple (*Malus domestica* Borkh.) fruit. *J. Agric. Food Chem.* **2005**, *53*, 7198–7203.

20. Villatoro, C.; Altisent, R.; Echeverria, G.; Graell, J.; Lopez, M.L.; Lara, I. Changes in biosynthesis of aroma volatile compounds during on-tree maturation of "Pink Lady" apples. *Postharvest Biol. Technol.* **2008**, *47*, 286–295.

21. González-Agüero, M.; Troncoso, S.; Gudenschwager, O.; Campos-Vargas, R.; Moya-León, M.A.; Defilippi, B.G. Differential expression levels of aroma-related genes during ripening of apricot (*Prunus armeniaca* L.). *Plant Physiol. Biochem.* **2009**, *47*, 435–440.

22. Aubert, C.; Chanforan, C. Postharvest changes in physicochemical properties and volatile constituents of apricot (*Prunus armeniaca* L.). Characterization of 28 cultivars. *J. Agric. Food Chem.* **2007**, *55*, 3074–3082.

23. Wendakoon, S.K.; Ueda, Y.; Imahori, Y.; Ishimaru, M. Effect of short-term anaerobic conditions on the production of volatiles, activity of alcohol acetyltransferase and other quality traits of ripened bananas. *J. Sci. Food Agric.* **2006**, *86*, 1475–1480.

24. Bouwmeester, H.J.; Gershenzon, J.; Konings, M.C.J.M.; Croteau, R. Biosynthesis of the monoterpenes limonene and carvone in the fruit of caraway—I. Demonstration of enzyme activities and their changes with development. *Plant Physiol.* **1998**, *117*, 901–912.

25. Bouwmeester, H.J.; Konings, M.C.J.M.; Gershenzon, J.; Karp, F.; Croteau, R. Cytochrome P-450 dependent (+)-limonene-6-hydroxylation in fruits of caraway (*Carum carvi*). *Phytochemistry* **1999**, *50*, 243–248.

26. Song, J.; Forney, C.F. Flavour volatile production and regulation in fruit. *Can. J. Plant Sci.* **2008**, *88*, 537–550.

27. Carcia, C.V.; Stevenson, R.J.; Atkinson, R.G.; Winz, R.A.; Yong Quik, S. Changes in the bound aroma profiles of "Hayward" and "Hort16A" kiwifruit (*Actinidia* spp.) during ripening and GC-olfactometry analysis. *Food Chem.* **2013**, *137*, 45–54.

28. Tressl, R.; Albrecht, W. Biogenesis of aroma compounds through acyl pathways. In *Biogeneration of Aromas*; Parliament, T.H., Croteau, R., Eds.; ACS: Washington, DC, USA, 1986; pp. 114–133.

29. Eduardo, I.; Chietera, G.; Bassi, D.; Rossini, L.; Vecchietti, A. Identification of key odor volatile compounds in the essential oil of nine peach accessions. *J. Sci. Food Agric.* **2010**, *90*, 1146–1154.

30. Tokitomo, Y.; Steinhaus, M.; Buttner, A.; Schieberle, P. Odor-active constituents in fresh pineapple (*Ananas comosus* [L.] Merr.) by quantitative and sensory evaluation. *Biosci. Biotechnol. Biochem.* **2005**, *69*, 1323–1330.

31. Klesk, K.; Qian, M.; Martin, R. Aroma extract dilution analysis of cv. Meeker (*Rubus idaeus* L.) red raspberries from Oregon and Washington. *J. Agric. Food Chem.* **2004**, *52*, 5155–5161.

32. Malowicki, S.M.; Martin, R.; Qian, M.C. Volatile composition in raspberry cultivars grown in the Pacific Northwest determined by stir bar sorptive extraction-gas chromatography-mass spectrometry. *J. Agric. Food Chem.* **2008**, *56*, 4128–4133.

33. Auldridge, M.E.; McCarty, D.R.; Klee, H.J. Plant carotenoid cleavage oxygenases and their apocarotenoid products. *Curr. Opin. Plant Biol.* **2006**, *9*, 315–321.

34. Song, J.; Bangerth, F. Fatty acids as precursors for aroma volatile biosynthesis in pre-climacteric and climacteric apple fruit. *Postharvest Biol. Technol.* **2003**, *30*, 113–121.

35. Schwab, W.; Schreier, P. Enzymic formation of flavor volatiles from lipids. In *Lipid Biotechnology*; Kuo, T.M., Gardner, H.W., Eds.; Marcel Dekker: New York, NY, USA, 2002; pp. 293–318.

36. Newman, J.D.; Chappell, J. Isoprenoid biosynthesis in plants: Carbon partitioning within the cytoplasmic pathway. *Crit. Rev. Biochem. Mol. Biol.* **1999**, *34*, 95–106.

37. Lichtenthaler, H.K. The 1-deoxy-D-xylulose-5-phosphate pathway of isoprenoid biosynthesis in plants. *Annu. Rev. Plant Physiol. Plant Mol. Biol.* **1999**, *50*, 47–65.

38. Boudhrioua, N.; Giampaoli, P.; Bonazzi, C. Changes in aromatic components of banana during ripening and air-drying. *Lebensm Wiss Technol.* **2003**, *36*, 633–642.

39. Aurore, G.; Ginies, C.; Ganou-Parfait, B.; Renard, C.M.G.C.; Fahrasmane, L. Comparative study of free and glycoconjugated volatile compounds of three banana cultivars from French West Indies: Cavendish, Frayssinette and Plantain. *Food Chem.* **2011**, *129*, 28–34.

40. Jayanty, S.; Song, J.; Rubinstein, N.M.; Chong, A.; Beaudry, R.M. Temporal relationship between ester biosynthesis and ripening events in bananas. *J. Am. Soc. Hort. Sci.* **2002**, *127*, 998–1005.

41. Turemis, N.; Kafkas, E.; Kafkas, S.; Kurkcuoglu, M.; Baser, K.H.C. Determination of aroma compounds in blackberry by GC/MS analysis. *Chem. Nat. Comp.* **2003**, *39*, 174–176.

42. Du, X.F.; Finn, C.E.; Qian, M.C. Volatile composition and odour-activity value of thornless "Black Diamond" and "Marion" blackberries. *Food Chem.* **2010**, *119*, 1127–1134.

43. Du, X.F.; Plotto, A.; Song, M.; Olmstead, J.; Rouseff, R. Blueberry volatile composition of four southern highbush cultivars and effect of growing location and harvest date. *J. Agric. Food Chem.* **2011**, *59*, 8347–8357.

44. Rosillo, L.; Salinas, M.R.; Garijo, J.; Alonso, G.L. Study of volatiles in grapes by dynamic headspace analysis Application to the differentiation of some *Vitis vinifera* varieties. *J. Chromatogr. A* **1999**, *847*, 155–159.

45. Bellincontro, A.; Nicoletti, I.; Valentini, M.; Tomas, A.; de Santis, D.; Corradini, D.; Mencarelli, F. Integration of nondestructive techniques with destructive analyses to study postharvest water stress of winegrapes. *Am. J. Enol. Vitic.* **2009**, *60*, 57–65.

46. Fenoll, J.; Manso, A.; Hellin, P.; Ruiz, L.; Flores, P. Changes in the aromatic composition of the *Vitis vinifera* grape Muscat Hamburg during ripening. *Food Chem.* **2009**, *114*, 420–428.

47. Narain, N.; Bora, P.S.; Narain, R.; Shaw, P.E. Mango. In *Tropical and Subtropical Fruits*; Shaw, P.E., Chan, H.T., Nagy, S., Eds.; AgScience: Auburndale, FL, USA, 1997; pp. 1–77.

48. Chang, B.W.; Sheng, H.L.; Yu, G.L.; Ling, L.L.; Wen, X.Y.; Guang, M.S. Characteristic aroma compounds from different pineapple parts. *Molecules* **2011**, *16*, 5104–5112.

49. Zhang, X.M.; Du, L.Q.; Sun, G.M.; Wei, C.B.; Liu, S.H.; Xie, J.H. Changes of aroma components in Yellow Mauritius pineapple during fruit development (In Chinese). *J. Fruit Sci.* **2009**, *26*, 245–249.

50. Liang, Y.Z.; Guang, M.S.; Yu, G.L.; Ling, L.L.; Wen, X.Y.; Wei, F.Z.; Chang, B.W. Aroma volatile compounds from two fresh pineapple varieties in China. *Int. J. Mol. Sci.* **2012**, *13*, 7383–7392.

51. Nijssen, L.M.; van Ingen-Visscher, C.A.; Donders, J.J.H. *VCF Volatile Compounds in Food: Database (Version 13.1.)*; TNO Triskelion Recuperato da: Zeist, The Netherlands, 2011.

52. Dixon, J.; Hewett, E.W. Factors affecting apple aroma/flavour volatile concentration: A review. *N. Z. J. Crop Hortic. Sci.* **2000**, *28*, 155–173.

53. Rudell, D.R.; Mattinson, D.S.; Mattheis, J.P.; Wyllie, S.G.; Fellman, J.K. Investigations of aroma volatile biosynthesis under anoxic conditions and in different tissues of "Redchief Dilicious" apple fruit (*Malus domestica* Borkh.). *J. Agric. Food Chem.* **2002**, *50*, 2627–2632.

54. Guadagni, D.G.; Bomben, J.L.; Hudson, J.S. Factors influencing the development of aroma in apple peel. *J. Sci. Food Agric.* **1971**, *22*, 110–115.

55. Defilippi, B.G.; Dandekar, A.M.; Kader, A.A. Relationship of ethylene biosynthesis to volatile production, related enzymes and precursor availability in apple peel and flesh tissues. *J. Agric. Food Chem.* **2005**, *53*, 3133–3141.

56. Aubert, C.; Milhet, C. Distribution of the volatile compounds in the different parts of a white-fleshed peach (*Prunus persica* L. Batsch). *Food Chem.* **2007**, *102*, 375–384.

57. Wang, Y.J.; Yang, C.X.; Li, S.H.; Yang, L.; Wang, Y.N.; Zhao, J.B.; Jiang, Q. Volatile characteristics of 50 peaches and nectarines evaluated by HP-SPME with GC-MS. *Food Chem.* **2009**, *116*, 356–364.

58. Aubert, C.; Gunata, Z.; Ambid, C.; Baumes, R. Changes in physicochemical characteristics and volatile constituents of yellow- and white-fleshed nectarines during maturation and artificial ripening. *J. Agric. Food Chem.* **2003**, *51*, 3083–3091.

59. Cheng, L.; Xiao, L.; Chen, W.; Wang, J.; Che, F.; Wu, B. Identification of compounds characterizing the aroma of oblate-peach fruit during storage by GC-MS. *J. Stored Prod. Postharv. Res.* **2012**, *3*, 54–62.

60. Zhang, B.; Shen, J.Y.; Wei, W.W.; Xi, W.P.; Xu, C.J.; Ferguson, I.; Chen, K.S. Expression of genes associated with aroma formation derived from the fatty acid pathway during peach fruit ripening. *J. Agric. Food Chem.* **2010**, *58*, 6157–6165.

61. Zhang, B.; Xi, W.P.; Wei, W.W.; Shen, J.Y.; Ferguson, I.; Chen, K.S. Changes in aroma-related volatiles and gene expression during low temperature storage and subsequent shelf-life of peach fruit. *Postharvest Biol. Technol.* **2011**, *60*, 7–16.

62. Rapparini, F.; Predieri, S. Pear fruit volatiles. In *Horticultural Reviews*; Janick, J., Ed.; John Wiley & Sons: Hoboken, NJ, USA, 2003; pp. 237–324.

63. Kahle, K.; Preston, C.; Richling, E.; Heckel, F.; Schreier, P. On-line gas chromatography combustion/pyrolysis isotope ratio mass spectrometry (HRGC-C/P-IRMS) of major volatiles from pear fruit (*Pyrus communis*) and pear products. *Food Chem.* **2005**, *91*, 449–455.

64. Nijssen, L.M.; Visscher, C.A.; Maarse, H.; Willemsens, L.C.; Boelens, M.H. *Volatile Compounds in Foods Qualitative and Quantitative Data*; TNO Nutrition and Food Research Institute: Zeist, The Netherlands, 2007; pp. 15–16.

65. Gonza'lez-Mas, M.C.; Rambla, J.L.; Alamar, M.C.; Antonio, A.G. Comparative analysis of the volatile fraction of fruit juice from different Citrus species. *PLoS One*, **2011**, *6*, e22016. Available online: http://www.plosone.org/article/ info%3Adoi%2F10.1371%2Fjournal.pone.0022016 (accessed on 15 November 2014).

66. Kelebek, H.; Selli, S. Determination of volatile, phenolic, organic acid and sugar components in a Turkish cv. Dortyol (*Citrus sinensis* L. Osbeck) orange juice. *J. Sci. Food Agric.* **2011**, *91*, 1855–1862.

67. Tomiyama, K.; Aoki, H.; Oikawa, T.; Sakurai, K.; Kashara, Y.; Kawakami, Y. Characteristic volatile components of Japanese sour citrus fruits: Yuzu, Sudachi and Kabosu. *Flavour Fragr. J.* **2012**, *27*, 341–355.

68. Phi, N.T.L.; Nishiyama, C.; Choi, H.S.; Sawamura, M. Evaluation of characteristic aroma compounds of *Citrus natsudaidai* Hayata (Natsudaidai) cold-pressed peel oil. *Biosci. Biotechnol. Biochem.* **2006**, *70*, 1832–1838.

69. Obando-Ulloa, J.M.; Moreno, E.; Garc ia-Mas, J.; Nicolai, B.; Lammertync, J.; Monforte, J.A.; Fernandez-Trujillo, J.P. Climacteric or non-climacteric behavior in melon fruit 1. Aroma volatiles. *Postharvest Biol. Technol.* **2008**, *49*, 27–37.

70. Perry, P.L.; Wang, Y.; Lin, J.M. Analysis of honeydew melon (*Cucumis melo* var. Inodorus) flavor and GC/MS identification of (*E, Z*)-2,6-nonadienyl acetate. *Flav. Frag. J.* **2009**, *24*, 341–347.

71. Beaulieu, J.C.; Grimm, C.C. Identification of volatile compounds in cantaloupe at various developmental stages using solid phase microextraction. *J. Agric. Food Chem.* **2001**, *49*, 1345–1352.

72. Jordán, M.J.; Shaw, P.E.; Goodner, K.L. Volatile components in aqueous essence and fresh fruit of *Cucumis melo* cv. Athena (muskmelon) by GC-MS and GC-O. *J. Agric. Food Chem.* **2001**, *49*, 5929–5933.

73. Dieguez, S.C.; Lois, L.C.; Gomez, E.F.; de Ia Pena, M.L.G. Aromatic composition of the *Vitis vinifera* grape *Albariño*. *LWT-Food Sci. Technol.* **2003**, *36*, 585–590.

74. Shibamoto, T.; Tang, C.S. "Minor" tropical fruit mango, papaya, passion fruit, and guava. In *Food Flavours: Part C: The Flavour of Fruit*; Morton, I.D., MacLeod, A.J., Eds.; Elsevier: Amsterdam, The Netherlands, 1990; pp. 221–234.

75. Pino, J.A.; Mesa, J. Contribution of volatile compounds to mango (*Mangifera indica* L.) aroma. *Flav. Frag. J.* **2006**, *21*, 207–213.

76. Quijano, C.E.; Salamanca, G.; Jorge, A.; Pino, J.A. Aroma volatile constituents of Colombian varieties of mango (*Mangifera indica* L.). *Flav. Frag. J.* **2007**, *22*, 401–406.

77. Goswami, T.K.; Rekha, R.M. Determination of aroma volatiles in mango fruit cv. Amrapali as affected during storage under rapid control atmosphere conditions using liquid nitrogen. *Int. J. Agric. Environm. Biotech.* **2013**, *6*, 471–478.

78. Bood, K.G.; Zabetakis, I. The biosynthesis of strawberry flavor (II): Biosynthetic and molecular biology studies. *J. Food Sci.* **2002**, *67*, 2–8.

79. Wyllie, S.G.; Leach, D.N.; Wang, Y.; Shewfelt, R.L. Key aroma compounds in melons: Their development and cultivar dependence. In *Fruit Flavors: Biogenesis, Characterization, and Authentication*; Rouseff, R.L., Leahy, M.M., Eds.; American Chemical Society: Washington, DC, USA, 1995; pp. 248–257.

80. Yoshii, F.; Yamada, Y.; Hoshi, T.; Hagiwara, H. The creation of a database of odorous compounds focused on molecular rigidity and analysis of the molecular features of the compounds in the database. *Chem. Sens.* **2002**, *27*, 399–405.

81. Benady, M.; Simon, J.E.; Charles, D.J.; Miles, G.E. Fruit ripeness determination by electronic sensing or aromatic volatiles. *Trans. ASAE* **1995**, *38*, 251–257.

82. Simon, J.E.; Herzroni, A.; Bordelon, B.; Miles, G.E.; Charles, D.J. Electronic sensing of aromatic volatiles for quality sorting of blueberries. *J. Food Sci.* **1996**, *61*, 967–970.

83. Bai, J.; Baldwin, E.A.; Fortuny, R.C.S.; Mattheis, J.P.; Stanley, R.; Perera, C.; Brecht, J.K. Effect of pretreatment of intact 'Gala' apple with ethanol vapor, hear, or 1-methylcyclopropene on quality and shelf life of fresh-cut slices. *J. Am. Soc. Hortic. Sci.* **2004**, *129*, 583–593.

84. Baumgartner, D.; Gabioud, S.; Hohn, E.; Gasser, F.; Nising, A.B. Messung der Aromaentwicklung wahrend der Reifung von Golden Delicious. *Schweiz. Z. Für Obst- Weinbau.* **2009**, *145*, 8–11.

85. Brezmes, J.; Llobet, E.; Vilanova, X.; Saiz, G.; Correig, X. Fruit ripeness monitoring using and electronic nose. *Sens. Actuators B Chem.* **2000**, *69*, 223–229.

86. Brezmes, J.; Llobet, E.; Vilanova, X.; Orts, J.; Saiz, G.; Correig, X. Correlation between electronic nose signals and fruit quality indicators of shelf-life measurements with pink lady apples. *Sens. Actuators B Chem.* **2001**, *80*, 41–50.

87. Brezmes, J.; Fructuoso, L.; Llobet, E.; Vilanova, X.; Recasens, I.; Orts, J.; Saiz, G.; Correig, X. Evaluation of an electronic nose to assess fruit ripeness. *IEEE Sens. J.* **2005**, *5*, 97–108.

88. Hines, E.L.; Llobet, E.; Gardner, J.W. Neural network based electronic nose for apple ripeness determination. *Electron. Lett.* **1999**, *35*, 821–823.

89. Herrmann, U.; Jonischkeit, T.; Bargon, J.; Hahn, U.; Li, Q.; Schalley, C.A.; Vogel, E.; Vogtle, F. Monitoring apple flavor by use of quartz microbalances. *Anal. Bioanal. Chem.* **2002**, *372*, 611–614.

90. Hui, G.H.; Wu, Y.L.; Ye, D.D.; Ding, W.W. Fuji apple storage time predictive method using electronic nose. *Food Anal. Method.* **2013**, *6*, 82–88.

91. Li, C.; Heinemann, P.H.; Irudayaraj, J. Detection of apple deterioration using an electronic nose and zNose™. *Trans. ASABE* **2007**, *50*, 1417–1425.

92. Marrazzo, W.N.; Heinemann, P.H.; Crassweller, R.E.; Leblanc, E. Electronic nose chemical sensor feasibility study for the differentiation of apple cultivars. *Trans. ASAE* **2005**, *48*, 1995–2002.

93. Pathange, L.P.; Mallikarjunan, P.; Marini, R.P.; O'Keefe, S.; Vaughan, D. Non-destructive evaluation of apple maturity using an electronic nose system. *J. Food Eng.* **2006**, *77*, 1018–1023.

94. Pruteanu, E.M.; Duta, D.E.; Hincu, F.A.; Calu, M. Electronic nose for discrimination of Romanian apples. *Lucr. Stiintifice* **2009**, *SV*, 398–404.

95. Saevels, S.; Lammertyn, J.; Berna, A.Z.; Veraverbeke, E.A.; Di Natale, C.; Nicolai, B.M. Electronic nose as a non-destructive tool to evaluate the optimal harvest date of apples. *Postharvest Biol. Technol.* **2003**, *30*, 3–14.

96. Saevels, S.; Lammertyn, J.; Berna, A.Z.; Veraverbeke, E.A.; di Natale, C.; Nicolai, B.M. An electronic nose and a mass spectrometry-based electronic nose for assessing apple quality during shelf life. *Postharvest Biol. Technol.* **2004**, *31*, 9–19.

97. Zou, X.B.; Zhao, J.W.; Li, Y.X. Objective quality assessment of apples using machine vision, NIR spectrophotometer, and electronic nose. *Trans. ASABE* **2010**, *53*, 1351–1358.

98. Young, H.; Rossiter, K.; Wang, M.; Miller, M. Characterization of Royal Gala apple aroma using electronic nose technology-potential maturity indicator. *J. Agr. Food Chem.* **1999**, *47*, 5173–5177.

99. Defilippi, B.G.; San Juan, W.; Valdes, H.; Moya-Leon, M.A.; Infante, R.; Campos-Vargas, R. The aroma development during storage of Castlebrite apricots as evaluated by gas chromatography, electronic nose, and sensory analysis. *Postharvest Biol. Technol.* **2009**, *51*, 212–219.

100. Parpinello, G.P.; Fabbri, A.; Domenichelli, S.; Mesisca, V.; Cavicchi, L.; Versari, A. Discrimination of apricot cultivars by gas multisensor array using an artificial neural network. *Biosys. Eng.* **2007**, *97*, 371–378.

101. Solis-Solis, H.M.; Calderon-Santoyo, M.; Gutierrez-Martinez, P.; Schorr-Galindo, S.; Ragazzo-Sanchez, J.A. Discrimination of eight varieties of apricot (*Prunus armeniaca*) by electronic nose, LLE and SPME using GC-MS and multivariate analysis. *Sens. Actuators B Chem.* **2007**, *125*, 415–421.

102. Llobet, E.; Hines, E.L.; Gardner, J.W.; Franco, S. Non-destructive banana ripeness determination using a neural network-based electronic nose. *Meas. Sci. Technol.* **1999**, *10*, 538–548.

103. Rosso, F.; Zoppellari, F.; Sala, G.; Malusa, E.; Bardi, L.; Bergesio, B. A study to characterize quality and to identify geographical origin of local varieties of sweet pepper from Piedmont (Italy). *Acta Hortic.* **2012**, *936*, 401–409.

104. Russo, M.; Serra, D.; Suraci, F.; Postorino, S. Effectiveness of electronic nose systems to detect bergamot (*Citrus bergamia* Risso et Poiteau) essential oil quality and genuiness. *J. Essent. Oil Res.* **2012**, *24*, 137–151.

105. Bernal, L.J.; Melo, L.A.; Diaz-Moreno, C. Evaluation of the antioxidant properties and aromatic profile during maturation of the blackberry (*Rubus glaucus* Benth) and the bilberry (*Vaccinium meridional* e Swartz). *Rev. Fac. Nac. Agron. Medellin* **2014**, *67*, 7209–7218.

106. Bernal-Roa, L.J.; Melo, L.A.; Diaz-Moreno, C. Aromatic profile and antioxidant properties during blackberry (*Rubus glaucus* Benth.) fruit ripening. *Acta Hortic.* **2014**, *1016*, 39–45.

107. Zawirska-Wojtasiak, R.; Goslinski, M.; Szwacka, M.; Gajc-Wolska, J.; Mildner-Szkudlarz, S. Aroma evaluation of transgenic, Thaumatin II-producing cucumber fruit. *J. Food Sci.* **2009**, *74*, C204–C210.

108. Lebrun, M.; Billot, C.; Harrak, H.; Self, G. The electronic nose: A fast and efficient tool for characterizing dates. *Fruits* **2007**, *62*, 377–382.

109. Pokhum, C.; Chawengkijwanich, C.; Maolanon, R. Application of electronic-nose for identification of ripeness stage of durian. *Acta Hortic.* **2010**, *875*, 319–328.

110. Athamneh, A.I.; Zoecklein, B.W.; Mallikarjunan, K. Electronic nose evaluation of cabernet sauvignon fruit maturity. *J. Wine Res.* **2008**, *19*, 69–80.

111. Lopez de Lerma, N.; Bellincontro, A.; Mencarelli, F.; Moreno, J.; Peinado, R.A. Use of electronic nose, validated by GC-MS, to establish the optimum off-vine dehydration time of wine grapes. *Food Chem.* **2012**, *130*, 447–452.

112. Lopez de Lerma, N.; Moreno, J.; Peinado, R.A. Determination of the optimum sun-drying time for *Vitis vinifera* L. cv. Tempranillo grapes by e-nose analysis and characterization of their volatile composition. *Food Bioprocess Biotech.* **2014**, *7*, 732–740.

113. Santonico, M.; Bellincontro, A.; de Santis, D.; di Natale, C.; Mencarelli, F. Electronic nose to study postharvest dehydration of wine grapes. *Food Chem.* **2010**, *121*, 789–796.

114. Zoecklein, B.W.; Devarajan, Y.S.; Mallikarjunan, K.; Gardner, D.M. Monitoring effects of ethanol spray on Cabernet franc and Merlot grapes and wine volatiles using electronic nose systems. *Am. J. Enol. Vitic.* **2011**, *62*, 351–358.

115. Devarajan, Y.S.; Zoecklein, B.W.; Mallikarjunan, M.; Gardner, D.M. Electronic nose evaluation of the effects of canopy side on Cabernet franc (*Vitis vinifera* L.) grape and wine volatiles. *Am. J. Enol. Vitic.* **2011**, *62*, 73–80.

116. Alasalvar, C.; Odabasi, A.Z.; Demir, N.; Balaban, M.O.; Shahidi, F.; Cadwallader, K.R. Volatiles and flavor of five Turkish hazelnut varieties as evaluated by descriptive sensory analysis, electronic nose, and dynamic headspace analysis/gas chromatography-mass spectrometry. *J. Food Sci.* **2004**, *69*, SNQ99–SNQ106.

117. Alasalvar, C.; Pelvan, E.; Bahar, B.; Korel, F.; Olmez, H. Flavour of natural and roasted Turkish hazelnut varieties (*Corylus avellana* L.) by descriptive sensory analysis, electronic nose and chemometrics. *Int. J. Food Sci. Technol.* **2012**, *47*, 122–131.

118. Guarrasi, V.; Farina, V.; Germana, M.A.; San Biagio, P.L.; Mazzaglia, A. Fruit quality evaluation of four loquat cultivars grown in Sicily. *Acta Hortic.* **2011**, *887*, 299–304.

119. Hernandez Gomez, A.; Wang, J.; Hu, G.; Garcia Pereira, A. Electronic nose technique potential monitoring mandarin maturity. *Sens. Actuators B Chem.* **2006**, *113*, 347–353.

120. Hernandez Gomez, A.; Wang, J.; Hu, G.; Garcia Pereira, A. Discrimination of storage shelf-life for mandarin by electronic nose technique. *LWT* **2007**, *40*, 681–689.

121. Hernandez Gomez, A.; Wang, J.; Hu, G.; Garcia Pereira, A. Mandarin ripeness monitoring and quality attribute evaluation using an electronic nose technique. *Trans. ASABE* **2007**, *50*, 2137–2142.

122. Kitthawee, U.; Pathaveerat, S.; Slaughter, D.C.; Mitcham, E.J. Nondestructive sensing of maturity and ripeness in mango. *Acta Hortic.* **2012**, *943*, 287–296.

123. Lebrun, M.; Ducamp, M.N.; Plotto, A.; Goodner, D.; Baldwin, E. Development of electronic nose measurements for mango (*Mangifera indica*) homogenate and whole fruit. *Proc. Fla. State Hort. Soc.* **2004**, *117*, 421–425.

124. Lebrun, M.; Plotto, A.; Goodner, K.; Ducamp, M.; Baldwin, E. Discrimination of mango fruit maturity by volatiles using the electronic nose and gas chromatography. *Postharvest Biol. Technol.* **2008**, *48*, 122–131.

125. Zakaria, A.; Shakaff, A.Y.M.; Masnan, M.J.; Saad, F.S.A.; Adom, A.H.; Ahmad, M.N.; Jaafar, M.N.; Abdullah, A.H.; Kamarudin, L.M. Improved maturity and ripeness classifications of *Magnifera indica* cv. Harumanis mangoes through sensor fusion of an electronic nose and acoustic sensor. *Sensors* **2012**, *12*, 6023–6048.

126. Abbey, L.; Aked, J.; Joyce, D.C. Discrimination amongst Alliums using an electronic nose. *Ann. Appl. Biol.* **2001**, *139*, 337–342.

127. Abbey, L.; Joyce, D.C.; Aked, J.; Smith, B.; Marshall, C. Electronic nose evaluation of onion headspace volatiles and bulb quality as affected by nitrogen, sulphur and soil type. *Ann. Appl. Biol.* **2004**, *145*, 41–50.

128. Russo, M.; di Sanzo, R.; Cefaly, V.; Carabetta, S.; Serra, D.; Fuda, S. Non-destructive flavour evaluation of red onion (*Allium cepa* L.) ecotypes: An electronic-nose-based approach. *Food Chem.* **2013**, *141*, 896–899.

129. Di Natale, C.; Macagnano, A.; Martinelli, E.; Paolesse, R.; Proietti, E.; D'Amico, A. The evaluation of quality of post-harvest oranges and apples by means of an electronic nose. *Sens. Actuators B Chem.* **2001**, *78*, 26–31.

130. Oshita, S.; Shima, K.; Haruta, T.; Seo, Y.; Kawagoe, Y.; Nakayama, S.; Takahara, H. Discrimination of odors emanating from 'La France' pear by semi-conducting polymer sensors. *Comput. Electron. Agric.* **2000**, *26*, 209–216.

131. Zhang, H.; Wang, J.; Ye, S. Prediction of acidity, soluble solids and firmness of pear using electronic nose technique. *J. Food Eng.* **2008**, *86*, 370–378.

132. Zhang, H.; Wang, J.; Ye, S. Prediction of soluble solids content, firmness and pH of pear by signals of electronic nose sensors. *Anal. Chim. Acta* **2008**, *606*, 112–118.

133. Benedetti, S.; Buratti, S.; Spinardi, A.; Mannino, S.; Mignani, I. Electronic nose as a non-destructive tool to characterize peach cultivars and to monitor their ripening stage during shelf life. *Postharvest Biol. Technol.* **2008**, *47*, 181–188.

134. Di Natale, C.; Macagnano, A.; Martinelli, E.; Proietti, E.; Paolesse, R.; Castellari, L.; Campani, S.; D'Amico, A. Electronic nose based investigation of the sensorial properties of peaches and nectarines. *Sens. Actuators B Chem.* **2001**, *77*, 561–566.

135. Di Natale, C.; Zude-Sasse, M.; Macagnano, A.; Paolesse, R.; Herold, B.; D'Amico, A. Outer product analysis of electronic nose and visible spectra: Application to the measurement of peach fruit characteristics. *Anal. Chim. Acta* **2002**, *459*, 107–117.

136. Infante, R.; Farcuh, M.; Meneses, C. Monitoring the sensorial quality and aroma through an electronic nose in peaches during cold storage. *J. Sci. Food Agric.* **2008**, *88*, 2073–2078.

137. Infante, R.; Rubio, P.; Meneses, C.; Contador, L. Ripe nectarines segregated through sensory quality evaluation and electronic nose assessment. *Fruits* **2011**, *66*, 109–119.

138. Rizzolo, A.; Bianchi, G.; Vanoli, M.; Lurie, S.; Spinelli, L.; Torricelli, A. Electronic nose to detect volatile compound profile and quality changes in "Spring Belle" peach (*Prunus persica* L.) during cold storage in relation to fruit optical properties measured by time-resolved reflectance spectroscopy. *J. Agric. Food Chem.* **2013**, *61*, 1671–1685.

139. Su, M.; Zhang, B.; Ye, Z.; Shen, J.; Li, H.; Chen, K. Non-destructive detection of peach (*Prunus persica*) fruit volatiles using an electronic nose. *J. Fruit Sci.* **2012**, *29*, 809–813.

140. Su, M.; Zhang, B.; Ye, Z.; Chen, K.; Guo, J.; Gu, X.; Shen, J. Pulp volatiles measured by an electronic nose are related to harvest seasons, TSS concentration and TSS/TA ratio among 39 peaches and nectarines. *Sci. Hortic.* **2013**, *150*, 146–153.

141. Zhang, H.; Chang, M.; Wang, J.; Ye, S. Evaluation of peach quality indices using an electronic nose by MLR, QPST and BP network. *Sens. Actuators B Chem.* **2008**, *134*, 332–338.

142. Zhang, H.; Wang, J.; Ye, S.; Chang, M. Application of electronic nose and statistical analysis to predict quality indices of peach. *Food Bioprocess Technol.* **2012**, *5*, 65–72.

143. Liu, H.; Ceng, F.K.; Wang, Q.H.; Wu, H.S.; Tan, L.H. Studies on the chemical and flavor qualities of white pepper (*Piper nigrum* L.) derived from fine new genotypes. *Eur. J. Food Res. Technol.* **2013**, *273*, 245–251.

144. Garcia Breijo, E.; Guarrasi, V.; Masot Peris, R.; Alcaniz Fillol, M.; Olguin Pinatti, C. Odour sampling system with modifiable parameters applied to fruit classification. *J. Food Eng.* **2013**, *116*, 277–285.

145. Li, J.; Peng, Z.; Xue, Y.; Chen, S.; Zhang, P. Discrimination of maturity and storage life for 'Mopan' persimmon by electronic nose technique. *Acta Hortic.* **2013**, *996*, 385–390.

146. Torri, L.; Sinelli, N.; Limbo, S. Shelf life evaluation of fresh-cut pineapple by using an electronic nose. *Postharvest Biol. Technol.* **2010**, *56*, 239–245.

147. Nugroho, J.; Rahayoe, S.; Oka, A.A. Applications of electronic nose to determine the maturity of sapodilla using pattern recognition system. *Acta Hortic.* **2013**, *1011*, 87–93.

148. Supriyadi; Shimizu, K.; Suzuki, M.; Yoshida, K.; Muto, T.; Fujita, A.; Tomita, N.; Watanabe, N. Maturity discrimination of snake fruit (*Salacca edulis* Reinw.) cv. Pondoh based on volatiles analysis using an electronic nose device equipped with a sensor array and fingerprint mass spectrometry. *Flavour Fragr. J.* **2004**, *19*, 44–50.

149. Akakabe, Y.; Sakamoto, M.; Ikeda, Y.; Tanaka, M. Identification and characterization of volatile components of the Japanese sour citrus fruit *Citrus nagato-yuzukichi* Tanaka. *Biosci. Biotechnol. Biochem.* **2008**, *72*, 1965–1968.

150. Marquez Cardozo, C.J.; Cartagena Valezuela, J.R.; Correa Londono, G.A. Determination of Soursop (*Annona muricata* L. cv. Elita) fruit volatiles during ripening by electronic nose and gas chromatography coupled to mass spectroscopy. *Rev. Fac. Nac. Agron. Med.* **2013**, *66*, 7117–7128.

151. Agulheiro-Santos, A.C. Quality of strawberry 'Camarosa' with different levels of nitrogen fertilization. *Acta Hortic.* **2009**, *842*, 907–910.

152. Du, X.; Bai, J.; Plotto, A.; Baldwin, E.; Whitaker, V.; Rouseff, R. Electronic nose for detecting strawberry fruit maturity. *Proc. Fla. State Hort. Soc.* **2010**, *123*, 259–263.

153. Qiu, S.; Wang, J.; Gao, L. Discrimination and characterization of strawberry juice based on electronic nose and tongue: Comparison of different juice processing approaches by LDA, PLSR, RF, and SVM. *J. Agric. Food Chem.* **2014**, *62*, 6426–6434.

154. Berna, A.Z.; Buysens, S.; Di Natale, C.; Grun, I.U.; Lammertyn, J.; Nicolai, B.M. Relating sensory analysis with electronic nose and headspace fingerprint MS for tomato aroma profiling. *Postharvest Biol. Technol.* **2005**, *36*, 143–155.

155. Berna, A.Z.; Lammertyn, J.; Saevels, S.; di Natale, C.; Nicolai, B.M. Electronic nose systems to study shelf life and cultivar effect on tomato aroma profile. *Sens. Actuators B Chem.* **2004**, *97*, 324–333.

156. Cheng, S.; Wang, J.; Wang, Y.; Wei, Z. Discrimination of tomato plant with different levels of mechanical damage by electronic nose. *Trans. Chin. Soc. Agric. Eng.* **2012**, *28*, 102–106.

157. Hernandez Gomez, A.; Hu, G.; Wang, J.; Garcia Pereira, A. Evaluation of tomato maturity by electronic nose. *Comput. Electron. Agric.* **2006**, *54*, 44–52.

158. Hernandez Gomez, A.; Wang, J.; Hu, G.; Garcia Pereira, A. Monitoring storage shelf life of tomato using electronic nose technique. *J. Food Eng.* **2008**, *85*, 625–631.

159. Hong, X.; Wang, J. Application of e-nose and e-tongue to measure the freshness of cherry tomatoes squeezed for juice consumption. *Anal. Methods* **2014**, *6*, 3133–3138.

160. Maul, F.; Sargent, S.A.; Huber, D.J. Non-destructive quality screening of tomato fruit using "electronic nose" technology. *Proc. Fla State Hortic. Soc.* **1997**, *110*, 188–194.

161. Maul, F.; Sargent, S.A.; Balaban, M.O. Aroma volatile profiles from ripe tomatoes are influenced by physiological maturity at harvest: An application for electronic nose technology. *J. Am. Soc. Hort. Sci.* **1998**, *126*, 1094–1101.

162. Sinesio, F.; di Natale, C.; Quaglia, G.B.; Bucarelli, F.M.; Moneta, E.; Macagnano, A.; Paolesse, R.; D'Amico, A. Use of electronic nose and trained sensory panel in the evaluation of tomato quality. *J. Sci. Food Agric.* **2000**, *80*, 63–71.

163. Wang, J.; Zhou, Y. Electronic-nose technique: Potential for monitoring maturity and shelf life of tomatoes. *N. Z. J. Agric. Res.* **2007**, *50*, 1219–1228.

164. Li, G.; Jia, H.; Wang, Q.; Zhang, M.; Teng, Y.W. Changes of aromatic composition in Youhongli (*Pyrus ussuriensis*) fruit during fruit ripening. *J. Fruit Sci.* **2012**, *29*, 11–16.

165. Guohua, H.; Yuling, W.; Dandan, Y.; Wenwen, D.; Linshan, Z.; Lvye, W. Study of peach freshness predictive method based on electronic nose. *Food Control* **2012**, *28*, 25–32.

166. Zhang, H.; Wang, J. Evaluation of peach quality attribute using an electronic nose. *Sens. Mater.* **2009**, *21*, 419–431.

167. Chatonnet, P.; Dubourdieu, D. Using electronic odor sensor to discriminate among oak barrel toasting levels. *J. Agric. Food Chem.* **1999**, *47*, 4319–4322.

168. Di Natale, C.; Paolesse, R.; Burgio, M.; Martinelli, E.; Pennazza, G.; D'Amico, A. Application of metalloporphyrins-based gas and liquid sensor arrays to the analysis of red wine. *Anal. Chim. Acta* **2004**, *513*, 49–56.

169. Buratti, S.; Benedetti, S.; Scampicchio, M.; Pangerod, E.C. Characterization and classification of Italian Barbera wines by using an electronic nose and an amperometric electronic tongue. *Anal. Chim. Acta* **2004**, *525*, 133–139.

170. Penza, M.; Cassano, G. Recognition of adulteration of Italian wines by thin-film multisensory array and artificial neural networks. *Anal. Chim. Acta* **2004**, *509*, 159–177.

171. McKellar, R.C.; Rupasinghe, H.P.V.; Lu, X.; Knight, K.P. The electronic nose as a tool for the classification of fruit and grape wines from different Ontario wineries. *J. Food Sci. Agric.* **2005**, *85*, 2391–2396.

172. Lozano, J.; Santos, J.P.; Horrillo, M.C. Classification of white wine aromas with an electronic nose. *Talanta* **2005**, *67*, 610–616.

173. Martin, A.; Mallikarjunan, K.; Zoecklein, B.W. Discrimination of wines produced from Cabernet Sauvignon grapes treated with aqueous ethanol post-bloom using an electronic nose. *Int. J. Food Eng.* **2008**, *4*.

174. Guarrasi, V.; Farina, V.; San Biagio, P.L.; Germana, M.A. Naso elettronico con sensori MOS per discriminare 3 cultivar di Eriobotrya japonica Lindl. *Italus Hortus* **2010**, *17*, 103–108.

175. Guohua, H.; Jiaojiao, J.; Shanggui, D.; Xiao, Y.; Mengtian, Z.; Minmin, W.; Dandan, Y. Winter jujube (*Zizyphus jujube* Mill.) quality forecasting method based on electronic nose. *Food Chem.* **2015**, *170*, 484–491.

176. Lewis, R.A. *CRC Dictionary of Agricultural Sciences*; CRC Press: Boca Raton, FL, USA, 2002; p. 375.

177. Mamatha, B.Z.; Prakash, M. Studies on Pepper (*Piper nigrum* L.) cultivars by sensory and instrumental techniques. *Z. Arznei- Gewurzpflanzen* **2011**, *16*, 176–180.

178. Wilson, A.D. Development of an electronic-nose technology for the rapid detection of agricultural pesticide residues. Proceedings of the annual meeting of American Phytopathological Society, Providence, RI, USA, 4–8 August 2012.

179. Wilson, A.D. Identification of insecticide residues with a conducting-polymer electronic nose. *Chem. Sens.* **2014**, *4*, 1–10.

180. Wilson, A.D. Fungicide residue identification and discrimination using a conducting polymer electronic-nose. Proceedings of the Fourth International Conference on Sensor Device Technologies and Applications, Barcelona, Spain, 25–31 August 2014; Xpert Publishing Services: Wilmington, DE, USA, 2013; pp. 116–121.

Assessment of Taste Attributes of Peanut Meal Enzymatic-Hydrolysis Hydrolysates Using an Electronic Tongue

Li Wang, Qunfeng Niu, Yanbo Hui, Huali Jin and Shengsheng Chen

Abstract: Peanut meal is the byproduct of high-temperature peanut oil extraction; it is mainly composed of proteins, which have complex tastes after enzymatic hydrolysis to free amino acids and small peptides. The enzymatic hydrolysis method was adopted by using two compound proteases of trypsin and flavorzyme to hydrolyze peanut meal aiming to provide a flavor base. Hence, it is necessary to assess the taste attributes and assign definite taste scores of peanut meal double enzymatic hydrolysis hydrolysates (DEH). Conventionally, sensory analysis is used to assess taste intensity in DEH. However, it has disadvantages because it is expensive and laborious. Hence, in this study, both taste attributes and taste scores of peanut meal DEH were evaluated using an electronic tongue. In this regard, the response characteristics of the electronic tongue to the DEH samples and standard five taste samples were researched to qualitatively assess the taste attributes using PCA and DFA. PLS and RBF neural network (RBFNN) quantitative prediction models were employed to compare predictive abilities and to correlate results obtained from the electronic tongue and sensory analysis, respectively. The results showed that all prediction models had good correlations between the predicted scores from electronic tongue and those obtained from sensory analysis. The PLS and RBFNN prediction models constructed using the voltage response values from the sensors exhibited higher correlation and prediction ability than that of principal components. As compared with the taste performance by PLS model, that of RBFNN models was better. This study exhibits potential advantages and a concise objective taste assessment tool using the electronic tongue in the assessment of DEH taste attributes in the food industry.

Reprinted from *Sensors*. Cite as: Wang, L.; Niu, Q.; Hui, Y.; Jin, H.; Chen, S. Assessment of Taste Attributes of Peanut Meal Enzymatic-Hydrolysis Hydrolysates Using an Electronic Tongue. *Sensors* **2015**, *15*, 11169–11188.

1. Introduction

Peanut meal, a good protein raw material, is the peanut byproduct obtained after high-temperature oil extraction; it is a plant-derived protein with a high nutritional value, the content of which can range from 40.1% to 50.9% [1,2]. Nevertheless, after extraction using high temperatures and organic solvents, this peanut meal protein is

35

highly denatured, and its nutritional value and functionality (flavor base) decrease significantly, thereby limiting its application in the food industry [3,4]. Protein enzymolysis technology has become one of the effective methods for preparing flavor bases because it is not time-consuming; it is also harmless when directly employed to cure indigestion [5]. The peanut meal protein enzymolysis solution contains several free amino acids and small-molecule peptides, which give the enzymolysis liquid a complex taste; different free amino acid and peptide compositions render different tastes. As compared to the single enzymolysis of peanut meal, double enzymolysis can reduce the bitter taste value of hydrolysates obtained via enzymatic hydrolysis and improve the utilization rate of protein, thereby providing a new method for preparing flavor bases using protein hydrolysates.

Conventionally, sensory analysis using trained panelists has been employed to assess taste attributes in food; it is the only method that directly measures the perceived food taste intensity. However, there are some disadvantages of using subjective human sensory organs to evaluate food taste characteristics. For example, the results of sensory evaluation are influenced by subjective and objective factors and with a certain degree of ambiguity and uncertainty. Furthermore, it is expensive as panelists have to be paid for their time and effort. Moreover, it is also time-consuming to organize the training. For all these reasons, there has been increasing research into alternative objective evaluation methods, such as the use of an electronic tongue, which is based on biosensors.

Since the late 1990s, electronic tongues which use an array of multi-channel taste sensors which measure response signals characteristic of the sample solution, coupled with a signal processing unit based on pattern recognition and/or multivariate data analysis algorithms have been studied as objective taste assessment devices for the qualitative and/or quantitative characterization of compounds [6–8]. In principle, the electronic tongue system works in a manner similar to that of a human gustatory system. The arrays of low-selectivity taste sensors mimic the human tongue to sense the different tastes instead of using special sensors to obtain single information. To identify different tastes, the signal processing unit mimics the human nervous system to collect excited sensory signals and process the data using software. Therefore, as compared to conventional sensory evaluation, the electronic tongue has the advantages of good repeatability, high resolution, as well as rapid and facile operation. To collect comprehensive signals characteristic of the sample solution, various electronic tongue systems use different electrochemical measurement methods such as potentiometry, cyclic voltammetry, and impedance spectroscopy. After over ten years of development, the electronic tongue has already been applied widely in the food and beverage industry in applications ranging from the distinction of food varieties as well as food freshness to the prediction of food ingredients and to classification of food quality in water [9–11], beverages [12–18], wine [19–21], milk [22,23], or oil [24–26]. Moreover, electronic tongues have been shown to exhibit the potential to mimic the human tasting process and the response of a sensory panel to assess samples taste [27–29]. They have also been used to predict the taste intensity in pharmaceutical formulations [30–37] and hydrolysates, which are mainly bitter [38,39], and the relationship between the amount of bitter substances adsorbed onto membranes and taste sensors [40]. For example, Legin [32] has applied the electronic tongue for pharmaceutical analysis, where it could discriminate between different taste modalities of substances and its masking efficiency was found to be consistent with that obtained from a human taste panel. Rachid [33] has used an Alpha M.O.S. Astree electronic tongue to evaluate the masking efficacy of sweetening and/or flavoring agents on the bitter taste of epinephrine. First, a bitterness model was constructed with six standard pharmaceutical ingredients and then the bitterness score of different E bitartrate (EB) and EB + NMI solutions was predicted. However, thus far, relatively few studies exist on the taste assessment of protein-rich samples using electronic tongues. Newman [38,39] has used the electronic tongue to assess bitter dairy protein hydrolysates and investigated the correlation between the electronic tongue and sensory panel results (R^2 of 0.98). Prediction models built using sensory, chromatographic, and electronic tongues were compared; strong correlations between these models were studied, showing an R^2 from 0.78 to 0.93. Multivariate data analysis and pattern recognition methods such as principal component analysis (PCA), linear discriminate analysis (LDA), and partial least-square regression (PLS) have been increasingly applied in the abovementioned taste assessment studies. In particular, PLS regression has been widely applied for the construction of numerical prediction models from using chromatographic data to electronic tongue data in the taste assessment.

These previous studies indicate that the use of an electronic tongue may be suitable for assessing taste characteristics of peanut meal double enzymatic-hydrolysis hydrolysates (DEH). However, until now, qualitative analysis and judgment of taste characteristics in DEH have predominantly been conducted by analyzing the composition of the free amino acids and peptide species as well as by molecular weight distribution. For example, acidic amino acids or a short peptide containing acidic amino acid residues tastes umami. Glutamic acid, aspartic acid, and its amide are sour, short peptides with molecular weight less than 1000 Da are slightly salty; some short peptides are bitter, while amino acids such as Gly, Ser, Thr, Ala, and Pr are sweet. Sensory analysis is the only method to quantify the perceived different taste intensity of the hydrolysates obtained via enzymatic hydrolysis. Thus far, few studies have been conducted to demonstrate the ability of the electronic tongue to assess the taste characteristics of peanut meal DEH, which has complex different tastes.

This study aims to investigate the ability of an electronic tongue to assess the taste attributes of peanut meal DEH. The response characteristics of the electronic tongue to the six types of peanut meal DEH samples and five standard taste samples were investigated by PCA and DFA. PLS and RBF neural network (RBFNN) quantitative umami and saltiness prediction models were employed to compare the predictive abilities of the intensity of umami and saltiness and to correlate results obtained from both the electronic tongue and sensory analysis, respectively.

2. Materials and Methods

2.1. Materials

2.1.1. Peanut Meal

Peanut meal samples (produced between 2013 and 2014) were commercially purchased at random from five different manufacturers in a local market. The peanut meal sample's nutritional ingredients are 44.70% protein, 8.28% water, 5.60% ash and 1.61% fat. The protein composition of the sample was determined by the micro-Kjeldahl method (GB 5009.5-2010, $F = 5.46$). The water and ash content was determined by the constant weight method (GB5497-85) and by combustion (GB 5009.3-2010), respectively, and the fat content was determined by Soxhlet extraction (GB5512-85).

2.1.2. Chemicals

Flavor protease and trypsin used for the enzymolysis of peanut meal were purchased from Novozymes (Bagsvaerd, Denmark). Acetonitrile (chromatographically pure) used for gel exclusion chromatography (GEC) was purchased from TEDIA (Fairfield, OH, America). Monosodium glutamate (MSG), citric acid, tannic acid, sugar, and salt used as the five standard taste samples were obtained from commercial suppliers. Standard chemicals such as 1 mol/L hydrochloric acid, 0.1 MSG, and sodium chloride were supplied by Alpha M.O.S. (Toulouse, France), which had to be diluted with distilled water prior to use.

2.2. Methods

2.2.1. Sample Preparation

First, a 5% peanut meal solution was preheated for 10 min. Second, when the temperature of the solution reached 50 °C, 2000 μ/g of the composite enzyme (800 μ/g trypsin and 1200 μ/g flavourzyme) was added to start the enzymolysis; third, after a certain time, enzyme deactivation was performed by placing the solution mixture in a boiling water bath for 10 min. Next, after cooling to room temperature, the solution was centrifuged at 4000 r/min for 20 min; the supernatant obtained was frozen, and the solid obtained after drying was stored at $-20\,^{\circ}\text{C}$. Enzymolysis was conducted for 10 min, 1 h, 3 h, 8 h, 12 h, and 24 h. For sample analysis by the sensory panel and electronic tongue, 2% peanut meal DEH solutions of the six peanut samples were used at different enzymolysis times. The solid samples were solubilized in distilled water before testing, while samples for GEC analysis were not solubilized. Before the test, MSG, citric acid, tannic acid, sugar, and salt were solubilized in distilled water and prepared using different concentration gradients (Table 1). These standard sample solutions were selected for investigating the predominant taste observed in peanut meal DEH via analysis using the electronic tongue.

2.2.2. Sensory Analysis

The sensory assess panel was organized with 30 panelists who have a professional food testing background. Their personal attributes were checked according to sensory standards (no smoking and no beverage drinking) and they were trained before providing the assessment score results. The training contents involved a questionnaire screening, taste discrimination of five standards, and a 5-point intensity scale assessment using the five basic tastes (Table 1) to aid scaling. The 5-point intensity scale is from 0 to 5, where 0 and 5 denote the least intense and most intense taste perception, respectively. Then, all peanut meal DEH samples under different enzymolysis times were provided to the panelists for sensory assessment at a room temperature at approximately $25\,^{\circ}\text{C}$. Water was provided for cleansing the palate between tasting of different samples. All tests were performed in triplicate.

Table 1. Different concentration gradients of the five standard taste samples.

Standard	Concentration Gradient					
Umami (MSG)	0.1%	0.2%	0.4%	0.6%	0.8%	1.0%
Saltiness (salt)	0.001%	0.005%	0.01%	0.05%	0.1%	0.2%
Sourness (citric acid)	0.02%	0.04%	0.08%	0.12%	0.16%	0.2%
Bitterness (tannic acid)	0.025%	0.05%	0.1%	0.15%	0.2%	0.25%
Sweetness (sugar)	0.25%	0.5%	1%	1.5%	2%	2.5%

2.2.3. Gel Exclusion Chromatography Analysis of DEH Samples

There is a typical relationship between the molecular weight distribution of protein DEH and its taste attributes. GEC was employed to determine the molecular weight of DEH using an Agilent protein purification system (sample volume: 20 µL, detection wavelength: 220 nm, mobile phase: 20% acetonitrile, elution rate: 0.5 mL/min); the relationship between molecular weight and retention volume is expressed as y = 6.699x − 0.393, R^2 = 0.99, where y is the standard molecular weight logarithm of the peptide, and x is sample elution volume.

2.2.4. Electronic Tongue

The Astree Electronic Tongue Analyzer (Alpha M.S.O, Toulouse, France) was used for measurement. It consists of a 16-position autosampler for automatic sampling with 120 mL beakers as sample containers, a sensor array of seven different lipid membrane sensors mounted around a Ag/AgCl reference electrode, and an electronic unit for data acquisition and autosampler control. The seven sensors used to detect chemically dissolved compounds and acquire data are ZZ, JE, BB, CA, GA, HA and JB, respectively. The working principle of the electronic tongue system is as follows: when certain samples pass through the lipid membrane, they will cause a change in the membrane potential, and then the ions of samples are detected. The different sensors are composed of different lipid membranes; hence, they exhibit different sensor selectivity and potentials. The electronic unit measures all potentials between each sensor and the reference electrode and investigates the taste characteristics of the sample by analyzing the difference in potential. Before sample analysis with the electronic tongue, it is necessary to finish the start-up procedure, which consists of the three tests using each of the 80 mL standard chemical samples: conditioning, calibration, and diagnostic.

Two kinds of samples were supplied for electronic tongue analysis: 2% peanut meal DEH solutions at different enzymolysis time and five standard taste sample solutions at different concentration gradients. For electronic tongue analysis, 80 mL of the sample solutions was poured into a 120 mL beaker and placed into the electronic tongue automatic sampler. Then, the solutions were tested according to numerical order. Cleaning fluid (distilled water) and samples were placed alternately.

To ensure the accuracy and stability of the response signals by the electronic tongue sensor, the time to acquire data for each solution was 120 s, and the cleaning time was 20 s after the measurement of each solution. Data were collected every 1 s, and measurement data obtained for each solution was taken as the average of the last 5 s. To reduce the measurement error, each solution was repeatedly measured 10 times, and the last three measured values of each sensor were considered as reliable data used as input for subsequent analysis. Five duplicates of each peanut meal DEH sample were prepared from different peanut meal samples. Hence, a dataset of 90 samples for peanut meal DEH samples was supplied for analysis. Once the peanut meal DEH samples which were dried and stored at −20 °C were removed and opened, simultaneous measurement was followed by sensory analysis, GEC, and electronic tongue to ensure consistency in sample data.

2.2.5. Data Processing

One-way ANOVA was conducted using SPSS 14.0 statistical analysis software with a significance difference ($p < 0.05$) for sensory assessment. PCA, DFA, and PLS were applied (SPSS Inc., Chicago, IL, USA) using the Alpha M.S.O data statistical software. RBFNN was conducted in MATLAB 7.1. Concretely, the response characteristics of the electronic tongue to DEH samples and five standard taste samples were investigated, while qualitative assessment of the taste attributes of DEH samples was performed by PCA and DFA. PLS and RBFNN quantitative prediction models were employed to predict taste intensity scores of peanut meal DEH and study the correlation between the results obtained by electronic tongue and sensory analysis, respectively.

As explained in Section 2.2.4, the total data set for the qualitative analysis of the DEH samples was 7 (sensors) × 90 (samples) matrix. Taking into account the construction of the prediction models, the reduction of the input data was important and necessary for reducing complexity, which in turn can avoid data redundancy, decrease model training time, and obtain a better prediction ability model. PCA can be used for the effective compression of data with less information loss. Two data sets were used in the PLS and RBFNN quantitative prediction models. One data set was a matrix of 7 (sensors) × 90 (samples) as before, while the other was a matrix of 4 (the first four principal component, PC, values) × 90 (samples).

To estimate the predictive ability of the PLS and RBFNN models, 3-fold cross-validation was performed, where the original 90 samples were randomly partitioned into three equal-sized subsamples. Of the three subsamples, a single subsample (30 samples) was retained as the validation data for testing the model, while the remaining two subsamples were used as training data. Cross-validation was then repeated three times with each of the three subsamples used exactly once as the validation data. The root-mean-square errors (RMSEs) from the folds were averaged.

In the cross-validation of PLS, R^2 (Test) and R^2 (Prediction) need to be calculated. R^2 (Test) provides a variation ratio from the explanation of the prediction variable in each response, which determines the goodness-of-fit of each model with sample data. On the other hand, R^2 (Prediction) indicates how fine each predicted response was from the calculated models; it was only calculated using cross-validation. If the two R^2 values are close, a fine model is built; however, if R^2 (Test) is significantly lower than R^2 (Prediction), prediction results are overoptimistic.

RBFNN is a three-layer forward network using the radial basis function as the activation function. It is based on the k-means clustering algorithm with two main parameters of overlap and hidden layer number, which affect network performance. After optimization of overlap and the hidden layer number with training, two types of structures in the RBF neural network were designed. One was composed of 7-8-1, where response voltages of seven sensors obtained by the electronic tongue were used as the model input neurons, eight neurons in the hidden layer, and one neuron for the prediction score of taste intensity by the electronic tongue. The other was composed of 4-6-1, where 4 denotes the first four PC values selected as the model input neurons, six neurons in the hidden layer, and one neuron as before.

3. Results and Discussion

3.1. Sensory Analysis

The quantitative sensory evaluation of the five taste attributes (umami, saltiness, sourness, sweetness, bitterness) of six peanut meal DEH samples (denoted as A1–A6) at different enzymolysis times (10 min, 1 h, 3 h, 8 h, 12 h, 24 h) was performed (Figure 1 and Table 2). Rather than a single taste, five complex tastes including umami, saltiness, sourness, sweetness, and bitterness were observed for the DEH samples. Among these tastes, the intensity of umami was the maximum (score 1.3–4.2), followed by that of saltiness (score 0.6–3.4); the intensities for both sourness and sweetness were weak, while the intensity of bitterness was the weakest (score 0.7–1.8).

As shown in Table 2, the trends of umami and saltiness gradually increased within 24 h with the progression of enzymolysis time. Sensory analysis showed that the umami score was the highest for samples A1–A6, with umami being predominant.

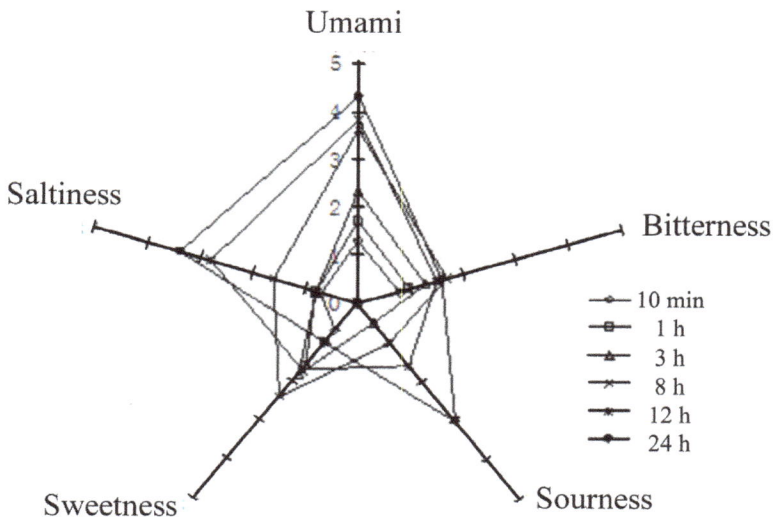

Figure 1. Radar distribution results of sensory analysis.

Table 2. Characteristics and taste intensity scores of peanut meal DEH, which were used to compare the taste attributes obtained by the electronic tongue.

Sample	Enzymolysis Time (min)	Degree of Hydrolysis (DH) (%)	Protein Extraction Rate (%)	Ratio of Different Relative Molecular Weight Peptides and the Total Peptides % (Da)			Taste Intensity Score	
				<1 k	1–3 k	>20 k	Umami	Saltiness
A1	10	4.8	61	6.38	11.64	37.72	1.3	0.6
A2	60	8.2	63	13.27	14.22	25.98	1.7	0.7
A3	180	11.0	78	14.6	16.8	21.96	2.2	0.9
A4	480	15.2	72	20.79	16.86	20.73	3.6	1.6
A5	720	17.1	71	24.02	16.77	15.4	3.8	2.8
A6	1440	20.4	68	26.34	20.36	10.23	4.2	3.4

3.2. Analysis of Taste Attributes by Gel Exclusion Chromatography

GEC was conducted to obtain the molecular weight distribution of peanut meal DEH (Table 2). The peptides with molecular weight greater than 20 kDa rapidly reduced as enzymolysis time progressed, while smaller-molecular-weight peptides (<1 kDa) gradually increased. For instance, at an enzymolysis time of 1440 min (24 h), the content of the peptides with molecular weight greater than 20 KDa decreased from 46.88% to 10.23%, while the content of peptides with molecular weight less than 1 KDa gradually increased from the original 0.99% to 26.34%, because trypsin (endonuclease) rapidly hydrolyzed the protein to macromolecular peptide, and the compound flavor enzyme (including endonuclease and exonuclease) continuously hydrolyzed the macromolecular peptide to small-molecule peptides or free amino acids during enzymolysis. The small peptides with molecular weight less than 1 kDa clearly exhibited umami and saltiness (small peptides with molecular weight 1 k–3 kDa have umami-enhancing effect). Hence, from the analysis of the taste attributes by GEC, peanut meal DEH predominantly exhibits umami, and with increasing enzymolysis time, the umami taste increases. These results are consistent with those reported by *Yamada and Nishimura* as well as with former sensory analysis.

3.3. Representation of Sensor Response of the Electronic Tongue

Further objective analysis of the DEH samples was performed using the electronic tongue. Figure 2 shows the response curves of ZZ, JE, BB, CA, GA, HA, and JB to sample A6 (enzymolysis time 24 h). The X- and Y-axes denote the data acquisition time (120 s) and response voltage values of the seven sensors, respectively. The sensors exhibited good stability and repeatability with respect to the measured signal, except for the variation of the acquired data before 30 s.

Figure 3 shows bar graphs of response intensity of the seven lipid membrane sensors to samples A1–A6. The X- and Y-axes denote each sensor and the final stable voltage response values of the sensors, respectively. The different colors represent different samples. Each sample was measured five times. Sensors ZZ, BB, and JB exhibited a higher response intensity, while CA exhibited the lowest response intensity. Moreover, JE, GA, and HA exhibited a similar response intensity.

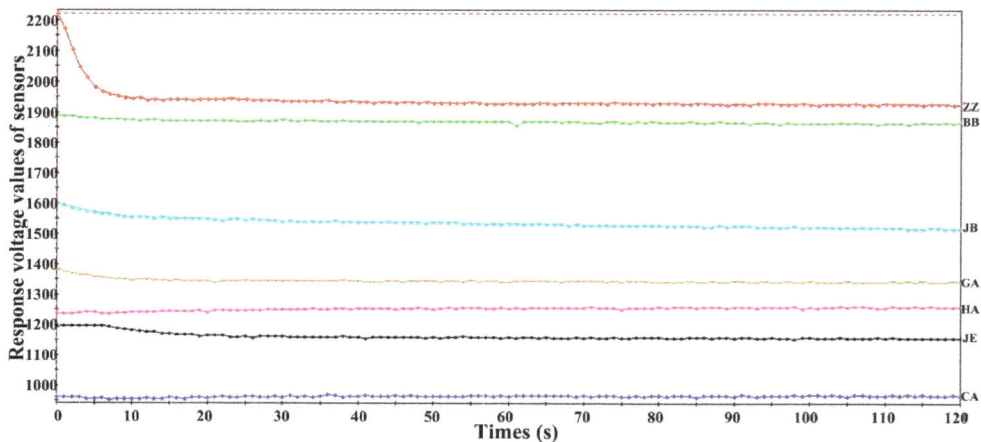

Figure 2. Response curves of the seven sensors to Sample A6.

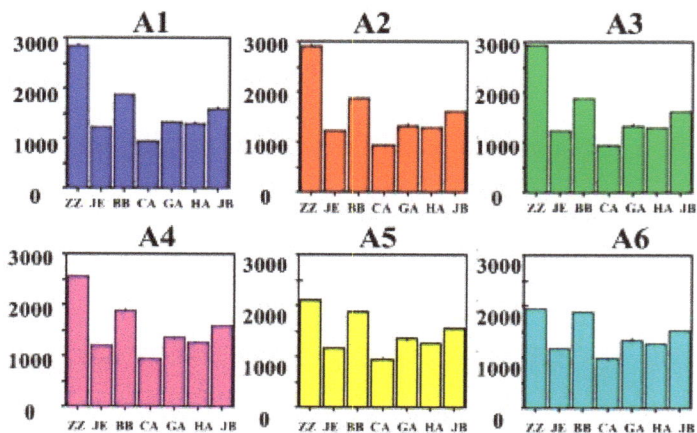

Figure 3. Bar graphs of the response values of sensors to all samples.

The relative standard deviation (RSD) values of the seven sensors in the DEH samples and the five standard taste samples were less than 6%, which indicates that the difference in experimental data obtained by the electronic tongue was small and exhibits good reproducibility.

3.4. Analysis of Five Taste Attributes of DEH Samples

DFA can be a useful tool for differentiating the individual to which each colony is affiliated; DFA was mapped using the response voltage values of the seven sensors for each sample as data for calculating a set of new variables called discriminant factors.

44

Each sensor response voltage value was decided according to the average of the last three measurements of the sensor of each sample. From Figure 4, DFA showed a total contribution of 99.56% with the contribution from the first two discriminant factors being 57.72% and 41.839%, respectively, with a Discrimination Index (DI) value of 98 All samples were separated into six distinct clusters. Five standard tastes (umami, saltiness, sourness, bitterness, and sweetness) were successfully discriminated. The space within each sample was close, while the distance between the two groups was large, indicating good data repeatability and a clear distinction between different samples. All six peanut meal DEH samples as unknown samples were projected into five tastes in the DFA map. From the cluster formations in the DFA map, the position of the peanut meal DEH samples cluster (unknown cluster) was relatively close to the umami and saltiness clusters, with the position being closer to the umami cluster. The unknown cluster was at a farther distance to the sourness cluster, and the distance between the sweetness and bitterness clusters was the farthest.

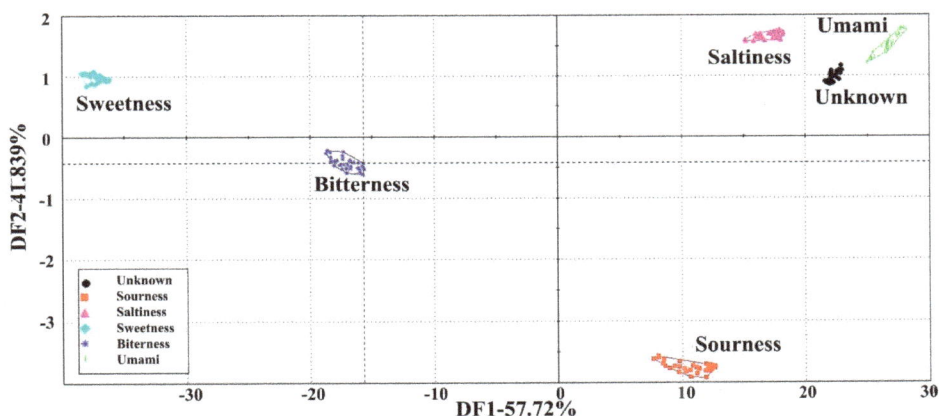

Figure 4. Analysis of the taste attributes of the five taste samples and peanut meal DEH samples (unknown samples was peanut meal DEH at enzymolysis times of 10 min, 1 h, 3 h, 8 h, 12 h, 24 h. Umami: 0.1%–1.0% MSG; saltiness: 0.001%–0.2% salt; sourness: 0.02%–0.2% citric acid; bitterness: 0.025%–0.25% tannic acid; sweetness: 0.25%–2.5% sugar). Data were processed by DFA.

Each point in each cluster represents each replicate sample measurement by the sensor array, indicating that peanut meal DEH samples exhibited characteristics similar to umami and saltiness. This result is in agreement with those obtained from sensory analysis and GEC. Samples A1–A6 mainly exhibited umami and saltiness, followed by sourness, while the intensities of both sweetness and bitterness were the weakest. Umami was the most prominent taste of peanut meal DEH samples.

The DFA map from the electronic tongue showed that it can assist and even replace human sensory analysis.

In addition, the lack of discrimination among samples of different concentration gradients in the umami and saltiness clusters as well as from samples of peanut meal DEH indicated that further research is required for the discrimination of these three taste samples. The discrimination of the three taste samples with different taste attributes was also mapped with DFA using the same data extraction method as shown in Figure 4. DFA (Figure 5) showed a complete 100% contribution with the first discriminant factor (DF1) with a DI value of 100. It indicated that the DFA map can represent the response data of all samples using the electronic tongue. All samples were separated into three distinct clusters. Three taste samples—umami, saltiness and peanut meal DEH samples—were successfully discriminated. The saltiness cluster was located at the left edge of the map, the umami cluster was located at the right edge of the map, and the peanut meal DEH (unknown cluster) was located between them, albeit closer to the umami cluster. Thus, the relationship between their relative positions on this map demonstrated a strong correlation with their taste attributes perceived by human. From the saltiness cluster, the six samples exhibiting saltiness (0.001%, 0.005%, 0.01%, 0.05%, 0.1%, and 0.2%) were separated from each other, indicating that the electronic tongue can discriminate different concentrations with the same taste. DF2 in the map approximately coincided with the direction of intensity change of saltiness in samples. The arrow on the map represents the increasing direction of the intensity of saltiness. Similar results were observed for the umami cluster. Six umami samples (0.1%, 0.2%, 0.4%, 0.6%, 0.8% and 1.0%) were separated, and the direction of the arrow on the map represents the increasing intensity of umami. For peanut meal DEH samples as projected into the map as unknown samples exhibited mainly two taste attributes, it located in a direction shown with an arrow direction representing the increasing enzymolysis time. The arrow direction is also called taste intensity increase direction because according to the previous analysis, the taste intensity increased with enzymolysis time. The first three samples of peanut meal DEH were not distinguished probably because of the low concentration of the prepared DEH solution. The lack of discrimination among these three samples indicated the necessity to search for more supervised methods for the quantitative analysis of peanut meal DEH samples, such as PLS or ANN.

The results from the above analysis demonstrated that the electric tongue has the potential to qualitatively assess different taste samples. The location of the different samples indicates the taste similarity degree. In the same cluster, the location can reflect different taste intensities and can regularly change with changing concentration.

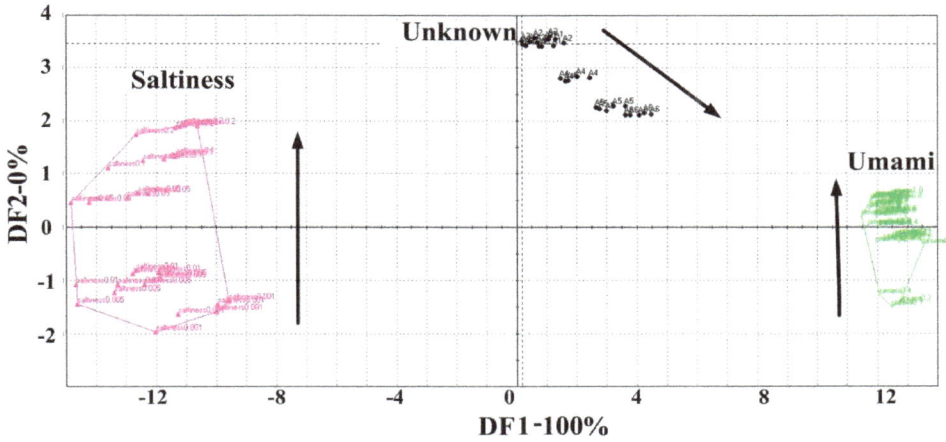

Figure 5. Evaluation of taste attributes of the three samples (Unknown peanut meal DEH samples under 10 min, 1 h, 3 h, 8 h, 12 h, 24 h; Umami: 0.1%–1.0% MSG; saltiness: 0.001%–0.2% salt).

3.5. Discrimination of Different DEH Samples by PCA and DFA

Because of the satisfactory taste trend observed in the previous analysis, the whole set of peanut meal DEH samples was further analyzed using only the electronic tongue, and data were processed by PCA and DFA to obtain consistent taste assessment results using different data processing methods (Figure 6). Figure 6a shows the distribution of the peanut meal DEH samples along the first two new coordinates (PC1 and PC2): an accumulated contribution of 99.94% was observed with a DI value of 92. The high contribution value indicated that nearly all the data from the original sensors are presented only by the two new functions. As can be seen, except samples A1 and A2, all other samples were successfully discriminated by analysis with PCA coordinates. The arrow direction on the map represents the increase in taste intensity. Figure 6b shows similar results by DFA analysis. It shows the distribution of the peanut meal DEH samples along the first two new coordinates (DF1 and DF2): an accumulated contribution of 99.96% was observed with a DI value of 94. As can be seen, except samples A1 and A2, all other samples were discriminated by analysis of DFA coordinates. The arrow direction on the map also represents the increase in taste intensity. DFA is significantly better than PCA because it can leave some samples as unknown, as compared to the expected ones, for projection into the built model to predict classes rather than use all samples to perform classification. Hence, the actual performance of the DFA classification model can be assessed.

The analysis result showed that the electronic tongue demonstrates potential for the qualitative assessment of the taste intensity of the different peanut meal DEH samples. The samples regularly varied according to the PC1/DF1 direction, which was the direction of increasing enzymolysis time. The two methods of the electronic tongue showed consistent taste assessment results. It was also in agreement with the results obtained from sensory analysis and GEC. The intensities of both umami and saltiness of the peanut meal DEH samples (A1–A6) increased with enzymolysis time, thereby resulting in the increase of the total taste intensity.

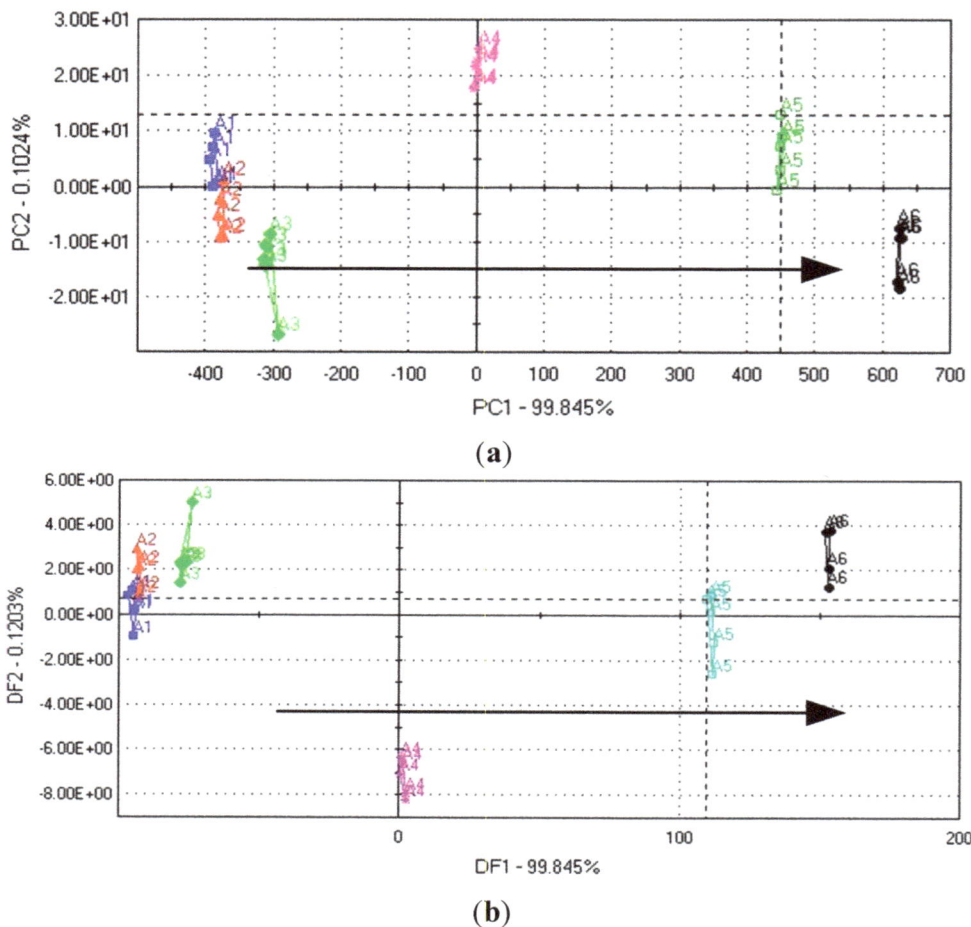

(a)

(b)

Figure 6. Taste discrimination of different peanut meal DEH samples (A1–A6) (a) score map obtained by PCA; (b) score map obtained by DFA.

48

3.6. Quantitative Taste Prediction of DEH

The taste attributes of peanut meal DEH were complex because a large amount of free amino acids and small-molecule peptides was present in the protein enzymolysis solution. Because of the different composition of free amino acids and peptides, the enzymolysis solution exhibits different tastes. From the information stated above, substances that were soluble in water exhibit umami and saltiness, which was easily detected by sensors; they were the main two taste attributes; also, their intensity scores for peanut meal DEH were predicted by the electronic tongue and were found to be comparable to those obtained by sensory analysis.

To assess whether the electronic tongue was suitable for the prediction of umami and saltiness in peanut meal DEH, it is essential to correlate between prediction scores obtained by the electronic tongue and actual scores obtained by sensory analysis. PLS and RBFNN prediction models were employed. Unlike previous studies that focus on the extraction of qualitative information, herein two quantitative models were built. In each model, two input datasets were extracted for model construction. One type of input variable was selected with the response voltage values obtained from the seven sensors of the electronic tongue, and the other type was the first four PC values. That is, the scores assigned for peanut meal DEH by sensory analysis were modeled from the data of sensors voltage response or from the data of the first four PC values.

3.6.1. PLS Taste Prediction Model

Figure 7a,b show the correlation between the predicted scores by PLS using the electronic tongue and those obtained by sensory analysis with respect to the intensity of umami in peanut meal DEH samples. PLS regression was built by using the response voltage obtained by the seven sensors of the electronic tongue for the peanut meal DEH samples and the scores obtained by sensory analysis. As stated before, the average of last three measurements of each sample was used as the response voltage value of each sensor. Table 3 lists the prediction formula, and Figure 7a shows the PLS regression results. A good correlation was observed between the intensity scores of umami predicted by the electronic tongue and the actual scores obtained by sensory analysis with a high R^2 (Test) value of 0.9805 and R^2 (Prediction) value of 0.9654 in the cross-validation test. The two R^2 values are similar, indicating the effectiveness of the taste prediction model: a satisfactory trend was obtained with the regression line being close to the theoretical line. The RMSE of prediction was 0.72374. Then, PLS regression was built by using the first four PC values as independent variables and the scores obtained from sensory analysis. A better trend was observed between the predicted intensity scores of umami and the actual scores obtained by sensory analysis with a high R^2 (Test) value of 0.9723 and R^2 (Prediction) value of 0.9481, albeit a relatively high RMSE of 1.024. From the above analysis, the PLS prediction model using the response voltage from the seven sensors of the electronic tongue as model data input was better than by using the first four PC values. This indicated that the model constructed by using the first four PCs is disadvantageous in that compressed data information obtained by sensor detection.

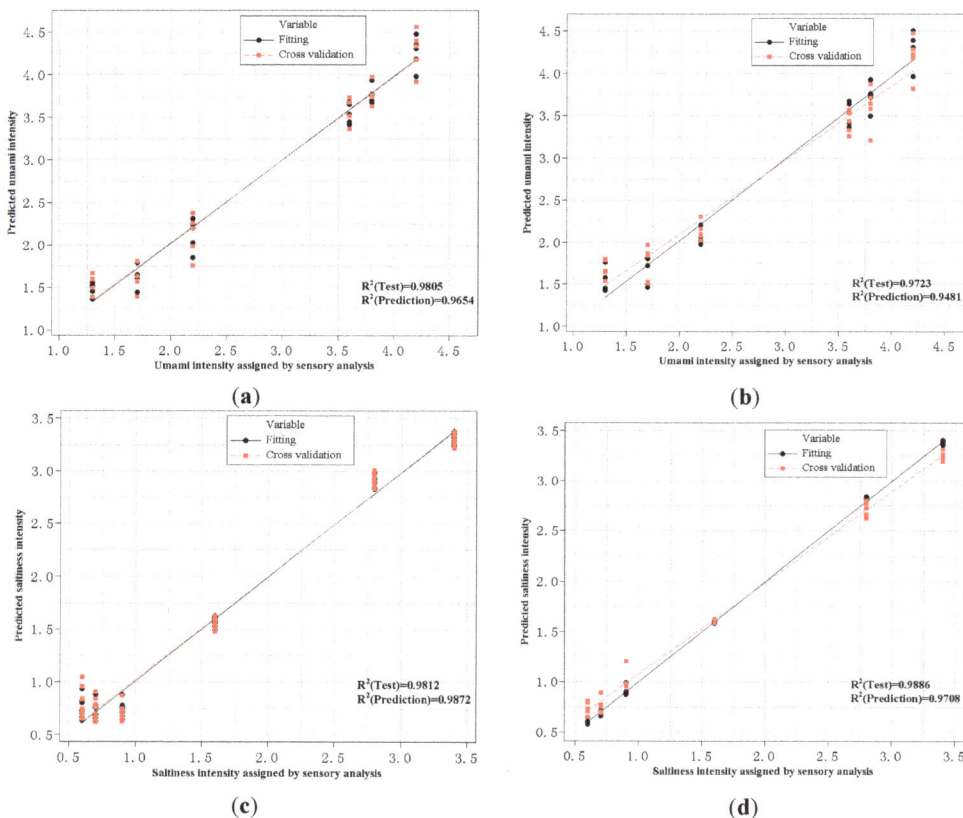

Figure 7. PLS model prediction of intensity scores by the electronic tongue against actual scores as scored by sensory analysis. (**a**) umami intensity scores using sensors voltage response values as model input; (**b**) umami scores using the first four PC values as the model input; (**c**) saltiness scores using sensors voltage response values as the model input; (**d**) saltiness scores using the first 4 PC values as the model input.

The same experimental procedure was performed to predict the intensity of saltiness in peanut meal DEH by using the PLS model. Table 3 lists the prediction formula for saltiness, and Figure 7c,d show the resultant PLS using input data obtained from the response voltage from the seven sensors of the electronic tongue or the first four PC values. A strong correlation was observed between the predicted scores by the electronic tongue, obtained from the voltage response data, and the actual scores for saltiness obtained by sensory analysis with R^2 (prediction) and R^2 (Test) of 0.9812, 0.9872 and a relatively low RMSE of 0.32548 (Figure 7c). The resultant PLS regression using the first four PCs as model data input exhibited a relatively worse correlation than the former using the voltage response data with two R^2 of 0.9886, 0.9708 and an RMSE of 0.45234 (Figure 7d).

Table 3. PLS models for prediction of taste intensity scores of DEH.

Taste	Input Data	Prediction Formula
umami	voltage response	Y = 76.2667 + 0.0029x1 − 0.0412x2 + 0.0303x3 − 0.0010x4 − 0.0207x5 − 0.0121x6 − 0.0280x7
	first 4 PC values	Y = 2.8 + 0.00256PC1 + 0.01402PC2 − 0.03365PC3 − 0.03814PC4
saltiness	voltage response	Y = −29.9924 − 0.0065x1 + 0.0302x2 − 0.0051x3 + 0.0009x4 − 0.016x5 + 0.0245x6 + 0.0068x7
	first 4 PC values	Y = 1.5 + 0.00266PC1 − 0.0246PC2 + 0.02367PC3 + 0022PC4

x1–x7 represented response voltage values from sensor1–7, respectively (x1: ZZ; x2: JE; x3: BB; x4: CA; x5: GA; x6: HA; x7: JB).

From the above analysis of taste prediction models, all four PLS models for taste intensity scores prediction of peanut meal DEH exhibited a good correlation with actual scores obtained by sensory analysis. The two PLS models used to predict the intensity scores for saltiness had stronger predictive capabilities with lower RMSE than those for the intensity scores for umami. In the same prediction of taste intensity scores, the PLS model for the prediction of taste intensity constructed with data from the response voltage values of the seven sensors was better than that constructed using data from the first four PCs with both higher correlation coefficients and relative lower RMSE as no detected data was lost.

3.6.2. RBFNN Taste Prediction Model

RBPNN was applied to further predict the taste intensity scores by the electronic tongue as compared to linear methods caused by its high performance. The two types of input data for RBFNN were selected to be the same as those for PLS. First, the k-means clustering algorithm was adopted to determine a proper data center for the radial basis function of the hidden layer nodes and then an appropriate overlap value and the number of hidden layer nodes were selected using different combination of the two parameters. The optimization range of overlap was 1–3, while the number of hidden layer nodes was 4–10. The optimization goal was the minimum RMSE value of RBFNN. The prediction error results showed that the error of the models decreases with increasing number of hidden layer nodes. However, after the hidden layer nodes increased to a certain number, "overfitting" was observed, leading to an increase in error. Hence, the optimum number of hidden layer nodes for the two types of constructed RBFNN models was determined according to the average minimum RMSE value after three times validation. In one construction of 7-8-1, the lowest average RMSEs of 0.139 and 0.1209 were predicted for umami and saltiness, respectively, with an optimized overlap of 2; the optimum number of the hidden layer nodes was 8. On the other hand, for the other construction of 4-6-1, the lowest average RMSEs of 0.1533 and 0.1793 were predicted for umami and saltiness, respectively, with an overlap of 2, and the optimum number of hidden layer nodes was 6.

Table 4 summarizes the prediction results. It lists the RMSE, maximum error, mean absolute error (MAE), and R^2, and rows having different taste attributes and different input data (using sensor voltage responses and four PCs). It also shows PLS results for comparison.

The prediction ability of the RBFNN model can be seen directly in Table 4. It shows a comparison of the prediction error for umami or saltiness assigned by using the response voltage values from the seven sensors and the first 4 PCs values; an R^2 of 0.994 and RMSE, Maximum error and MAE of 0.139, 0.1531 and 0.13, respectively, was better than that by the first 4 PCs values, for which an R^2 of 0.992 and RMSE, Maximum error and MAE of 0.1533, 0.2738 and 0.14, respectively. Similarly, the model prediction results for the intensity of saltiness were shown. A strong correlation was observed between the scores predicted by the electronic tongue using the response voltage data and the actual saltiness scores obtained from sensory analysis with an R^2 of 0.998 and RMSE, Maximum error and MAE of 0.1209, 0.2162 and 0.115, respectively. The latter also had an excellent correlation by using the first 4 PCs as model data input with an R^2 of 0.996 but with a relative higher error of 0.1793, 0.2275 and 0.169, respectively.

Table 4. Comparison of the model for the ability to predict different errors and R^2 for DEH.

Models	Taste Prediction	Input Data	RMSE	Maximum Error	MAE	R^2
PLS	umami	voltage response	0.7237	0.5056	0.222	0.9805/0.96544
		the first 4 PCs	1.024	0.7044	0.246	0.9723/0.9481
	saltiness	voltage response	0.3255	0.241	0.117	0.9812/0.9872
		the first 4 PCs	0.4523	0.4965	0.119	0.9886/0.9708
RBFNN	umami	voltage response	0.139	0.1531	0.13	0.994
		the first 4 PCs	0.1533	0.2738	0.14	0.992
	saltiness	voltage response	0.1209	0.2162	0.115	0.998
		the first 4 PCs	0.1793	0.2275	0.169	0.996

The assessment results using RBPNN model to predict taste showed that the results obtained from all four RBFNN prediction models well correlated with those obtained from the actual sensory analysis scores. In the same prediction of taste intensity, RBFNN with model input data from the response voltage values of seven sensors exhibited better ability to predict scores than that with model data from the first 4 PCs, caused by the relatively lower RMSE. In the study, RBFNN exhibited the best ability to predict the intensity for saltiness using the input data from the response voltage value of seven sensors, with the lowest RMSE of 0.1209.

A comparison of the taste assessment by the RBPNN and PLS prediction models shown in Table 4 indicated that RBPNN exhibited a stronger correlation with higher correlation coefficients and better taste prediction ability with a relatively lower error than those of PLS. For the construction of the two taste prediction models, using the response voltage values of the sensors of the electronic tongue, as compared to the first four PCs, was better.

However, despite the good prediction results obtained with both PLS and RBFNN, it is necessary to highlight that prediction results may be overoptimistic if some of the measurements in similar hydrolyzate solutions prepared from similar peanut meal samples were used both for calibration and validation.

4. Conclusions

Qualitative and quantitative assessment of taste attributes of peanut meal DEH samples using an electronic tongue was attempted. Concretely, the electronic tongue was used for the qualitative assessment of the taste attributes of peanut meal DEH using PCA and DFA, and the quantitative taste was evaluated by the scores predicted by employing PLS and RBFNN models. The two taste scores from the prediction models were compared on aspects of predictive abilities of umami and saltiness intensity and the correlation between electronic tongue and sensory analysis. The results showed the electronic tongue could distinguish five tastes with different concentrations and different peanut meal DEH, which is an easy and visual indicator of the taste attributes shown in the map. Both prediction models used for quantitative assessment can evaluate taste scores in peanut meal DEH. As compared to the PLS prediction model, the RBPNN prediction model showed a stronger correlation with higher correlation coefficients and better ability to predict taste with a relatively lower RMSE. Moreover, the results further proved good consistency with those obtained by sensory analysis and gel exclusion chromatography, indicating that the electronic tongue is an effective tool for the assessment of complex taste attributes as it involves a less sample preparation time and short analysis time. Furthermore, this study demonstrated applications of the electronic tongue in the assessment of hydrolysates with high bitterness or unpleasant tastes such as single enzymatic-hydrolysis hydrolysates. The study provides a new taste assessment tool for the preparation of flavor base with peanut meal DEH in the food industry. Finally, future efforts may involve further construction of more models, validation and correlation between the electronic tongue, and sensory analysis or other chemical methods with increasing various peanut meal DEH samples under different types of enzymolysis, as well as taste assessment to mimic the human tongue as better as possible.

Acknowledgments: This study is funded by the National Technology Support Program of China (2012BAF12B13) as well as by the Fundamental Research Funds for the Henan Provincial Colleges and Universities in Henan University of Technology (2014YWJC03).

Author Contributions: Li Wang, Qunfeng Niu and Huali Jin conceived and designed the experiments; Qunfeng Niu and Yanbo Hui performed the experiments; Li Wang and Yanbo Hui analyzed the data; Huali Jin contributed reagents, materials and sensory analysis tools; Li Wang and Shengsheng Chen wrote the paper.

Conflicts of Interest: The authors declare no conflict of interest.

References

1. Batal, A.; Dale, N.; Café, M. Nutrient composition of peanut meal. *J. Appl. Poult. Res.* **2005**, *14*, 254–257.

2. Natarajan, K. Peanut protein ingredients: Preparation, properties and food uses. *Adv. Food Res.* **1980**, *26*, 215–220.

3. Costa, E.; Miller, B.; Pesti, G.; Bakalli, R.; Ewing, H. Studies on feeding peanut meal as a protein source for broiler chickens. *Poult. Sci.* **2001**, *80*, 306–313.

4. Su, G.; Ren, J.; Yang, B.; Cui, C.; Zhao, M. Comparison of hydrolysis characteristics on defatted peanut meal proteins between a protease extract from Aspergillus oryzae and commercial proteases. *Food Chem.* **2011**, *126*, 1306–1311.

5. Wu, H.; Wang, Q.; Ma, T.; Ren, J. Comparative studies on the functional properties of various protein concentrate preparations of peanut protein. *Food Res. Int.* **2009**, *42*, 343–348.

6. Toko, K.; Masaru, T.; Satoru, L. Self-sustained oscillations of electric potential in a model membrane. *Biophys. Chem.* **1986**, *23*, 201–210.

7. Winquist, F.; Wide, P.; Lundström, I. An electronic tongue based on voltammetry. *Anal. Chim. Acta* **1997**, *357*, 21–31.

8. Legin, A.; Rudnitskaya, A.; Vlasov, Y. Application of electronic tongue for qualitative and quantitative analysis of complex liquid media. *Sens. Actuators B Chem.* **2011**, *65*, 232–234.

9. Winquist, F.; Olsson, J.; Eriksson, M. Multicomponent analysis of drinking water by a voltammetric electronic tongue. *Anal. Chim. Acta* **2011**, *683*, 192–197.

10. Kirsanov, D.; Zadorozhnaya, O.; Krasheninnikov, A.; Komarova, N.; Popov, A.; Legin, A. Water toxicity evaluation in terms of bioassay with an Electronic Tongue. *Sens. Actuators B Chem.* **2013**, *179*, 282–286.

11. Ariza-Avidada, M.; Cuellarb, M.P.; Salinas-Castillo, A. Feasibility of the use of disposable optical tongue based on neural networks for heavy metal identification and determination. *Anal. Chim. Acta* **2013**, *783*, 56–64.

12. Liu, M.; Wang, M.J.; Wang, J.; Li, D. Comparison of random forest, support vector machine and back propagation neural network for electronic tongue data classification: Application to the recognition of orange beverage and Chinese vinegar. *Sens. Actuators B Chem.* **2013**, *177*, 970–980.

13. Escriche, I.; Kadar, M.; Domenech, E.; Gil-Sanchez, L. A potentiometric electronic tongue for the discrimination of honey according to the botanical origin. Comparison with traditional methodologies: Physicochemical parameters and volatile profile. *J. Food Eng.* **2012**, *109*, 449–456.

14. Tiwari, K.; Tudu, B.; Bandyopadhyay, R.; Chatterjee, A. Identification of monofloral honey using voltammetric electronic tongue. *J. Food Eng.* **2013**, *117*, 205–210.

15. Ghosh, A.; Tamuly, P.; Bhattacharyya, N. Estimation of theaflavin content in black tea using electronic tongue. *J. Food Eng.* **2012**, *110*, 71–79.

16. Ivarsson, P.; Holmin, S.; Höjer, N.E.; Krantz-Rülcker, C.; Winquist, F. Discrimination of tea by means of a voltammetric electronic tongue and different applied waveforms. *Sens. Actuators B Chem.* **2001**, *76*, 449–454.

17. Berenice Domínguez, R.; Moreno-Barón, L.; Muñoz, R.; Manuel Gutiérrez, J. Voltammetric Electronic Tongue and Support Vector Machines for Identification of Selected Features in Mexican Coffee. *Sensors* **2014**, *14*, 17770–17785.

18. Zakaria, N.Z.I.; Masnan, M.J.; Zakaria, A.; Shakaff, A.Y.M. A Bio-Inspired Herbal Tea Flavour Assessment Technique. *Sensors* **2014**, *14*, 12233–12255.

19. Legin, A.; Rudnitskaya, A.; Seleznev, B.; Vlasov, Y. Electronic tongue for quality assessment of ethanol, vodka and eau-de-vie. *Anal. Chim. Acta* **2005**, *534*, 129–135.

20. Di Natale, C.; Paolesse, R.; Macagnano, A. Application of Combined Artificial Olfaction and Taste System to the Qyantification of Relevant Compounds in Red Wine. *Sens. Actuators B Chem* **2009**, *69*, 342–347.

21. Wei, Z.; Wang, J.; Ye, L. Classification and prediction of rice wines with different marked ages by using a voltammetric electronic tongue. *Biosens. Bioelectron.* **2011**, *26*, 4767–4773.

22. Wu, C.; Wang, J.; Wei, Z. Prediction of apparent viscosity of milk with different volume fraction using electronic tongue. *Trans. CSAE* **2010**, *26*, 226–230.

23. Mottram, T.; Rudnitskaya, A.; Legin, A.; Fitzpatrick, J.L.; Eckersall, P.D. Evaluation of a novel chemical sensor system to detect clinical mastitis in bovine milk. *Biosens. Bioelectron.* **2007**, *22*, 2689–2693.

24. Apetrei, C.; Rodriguez-Mendez, M.L.; de Saja, J.A. Modified carbon paste electrodes for discrimination of vegetable oils. *Sens. Actuators B Chem* **2005**, *111*, 403–409.

25. Rodriguez-Mendez, M.L.; Apetrei, C. Evaluation of the polyphenolic content of extra virgin olive oils using an array of voltammetric sensors. *Electrochim. Acta* **2008**, *53*, 5867–5872.

26. Oliveri, P.; Baldo, M.A.; Daniele, S.; Forina, M. Development of a voltammetric electronic tongue for discrimination of edible oils. *Anal. Bioanal. Chem.* **2009**, *395*, 1135–1143.

27. Ceó, X.; González-Calabuig, A.; Capdevila, J.; Puig-Pujol, A.; del Valle, M. Instrumental measurement of wine sensory descriptors using avoltammetric electronic tongue. *Sens. Actuators B Chem.* **2015**, *207*, 1053–1059.

28. Apetrei, C.; Gutierez, F.; Rodr'iguez-M'endez. Novel method based on carbon paste electrodes for the valuation of bitterness in extra virgin olive oils. *Sens. Actuators B Chem.* **2007**, *121*, 567–575.

29. Eckert, C.; Lutz, C.; Breitkreutz, J.; Woertz, K. Quality control of oral herbal products by an electronic tongue—Case study on sage lozenges. *Sens. Actuators B Chem.* **2011**, *156*, 204–212.

30. Woertz, K.; Tissen, C.; Kleinebudde, P. Performance qualification of an electronic tongue based ICH guideline Q2. *J. Pharm. Biomed. Anal.* **2010**, *51*, 497–506.

31. Apetrei, C.; Rodr guez-Mndez, M.L. Array of voltammetric sensors for the discrimination of bitter solutions. *Sens. Actuators B Chem.* **2004**, *103*, 145–152.

32. Legin, A.; Rudnitskaya, L.; Clapham, D.; Seleznev, B.; Lord, K.; Vlasov, Y. Electronic tongue for pharmaceutical analytics: Quantification of tastes and masking effects. *Anal. Bioanal. Chem.* **2004**, *380*, 36–45.

33. Rachid, O.; Simons, F.E.R.; Rawas-Qalaji, M.; Simons, K.J. An electronic tongue: Evaluation of the masking efficacy of sweetening and/or flavoring agents on the bitter taste of epinephrine. *AAPS PharmSciTechl.* **2010**, *11*, 550–557.

34. Pein, M.; Preis, M.; Eckert, C.; Kiene, F.E. Taste-masking assessment of solid oral dosage forms—A critical review. *Int. J. Pharm.* **2014**, *465*, 239–254.

35. Woertz, K.; Tissen, C.; Kleinebudde, P.; Breitkreutz, J. A comparative study on two electronic tongues for pharmaceutical formulation development. *J. Pharm. Biomed. Anal.* **2011**, *55*, 272–281.

36. Woertz, K.; Tissen, C.; Kleinebudde, P.; Breitkreutz, J. Rational development of taste asked oral liquids guided by an electronic tongue. *Int. J. Pharm.* **2010**, *400*, 114–123.

37. Harada, T.; Uchida, T.; Yoshida, M.; Kobayashi, Y.; Narazaki, R.; Ohwaki, T. A new method for evaluating the bitterness of medicines in development using a taste sensor and a disintegration testing apparatus. *Chem. Pharm. Bull.* **2010**, *58*, 1009–1014.

38. Newman, J.; Harbourne, N.; O'Riordan, D.; Jacquier, J.C.; O'Sullivan, M. Comparison of a trained sensory panel and an electronic tongue in the assessment of bitter dairy protein hydrolysates. *J. Food Eng.* **2014**, *128*, 127–131.

39. Newman, J.; Egan, T.; Harbourne, N.; O'Riordan, D.; Jacquier, J.C.; O'Sullivan, M. Correlation of sensory bitterness in dairy protein hydrolysates: Comparison of prediction models built using sensory, chromatographic and electronic tongue data. *Talanta* **2014**, *126*, 46–53.

40. Toko, K.; Hara, D.; Tahara, Y.; Yasuura, M.; Ikezaki, H. Relationship between the Amount of Bitter Substances Adsorbed onto Lipid/Polymer Membrane and the Electric Response of Taste Sensors. *Sensors* **2014**, *14*, 16274–16286.

Fruit Quality Evaluation Using Spectroscopy Technology: A Review

Hailong Wang, Jiyu Peng, Chuanqi Xie, Yidan Bao and Yong He

Abstract: An overview is presented with regard to applications of visible and near infrared (Vis/NIR) spectroscopy, multispectral imaging and hyperspectral imaging techniques for quality attributes measurement and variety discrimination of various fruit species, *i.e.*, apple, orange, kiwifruit, peach, grape, strawberry, grape, jujube, banana, mango and others. Some commonly utilized chemometrics including pretreatment methods, variable selection methods, discriminant methods and calibration methods are briefly introduced. The comprehensive review of applications, which concentrates primarily on Vis/NIR spectroscopy, are arranged according to fruit species. Most of the applications are focused on variety discrimination or the measurement of soluble solids content (SSC), acidity and firmness, but also some measurements involving dry matter, vitamin C, polyphenols and pigments have been reported. The feasibility of different spectral modes, *i.e.*, reflectance, interactance and transmittance, are discussed. Optimal variable selection methods and calibration methods for measuring different attributes of different fruit species are addressed. Special attention is paid to sample preparation and the influence of the environment. Areas where further investigation is needed and problems concerning model robustness and model transfer are identified.

Reprinted from *Sensors*. Cite as: Wang, H.; Peng, J.; Xie, C.; Bao, Y.; He, Y. Fruit Quality Evaluation Using Spectroscopy Technology: A Review. *Sensors* **2015**, *15*, 11889–11927.

1. Introduction

Over the last couple of decades, with the rapid development of the economy and improvement of living standards, fruit consumption has increased significantly. Meanwhile, consumers have higher expectations of fruit qualities such as ripeness, firmness, soluble solids content (SSC) and acidity. However, many fruit quality attributes affecting consumer acceptance and price are still tested using traditional approaches which are either subjective or time-consuming, so it should be a surprise that how to measure fruits' internal and external attributes nondestructively and rapidly has become a research hotspot. Researchers all over the world have investigated the potential of various technologies, including acoustic techniques, spectroscopic techniques, machine vision and electronic noses, for the assessment of fruit qualities. Among all these technologies, spectroscopic techniques have drawn great attention for their prominent advantages: (1) they are nondestructive methods which enable the acquisition of fruits' internal quality parameters without damaging their surfaces; (2) the measurement processes are simple and rapid, as no complex pretreatments or chemical reactions on fruit samples are needed; (3) they enable the detection of several fruit internal attributes simultaneously. As a disadvantage, however, the small point-source measurements which are commonly used in spectral assessment cannot provide spatial information, which is important in many fruit quality evaluation instances.

Imaging and spectroscopy are two important directions of conventional optical technology. Imaging techniques can obtain the images of fruits and acquire their spatial information while spectroscopy provides access to information about the chemical components and physical properties of fruits by obtaining optical information. Imaging spectral techniques enable the acquisition of fruit images and spectral information simultaneously, with the advantages of high spectral resolution and multiple wavebands. According to the spectral resolution, imaging spectroscopy can be divided into multispectral imaging, hyperspectral imaging and ultra-spectral imaging. Multispectral imaging and hyperspectral imaging are proved to be feasible for the measurement of fruit quality parameters. However, few papers concerning both imaging technique and spectroscopy technique can be found yet, so in this review, most attention was paid to spectroscopic techniques, rather than imaging techniques.

Visible and near infrared (Vis/NIR) radiation covers the range from 380–2500 nm in the electromagnetic spectrum. As the signals of almost all major structures and functional groups of organic compounds can be detected in the Vis/NIR spectrum with a considerably stable spectrogram, spectra in the Vis/NIR range are frequently used for analysis [1]. Wavebands which are commonly used in multispectral and hyperspectral imaging technologies to assess fruit quality are also in the Vis/NIR region [2–5]. When incident radiation hits a sample, it may be reflected, transmitted or absorbed. Correspondingly, a spectrum is obtained in the reflectance, transmittance or absorbance mode, each of which can reflect some physical attribute and chemical constitution of the sample.

After the spectrum is obtained, chemometric methods are applied to extract information concerning the quality attributes and to eliminate the interference of factors irrelevant to sample concentration. In general, chemometrics consist of two parts, spectral pretreatments and regression methods.

The objective of this review is to offer a comprehensive overview of the use of Vis/NIR spectroscopy, multispectral imaging and hyperspectral imaging techniques in the measurement of various fruit quality attributes. We will briefly introduce the chemometric methods commonly used, and pay extra attention to the identification of optimal methods for variable selection and quality measurement.

2. Chemometrics

Applications of spectroscopy as well as multispectral and hyperspectral imaging technologies to measure fruit quality attributes are usually carried out in the Vis/NIR region since spectra in this range incorporate abundant information concerning O-H, C-H and N-H vibration absorptions [6]. However, in this region, the spectrum is basically dominated by water which highly absorbs near infrared radiation [7]. Besides, the Vis/NIR spectrum has a low signal-to-noise ratio and high overlap of combination bands and overtones, not to mention the complex constitution of fruits, wavelength-dependent light scattering and instrumental noise. All these cause the convolution of the Vis/NIR spectrum. Therefore, chemometrics are applied for extracting information concerning certain quality attributes from the spectral data.

2.1. Pretreatment Methods

2.1.1. Smoothing

Smoothing is an effective approach for removing high-frequency noise from a spectrum and improving the signal-to-noise ratio. Its basic idea is to obtain an optimal estimation value through the "averaging" or "fitting" of several points in a window. The broader the window is, the lower the spectral resolution would be. Thus it is crucial to choose the window width properly. Based on different smoothing fit methods, smoothing could be divided into moving average smoothing, Gaussian filter smoothing, median filter smoothing and Savitzky-Golay smoothing (S-G smoothing). Sun *et al.* [8] proved that moving average smoothing was the most feasible pretreatment method for SSC prediction of navel oranges, and Roger and Bellon-Maurel [9] applied NIR spectra processed with moving average smoothing for the measurement of sugar content in cherry fruit. However, smoothing is usually used in combination with other pretreatment methods, such as Multiplicative Scatter Correction (MSC). Liu and Zhou [10] claimed that the combination of 1st derivative, MSC and smoothing was feasible to process Vis/NIR transmittance spectra for predicting SSC in apples.

2.1.2. Offset Correction

This is a centralized processing method which is realized by subtracting the average value of the first few wavelength points (for example five) from each spectrum. Offset correction only adjusts the baseline drift, leaving the spectrum shape unchanged. It is mainly used for weakening the influence of instrumental noise, optical distance and detection environment.

2.1.3. De-Trending

De-trending is an approach to eliminate the baseline drift in the spectrum. Firstly a trend line was derived from spectral values and wavelengths through least squares fitting, and then the trend line was subtracted from the original spectrum. De-trending is often used in combination with standard normal variate correction (SNV), which we could find in the studies of Sanchez *et al.* [11] who predicted firmness in strawberries and Paz *et al.* [12], who predicted SSC and firmness in plums.

2.1.4. Multiplicative Scatter Correction (MSC)

First proposed by Ilari *et al.* [13], MSC was used to compensate the effect of non-uniform scattering induced by diverse particle sizes, uneven distribution and other physical effects in the spectral data. MSC is performed by linearizing each spectrum to an "ideal" spectrum, which corresponds to the average spectrum of the calibration set. The linear relationship between each spectrum and the average spectrum is fitted through the method of least squares. This suggests that MSC is feasible for removing the 'ideal' linear scattering and effects well when the linear relationship between absorbance and sample concentration is good. The feasibility of MSC was already confirmed by Liu *et al.* [14] and Shao *et al.* [15].

2.1.5. Standard Normal Variate (SNV)

Basically the same as MSC, the objective of SNV is to eliminate the deviations caused by particle size and scattering [16]. The method assumes that the absorbance of each wavelength point in the spectrum meets some certain distribution such as a Gaussian distribution. Based on this hypothesis, each spectrum is calibrated. Firstly the average value of a spectrum is subtracted from the original spectrum, and then the result is divided by the standard deviation (*SD*). For SNV effects on each spectrum alone, the correction capability of SNV is usually stronger than that of MSC. In the model established by Shi *et al.* [17] to evaluate the firmness of apples, the relative standard deviation of prediction (*RSDP*) was reduced from 16.65% to 14.82% after SNV processing.

2.1.6. Derivative Correction

As a widely-used pretreatment method, first and second derivatives are applied to eliminate drifting and scattering, respectively. They can remove background interference, distinguish superposed peaks and enhance the spectral resolution and sensitivity. Two commonly-used spectral derivative approaches are direct finite difference and Savitzky-Golay (S-G) derivatives. Before derivatization, smoothing should be applied because derivatives may extract differences of adjacent wavelength points and amplify spectral noise. Pissard *et al.* [6] proved S-G 1st derivative processing was the best pretreatment method, while Liu *et al.* [18] claimed that the 2nd derivative was the best.

2.1.7. Wavelet Transformation (WT)

Introduced and applied in the study of Liu *et al.* [19], WT is an emerging signal and image processing method. In spectral analysis, WT is often used for data compression, smoothing and filtering, as well as the extraction of effective information. By applying a basis function, chemical signals can be decomposed into various scale compositions according to their different frequencies. Sampling windows of corresponding width are applied to scale compositions of different sizes, thus any part of the signal could be focused on. Narrow windows could be used to observe drastic changes while wide windows could be used to observe the overall features of the spectrum. Among the various wavelet functions, some were proved to be quite effective, such as Daubechies and Symlets. Xia *et al.* [20,21] chose Daubechies 3 while Shao and He [22] applied Daubechies 2.

2.1.8. Orthogonal Signal Correction (OSC)

When there's little correlation between the spectral matrix and the concentration matrix of certain quality attributes or the background noise is big, the first several principal components (PCs) selected by PLS or PCA contain very limited information about the concentration matrix. Eliminating these irrelevant signals before calibration through some orthogonal approaches could effectively reduce the number of PCs and enhance the prediction ability as well as stability of the calibration model. Therefore, OSC was introduced to calibration transfer by Sjoblom *et al.* [23]. In addition, OSC could also be applied to solve problems concerning model transfer and outlier detection. Its outstanding ability of improving the prediction ability was proved by Shi *et al.* [17], who applied direct OSC (DOSC) in his study.

2.1.9. Net Analyte Preprocessing (NAP)

Introduced by Goicoechea *et al.* [24], NAP is mainly used for extracting spectral information concerning a certain ingredient in the spectra of the mixture. The feasibility of NAP was proved by Lv *et al.* [25].

2.2. Variable Selection Methods

Due to the abundance of information and severe nonlinearity in the full spectrum, some processing methods have been applied on the original spectral data in the whole wavelength range to extract characteristic wavelengths (CWs) with the highest predictive ability. In this way the number of input variables is reduced and the calibration time is shortened.

2.2.1. Successive Projections Algorithm (SPA)

Introduced by Araujo *et al.* [26], SPA is a forward selection method searching for a group of variables with minimum redundant information and minimal collinearity through simple operations in a vector space. Starting with one wavelength, SPA incorporates a new wavelength during each iteration until a certain number is reached. The CWs extracted by SPA could represent the spectral information of most samples and avoid information overlap at the highest degree possible. In the study of Zhang *et al.* [27], the Least Squares Support Vector Machine (LS-SVM) model combined with SPA could yield a result better than that of the LS-SVM for full spectrum.

2.2.2. Regression Coefficient (RC)

RCs could be obtained during the calibration of partial least squares regression (PLS) [28]. The RC value corresponding to each wavelength point represents its ability to affect the predictive performance of the model. Thus based on the absolute value of RCs, CWs could be identified.

2.2.3. Loading Weights (LW)

Loading weights could also be obtained during the calibration of PLS. Under each latent variable, loading weights corresponding to wavelengths could be obtained and their absolute values illustrate the wavelengths' impact on the prediction model, so a wavelength with the maximum loading weight value is selected as the CW and the number of CWs is the same of the number of latent variables. Fernandez-Novales *et al.* [29] used CWs selected by the loading weights of latent variables to build a MLR model for sugar content prediction and obtained a satisfactory result.

2.2.4. Genetic Algorithm (GA)

As an effective global searching method, GA mimics the competitive mechanism of survival of the fittest in biological world. Based on a fitness function, GA is an iterative process starting from a population of randomly generated individuals and achieves optimal solutions through genetic operations including crossover, selection and mutation. When GA is applied for variable selection, the number of iterations is set, usually above 100 and root mean square error for cross validation (*RMSECV*) is often used as the fitness function. After the iteration, the variables are realigned based on the frequency they are selected. The variable with the highest frequency is used for calibration and one more variable is included each time sequentially. The optimal number of variables used is determined when a minimized *RMSECV* is achieved. Through such operation, irrelevant spectral information is eliminated and the number of spectral variables is reduced. Cao *et al.* [30] executed GA for the selection of CWs for predicting SSC in grapes and yielded a satisfactory result.

2.2.5. Competitive Adaptive Reweighted Sampling (CARS)

CARS is a novel wavelength selection method introduced by Li *et al.* [31]. Wavelengths with large absolute coefficients are sequentially identified as the CWs based on the adaptive reweighted sampling technique in a PLS model. A series of variable subsets are obtained and cross validation is employed to choose the optimal one with the lowest *RMSECV*. Sun *et al.* [8] validated that the CWs selected with CARS could yield the best result.

2.2.6. Uninformative Variables Elimination (UVE)

UVE was first proposed by Centner *et al.* [32] and its basic approach is to add some noise variables into the experimental variables and calibration models are built with the mixed variables. The importance of each variable is evaluated and the experimental variables with no more importance than the noise variables are eliminated. The UVE method was employed by Sun *et al.* [8].

Readers are referred to the corresponding references for details about many other effective variable selection methods that we do not introduce in detail here, including backward interval PLS (BiPLS) [33], synergy interval PLS (siPLS) [34,35], independent component analysis (ICA) [36], simulated annealing algorithm (SAA) [37] and stepwise regression analysis (SRA) [27].

2.3. Discriminant Methods

2.3.1. Principal Component Analysis (PCA)

With the application of PCA, a set of principal components (PCs) are obtained. The first PC contains the largest percentage of data variance and the variance decreases in the following PCs. These PCs are linear combinations of the original spectral data but uncorrelated with each other, endowing their ability to handle multicollinearity. PCA is often utilized in combination with other discriminant methods [38–41].

2.3.2. Partial Least Squares-Discriminant Analysis (PLS-DA)

Introduced by Liu *et al.* [42], PLS-DA is a method commonly applied for optimal classification. Based on PLS regression, PLS-DA uses dummy variables such as 1, 2, 3, *i.e.*, as variables of Y matrix instead of the concentration of some quality attributes. The optimal number of PLS components are decided by full cross-validation. The feasibility of PLS-DA was proved by Cen *et al.* [39] and Hao *et al.* [43].

2.3.3. Soft Independent Modeling of Class Analogy (SIMCA)

When SIMCA is applied, a PCA model is established for each class in a certain training data set. Then each observation is assigned to a class based on its residual distance from the model. However, each model is established independently without consideration of other classes. Due to the overlapping between classes, there's a chance of producing a non-optimized discriminant model. SIMCA was employed by Cao *et al.* [40], Baranowski *et al.* [44] and Hao *et al.* [43].

2.3.4. Linear Discriminant Analysis (LDA)

Introduced and used by Baranowski *et al.* [44], LDA is commonly used in machine learning to search for a linear combination of characteristics separating different classes of objects. It offers a linear transformation of n-dimensional feature vectors into an m-dimensional space (m < n). This linear combination could be used as a classifier or for data dimensionality reduction.

2.3.5. Support Vector Machine (SVM)

Like SIMCA and LDA, SVM is a pattern recognition method which is quite useful for supervised classification. It is feasible to both linear and nonlinear data, by using kernel function, which maps from the original space to the feature space and guarantees the ability to handle nonlinear classification. With the use of statistical learning, a hyperplane for optimal discrimination is determined. The feasibility of SVM has been proved by Baranowski *et al.* [44] and Guo *et al.* [45]. Other commonly applied discriminant methods include stepwise discriminant analysis (SDA) [46,47], and BP-ANN [15,48–50].

2.4. Calibration Methods

2.4.1. Multiple Linear Regressions (MLR)

MLR predicts the dependent variables by a linear combination of spectral values at each wavelength point. The error between predicted and measured values is minimized in a least squares sense. In spectral analysis, multicollinearity between the variables degrades the performance of MLR algorithms. MLR was successfully employed by Peiris *et al.* [51] and ElMasry *et al.* [52]. However, Jaiswal *et al.* [53] reported a big gap between r_c and r_p in the MLR model they built, indicating unstable prediction.

2.4.2. Principal Component Regression (PCR)

In PCR, a small number of principal components (PCs) are selected by a principal component analysis (PCA). These PCs are applied as predictors instead of the original spectral data and used to fit a MLR model. PCR was used for calibration in the studies of Park *et al.* [54] and Angra *et al.* [55], with the advantage of eliminating the multicollinearity for the PCs are uncorrelated. However, Hadi and Ling [56] pointed out the potential drawback that the PCs are decided only according to the variables and they may contain little information about the dependent variables.

2.4.3. Partial Least Squares Regression (PLS)

To overcome the drawback of PCR, PLS regression was introduced by Wold *et al.* [57]. It predicts the dependent variables by extracting a smallest possible set of orthogonal factors with greatest predictive abilities from the variables. These orthogonal factors, called latent variables (LVs) were arranged according to the relevance for predicting the dependent variables. Synthesizing the sense of principal component analysis (PCA) and multiple linear regressions (MLR), PLS regression is especially feasible in circumstances where multicollinearity exists between the variables and the number of latent variables is usually smaller than that in the PCR regression. The advantage toward PCA was confirmed by Liu *et al.* [58] and Lu *et al.* [59]. Lots of researchers have applied PLS in their studies, including Shan *et al.* [5] and Bureau *et al.* [60].

2.4.4. Least Squares Support Vector Machine (LS-SVM)

LS-SVM is an emerging statistic learning algorithm which improves the generalization ability of the learning machine based on the principle of structural risk minimization [42]. The computational complexity and quality of the support vector machine does not directly depend on the dimension of input data. Therefore, LS-SVM is widely applied in pattern recognition and function regression for the advantage of limited over-fitting, high predictive reliability and strong generalization ability. LS-SVM is especially feasible for circumstances of small sample space modeling. LS-SVM was applied as the best calibration method in the studies of Suykens and Vanderwalle [61], Zhang *et al.* [27], Liu and Zhou, Pissard *et al.* [10] and Liu *et al.* [62].

2.4.5. Artificial Neural Network (ANN)

ANN has been widely used in NIR calibration. Usually an ANN model consists of three layers of neurons, which are the input layer, the hidden layer and the output layer. Each neuron in the previous layer is connected to each neuron in the latter layer and every connecting line has a weight factor, the value of which is assessed based on a calibration set using cross validation and keeps changing with the influx of new information. The value of neurons in the hidden layer is decided by weighted sum of values of neurons in the input layer using a nonlinear function and the value of neurons in the output layer is decided by the values of neurons in the hidden layer similarly. In certain circumstances the predictive performance of the ANN model may be excellent, but it also faces drawbacks such as slow training speed, over-fitting and the visualization difficulty. ANN was proved effective in the studies of Liu *et al.* [63], Zhang *et al.* [64] and He *et al.* [65].

These mentioned above are the most commonly applied calibration methods. Other improved approaches such as Spline-PLS [18] and stepwise MLR [66] have been introduced and utilized by some researchers.

2.5. Model Evaluation

The prediction ability of a calibration model is mainly evaluated by the correlation coefficient (r) and root mean square error (*RMSEP*) between the predicted value and the measured value in validation set. The higher is the correlation coefficient and the lower is the *RMSEP*, the better is the prediction performance. When cross validation is employed, the prediction performance could also be assessed by the root mean square error for cross validation (*RMSECV*).

Other commonly used evaluation parameters include the standard error of prediction (*SEP*), the standard error of cross validation (*SECV*), the residual predictive deviation (*RPD*) and relative standard deviation (*RSD*). *RPD* is the ratio of standard deviation of the dependent variable to *RMSEP* or *RMSECV*. According to Nicolai *et al.* [7] and Pissard *et al.* [6], for a prediction model, when the *RPD* value is between 2 and 2.5 coarse prediction is possible, while an *RPD* value above 2.5 indicates good to excellent prediction. A similar standard was defined by Davey *et al.* [67], who proved that total carotenoids and β-carotene in banana could be measured accurately (*RPD* = 3.34, 2.74, respectively), α-carotene and c-carotene could be predicted coarsely (*RPD* = 1.68 and 1.96 respectively), and lutein could not be predicted (*RPD* = 1.16).

3. Quality Evaluation for Different Fruit Varieties

Spectra in the Vis/NIR range contain abundant information concerning O-H, C-H and N-H vibration absorptions [6], making the measurement of various quality attributes of fruits possible. Some wavebands contain typical absorption bands for some chemical groups. A brief overview was presented in Table 1 to give some guidance for waveband selection.

Table 1. Overview of wavebands containing typical absorption bands for certain chemical groups.

Quality Attribute	Chemical Group	Wavelength/nm	Ref.
Sugar	O-H	1190, 1400	[68]
SSC	C-H	910	[69]
	O-H	960, 1450	
	C-H and O-H	1210	[14]
	O-H	975	[70]
	O-H	960, 1180, 1450, 2000	
	C-H	963	[27]
	Combination bands of C-H and O-H	2000–2500	
	O-H and C-H	950–1075	[71]
Acidity	C-O from COOH	1607	
	O-H from carboxyl acids	1127	[72]
	C=O from saturated and unsaturated carboxyl acid	1437	
pH	C-H	768	[73]
	O-H	986	

3.1. Apples

Apples are among the most widely cultivated and eaten fruits all over the world. Important attributes affecting the taste of apples include firmness, sugar content and acidity. During the harvest and transport of apples, bruising is inevitable and could affect the quality attributes including appearance, water loss and enhance the risk of bacterial and fungal contamination. The discrimination of bruised apples from the intact ones could ensure the postharvest quality. Other important attributes such as vitamin and polyphenol content also have drawn some attention. A brief overview is presented in Table 2.

Table 2. Overview of the applications for measuring quality attributes in apple.

Quality Attribute	Cultivar	Method	Spectral Mode	Spectral Range/nm	Calibration Method	r_p	SEP	Ref.
SSC or sugar content	GA	Spectroscopy	Reflectance	800–1100	PCR	0.97	0.28	[54]
	RD			800–1100		0.98	0.34	
	GD	Hyperspectral imaging	Reflectance	500–1000	PLS	0.88	0.7	[4]
	JG					0.78	0.7	
	RD					0.66	0.9	
	Various	Spectroscopy	Reflectance	800–1600	PCR	Unknown	0.73–1.78	[55]
	FJ	Spectroscopy	Transmittance	505–1031	LS-SVM	0.98	0.29	[10]
	Unknown	Spectroscopy	Reflectance	482–1009	PLS	0.96	0.23 (RMSEP)	[33]
	FJ	Spectroscopy	Reflectance	833–2500	PLS	0.98	0.69–0.72 (RMSEP)	[68]
	FJ	Hyperspectral imaging	Reflectance	480–1016	PLS	0.92	0.67	[5]
	Various	Spectroscopy	Reflectance	400–2500	LS-SVM	0.97	0.37	[6]
	Various	Spectroscopy	Reflectance	5882–9900	PLS	≥0.98	1.9%–3.4% (RMSEP)	[60]
Titratable acidity	Various	Spectroscopy	Reflectance	5882–9900	PLS	0.98	6.0% (RMSEP)	[60]
Malic acid						0.98	4.7% (RMSEP)	
Citric acid						0.34	100% (RMSEP)	

Table 2. Cont.

Table 2. Cont.

Quality Attribute	Cultivar	Method	Spectral Mode	Spectral Range/nm	Calibration Method	r_p	SEP	Ref.
Firmness	GA	Spectroscopy	Reflectance	800–1100	PCR	0.47	4.9	[54]
	RD			400–1800		0.89	7.0	
	GD	Hyperspectral imaging	Reflectance	500–1000	PLS	0.87	5.9	[4]
	JG					0.95	7.1	
	RD					0.84	8.7	
	FJ	Spectroscopy	Reflectance	1000–2339	PLS	0.82	14.1 (RSD)	[17]
Total polyphenol	Various	Spectroscopy	Reflectance	400–2500	LS-SVM	0.97	140	[6]
	Various	Spectroscopy	Reflectance	6378–9900	PLS	0.98	87.1 (RMSEP)	[60]
Vitamin C	Various	Spectroscopy	Reflectance	400–2500	LS-SVM	0.90	0.049	[6]

GA: Gala; RD: Red Delicious; GD: Golden Delicious; JG: Jonagold; FJ: Fuji; IA: Indian apples.

3.1.1. Soluble Solids Content (SSC) or Sugar Content

Park *et al.* [54] found that by using spectra ranging from 800 to 1100 nm, SSC in GA and RD could both be predicted with excellent accuracies (r_p = 0.97, 0.96; SEP = 0.28, 0.34, respectively). Angra *et al.* [55] evaluated Brix values of Indian apple along with apples from other countries. Reflectance spectra were acquired at ten wavelengths in the range of 800–1600 nm and the change in spectral reflectance caused by apple shape was eliminated by normalizing the spectral reflectance against a non-absorbing wavelength. SSC values of all cultivars could be predicted with SEPs of 0.73–1.78. Liu and Zhou [10] built models with PLS and LS-SVM, respectively, using Vis/NIR transmittance spectra, and the LS-SVM model (r = 0.98, SEP = 0.29) outperformed the PLS one. The superiority of LS-SVM was proved by Pissard *et al.* [6], who predicted sugar content in more than 150 apple genotypes with Vis/NIR reflectance spectroscopy in the wavelength range of 400–2500 nm. S-G first derivative was proved to be the best data pretreatment and the LS-SVM yielded excellent result, with r_p of 0.97, SEP of 0.37°Brix and RPD of 4.3. Ouyang *et al.* [33] compared the performance of PLS combined with the backward interval partial least squares method (BiPLS-PLS), genetic algorithm (GA-PLS) and successive projection algorithm (SPA-PLS), respectively. GA-PLS performed the best, with r_p increased from 0.93 to 0.96 and RMSEP decreased from 0.30°Brix to 0.23°Brix compared to the PLS model built with full spectrum (482–1009 nm).

Wang *et al.* [74] found the fluctuation of temperature could influence the prediction accuracy in a nonlinear way. When no precautions were taken, the SEP of the SSC prediction model could reach as high as 2.55. They offered two methods to enhance the accuracy: a temperature variable-eliminating calibration model and a global robust calibration model, both of which performed well, with RMSEP of 0.72 and 0.69, respectively. Bureau *et al.* [60] monitored the change in sugar content during sample preparation. Results showed that different conditions of sample preparation could not affect the sugar concentration. Mid-infrared spectra in the range of 5882–9900 nm showed an excellent ability for predicting sugar content ($r_p \geqslant 0.98$ and $RMSEP \leqslant 3.4\%$).

Mendoza et al. [4] combined spectroscopy with image analysis, reducing the SEPs by 11.2, 2.8 and 3.0% ($r_p = 0.88, 0.78, 0.66$; $SEP = 0.7\%, 0.7\%, 0.9\%$) for GD, JG and RD, respectively. Shan et al. [5] observed a slightly better result with hyperspectral imaging in the range of 480–1016 nm for 'Fuji' apple. The PLS model based on spectra processed with MSC, 1st derivative and S-G smoothing sequentially yielded an r_p of 0.92 ($SEP = 0.67°$Brix). Zou et al. [75] combined a near-infrared spectrophotometer, a machine vision system, and an electronic nose system through ANN to classify 'Fuji' apples based on sugar content, making the classification error drop from around 17% when only NIR spectra were used to around 6%.

According to all these studies above, Spectroscopy combined with other measurements performs well in prediction of SSC in apples. However, they all have some drawbacks. NIRS could not obtain the spatial information of samples. The hyperspectral imaging technique had a relatively poor performance in SSC prediction compared with spectroscopy, and the presence of bruises seriously influences the prediction of SSC. MIR shows a good ability to estimate sugar content. However, MIR needs the crushing to sample preparation, which makes it comparatively time consuming. In all, SSC or sugar content of apple could be predicted with good performance though subtle differences existed among cultivars.

3.1.2. Acidity

Bureau et al. [60] employed mid-IR spectroscopy in the 5882–9900 nm range to measure organic acids. Through monitoring its quantitative changes, the concentration of organic acids was proved to be unaffected by different conditions of sample preparation, such as storage temperature, sample grinding and sample oxidation. Mid-infrared spectra were feasible for predicting organic acid contents ($r_p \geqslant 0.98$; $RMSEP \leqslant 4.7\%$), except for citric acid ($r_p = 0.75$), probably due to its very low content in apple fruit.

3.1.3. Firmness

As shown in Table 2, Park et al. [54] predicted the firmness using Vis/NIR diffuse reflectance spectra. According to the PCR models they built, a pretty good result could be obtained for RD using the full spectrum ($r_p = 0.89$, $SEP = 7.0$). However, for GA, the r_p and SEP could only reach 0.47 and 4.9, respectively. In the study of Shi et al. [17], after DOSC combined with first derivative was employed to filter out the background and extract useful information, the model was simplified and an acceptable result was obtained ($r_p = 0.82$; $RSDP = 14.08\%$). Mendoza et al. [4] employed critical spectral and image features extracted from hyperspectral scattering images in the wavelength range of 500–1000 nm to predict the firmness. Spectral scattering in combination with image features could significantly improve the prediction. The standard error of prediction (*SEP*) for GD, JG and RD apple was reduced by 6.6%, 16.1% and 13.7%, respectively (r_p = 0.87, 0.95, 0.84; *SEP* = 5.9%, 7.1%, 8.7%). These investigations indicated fluctuations in firmness prediction of apple fruit with regard to different varieties. By contrast, hyperspectral imaging shows better performance than spectroscopy in the firmness prediction, though less performance in the SSC prediction.

3.1.4. Total Polyphenols

Pissard *et al.* [6] established an LS-SVM model for the prediction of total polyphenol content in apples using spectra recorded in the 400–2500 nm region. S-G 1st derivative was proved to be the best pretreatment method. The model performed well with r_p of 0.97 and *SEP* of 140 mg/g. A similar result was obtained by Bureau *et al.* [60], who applied mid-infrared spectroscopy in the 6378-9900 nm range. The PLS model they established performed excellently, with r_p of 0.98 and *RMSEP* of 9.0%. They also found phenolic compounds, contrary to sugar content and organic acid, could be affected by sample oxidation, grinding and storage temperature in a descending order of degree. The two researches proved the feasibility of predicting total polyphenol levels in apple using spectroscopic technology.

3.1.5. Variety Discrimination

He *et al.* [48] applied Vis/NIR diffuse reflectance spectra to discriminate apple cultivars. Spectra of three cultivars including 'Fuji', 'Red Delicious' and 'Copefrut Royal Gala' in the wavelength range of 400–960 nm were obtained, processed with moving average and compressed using PCA. Based on the loading plots, wavebands of 650–690 nm and 550–565 nm were identified as being sensitive to varieties and used as input of a BP-ANN model. A discriminant accuracy of 100% could be achieved, with a residual error of 9.94×10^{-5}. Moreover, wavelet transformation (WT) based on Daubechies 5 could reduce the size of variables to 4% [48]. Guo *et al.* [45] applied hyperspectral images in the wavelength range of 400–1000 nm for discrimination according to origins. Three CWs around 576, 678 and 971 nm were selected by PCA and texture analysis based on gray level co-occurrence matrix (GLCM) and used to build a SVM model, yielding a discriminant accuracy of 89.86% for the predicting sets. The excellent discriminant performance of CWs and wavebands indicated the bright future of online classification and instrument development. In the variety discrimination, spectroscopic and hyperspectral imaging techniques both show excellent performance with appropriate multivariate calibration techniques.

3.1.6. Bruise Detection

Luo *et al.* [76] selected characteristic wavelengths (CWs) in the 380–1000 nm range for bruise detection. Each CW was considered as an independent classifier for bruise/normal identification and evaluated with receiver operating characteristic (ROC) analysis. The performance of the model based on CWs was compared with the PLS-DA model based on the full spectrum. The accuracies of both methods could exceed 95%. Similar accuracies was obtained by Baranowski *et al.* [44], who detected early bruises on six apple cultivars using hyperspectral imaging in the Vis/NIR region (400–2500 nm) and thermal imaging of emitted radiation in mid-wavelength infrared range (MWIR, 3500–5000 nm). The whole spectral range (400–5000 nm) was found useful. Minimum noise fraction (MNF) could yield accuracy rates of 87%–97%, better than PCA. The performance of linear discriminant analysis LDA, SVM and SIMCA were compared and the best one to distinguish bruised and intact apples was LDA, with a total success rate of 95%, while the best result for distinguishing deep and shallow bruised areas was obtained by SVM, with a total success rate of 77%. Huang *et al.* [77,78] employed two hyperspectral imaging systems in the 400–1000 nm and 1000–2500 nm range, respectively. PCA was used and CWs were determined based on the weighting coefficients plot of the best PC images. An overall classification accuracy of 90% and 97% could be obtained by the two systems, respectively.

These research works proved the feasibility of bruise discrimination based on spectra or hyperspectral images in the whole wavelength as well as characteristic variables selected, indicating the potential of instrument development and online detection. Although good performances were obtained based on spectroscopic and hyperspectral imaging techniques for bruise detection in the NIR region, subtle bruises could not be easily detected. The surface morphology and skin coloration can significantly affect the performance of bruise detection. Thus a chemometric-based hyperspectral imaging system is a good choice and is more appropriate for online applications.

3.1.7. Pigment

Pigment content was proved to link with quality attributes such as SSC and firmness. Zude *et al.* [79] found strong correlations between the peak absorbance of chlorophyll at 680 nm and harvest date ($r = 0.59$), background color ($r = 0.74$) and the starch index SI ($r = 0.64$). It also had some correlation with firmness ($r = 0.48$) and SSC ($r = 0.46$). Rutkowski *et al.* [80] found the index of anthocyanin (NAI), calculated as $(I_{780}I_{570})/(I_{780} + I_{570})$ significantly correlated with fruit firmness ($r = 0.86$) and titratable acid ($r = 0.81$) in 'Golden Delicious' apples. They claimed that NAI was the most suitable index to assess apple maturity, whereas Kuckenberg *et al.* [81] claimed

NDVI better. Combined with further investigation, it could be concluded that the most suitable index representing maturity of apple fruit was variety-specific.

3.1.8. Other Parameters

There are also some other attributes affecting the quality of apple that can be measured spectroscopically. Pissard *et al.* [6] found that vitamin C varied greatly among different cultivars. An overall model could just provide a coarse prediction ($r = 0.89$).

Table 3. Overview of applications for measuring quality attributes in oranges.

Quality Attribute	Cultivar	Method	Spectral Mode	Spectral Range/nm	Calibration Method	r_p	RMSEP	Ref.
SSC or sugar content	GN	Spectroscopy	Reflectance	361–2488	PLS	0.93	0.59	[58]
	GN	Spectroscopy	Transmittance	1100–2500	PLS	0.90	0.46	[59]
	GN	Spectroscopy	Reflectance	820–950	LS-SVM	0.92	0.32	[62]
	Mixed	Spectroscopy	Reflectance	350–1800	PCA-BPNN	0.90	0.70	[63]
	Mixed	Spectroscopy	Reflectance	500–2300	PLS	0.91	0.74	[82]
				1100–2300		0.89	0.68	
	GN	Spectroscopy	Reflectance	350–1800	PCA-BPNN	0.90	0.68	[14]
	GN	Spectroscopy	Transmittance	350–1000	CARS-PLS	0.92	0.39%	[8]
	GN	Spectroscopy	Transmittance	465–1150	PLS	0.88	0.49%	[83]
	ST	Spectroscopy	Reflectance	400–1000	PCA-PLS	0.84	0.29 (SEP)	[15]
	HY					0.87	0.30 (SEP)	
	GN	Spectroscopy	Reflectance	450–1750	Spline-PLS	0.87	0.47	[18]
	GN	Spectroscopy	Reflectance	700–934	LS-SVM	0.85	0.41	[84]
Acidity	GN	Spectroscopy	Transmittance	1100–2500	PLS	0.64	0.70	[59]
						0.65 (pH)	0.13	
	Mixed	Spectroscopy	Reflectance	500–2300	PLS	0.83	0.17	[82]
						0.88 (pH)	0.15	
				1100–2300		0.77	0.19	
						0.81 (pH)	0.16	
Vitamin C	Mixed	Spectroscopy	Reflectance	1333–1835	PLS	0.96	0.039	[20,21]

GN: Gannan navel orange; HM: Hamlin orange; ST: Shatangju; HY: Huangyanbendizao.

3.2. Oranges

Rich in Vitamin C and other nutrients, orange is another widespread and popular fruit. Due to its short maturity period, the supply and demand contradiction is sharp. To extend the sales period, better storage and preservation methods are required as the internal quality of oranges declines during storage. If there wasn't a rapid and simple approach to monitor their inner attributes, lots of oranges would decay and lose value. Most of the studies on the quality assessment of oranges have focused on attributes affecting its taste, including sugar content, acidity and Vitamin C. A brief overview is presented in Table 3.

3.2.1. SSC or Sugar Content

Liu *et al.* [58] applied Vis/NIR diffuse reflectance spectroscopy to predict SSC of navel oranges. PLS models were built with 2nd derivative spectra in four different wavebands strongly correlated with SSC, *i.e.*, 361–2488 nm, 530–690 nm, 940–1420 nm and 1630–2488 nm. The model based on 940–1420 nm spectra provided a good result, with r_p of 0.90 and *RMSEP* of 0.75, quite close to that obtained by the model based on the full spectrum. The r_p of the PLS and PCR models was 0.93 and 0.61, respectively, with *RMSEP* of 0.59 and 0.69. The superiority of PLS to PCR was confirmed by Lu *et al.* [59], while Liu *et al.* [62] found that LS-SVM outperformed PLS. This result was in accordance with Sun *et al.* [84]. In the experiment of Liu *et al.* [18], the best performance was achieved by Spline-PLS. Cayuela and Weiland [82] compared 500–2300 nm and 1100–2300 nm spectra. The 600–750 nm spectra were excluded as they were strongly affected by skin chlorophyll, whose absorbance band corresponds to 680 nm. The r_{cv} yielded was 0.91 and 0.89 respectively, with *RMSEP* of 0.74 and 0.68. Shao *et al.* [15] also found the 970–990 nm waveband to be particularly important while the 750–800 nm waveband made rather small contributions.

Moreover, Liu *et al.* [63] extracted characteristic wavelengths (CWs) by PCA, and used them as input of an ANN, yielding an r_p of 0.90 and a *RMSEP* of 0.70. The result was quite similar with that reported by Liu *et al.* [14], who utilized Vis/NIR diffuse reflectance spectra in the 350–1800 nm range. PLS and back propagation neural network based on PCA (PCA-BPNN) were compared and the best result was achieved by the PCA-BPNN model combined with MSC, with r_p of 0.90 and *RMSEP* of 0.68°Brix. In the study of Sun *et al.* [8], several variable selection methods including competitive adaptive reweighted sampling (CARS), uninformative variables elimination (UVE) and SPA were compared. The best result was achieved by the CARS-PLS model, with r_p of 0.92 and *RMSEP* of 0.39%.

Xu *et al.* [83] investigated the influence of placement position. According to the different angles between incident light and the line composed by orange stem and pit, spectra were obtained at three different positions: vertical (90°), parallel (0°) and random. The best result was yielded by the model built with vertical spectra, with r_p of 0.88, and *RMSEP* of 0.49%. Sun *et al.* [84] investigated the effect of three different reference points *i.e.*, λ_{cv}(max), λ_{cv}(min) and λ_{cv}(median), which were identified by the coefficients of variation (CVs) at different wavelengths. The best result was achieved when λ_{cv}(max) was used.

Taken together these results indicate that SSC or sugar content in oranges could be excellently predicted using spectroscopy. Spectra in the NIR range were particularly important for the prediction while spectra in the visible region could enhance the accuracy. Meanwhile, with regard to measurement modes, transmittance has slightly higher predictive outcomes than reflectance and interactance. Reflectance is the easiest mode to obtain measurements due to the relatively high light levels. LS-SVM was proved slightly better than PLS and PCA.

3.2.2. Acidity

Lu *et al.* [59] employed Vis/NIR transmittance spectra to predict titratable acidity and available acidity (pH) in 'Gannan' navel oranges. PLS models outperformed PCR ones, but only coarse prediction results were obtained, with r_p of 0.64 and 0.65 and *RMSEP* of 0.70 and 0.13, respectively. Cayuela and Weiland [82] got slightly better results with PLS models based on reflectance spectra in the 1100–2300 nm range. For pH, r_{cv} was 0.81, while *RMSEP* was 0.16. For titratable acidity, r_{cv} was 0.77 while *RMSEP* was 0.19. Better results were obtained using spectra in the 500–2300 nm range, with r_{cv} of 0.88 and 0.83, respectively, indicating the possibility of acidity measurement using Vis/NIR spectroscopy. More efforts should be made to enhance the prediction performance, because total acidity prediction by NIRS has been considered difficult to achieve, due to the relatively low levels of organic acids found in oranges.

3.2.3. Vitamin C

Xia *et al.* [20,21] predicted vitamin C content in oranges using spectra in the 833–2500 nm wavelength range. Several preprocessing algorithms were compared, including constant offset elimination (COE), vector normalization (VN), MSC, first and second derivative and Daubechies 3 WT with different decomposing levels. Daubechies 3 WT at level 4 was proved to be the best. The optimal waveband for prediction was 1333–1835 nm, and a PLS model built with spectra in this waveband performed excellent, with r_p of 0.96 and *RMSECV* of 0.039 mg/g, indicating the feasibility of NIR spectroscopy for this application.

3.2.4. Variety Discrimination

Shao *et al.* [15] employed diffuse reflectance spectra in the 400–1000 nm region to discriminate four cultivars, including 'Shatangju', 'Huangyanbendizao', 'Gongju' and 'Huangdigan'. By using the predicted error sum of squares (PRESS) as an indicator, Daubechies 1WT with a decomposition length of 5 was chosen. Afterward, a BP-ANN model was established and yielded a discriminant accuracy of 100%, with a residual error of 8.27×10^{-5}. Cen *et al.* [39] got a result at the same level using reflectance spectra in the 325–1075 nm wavelength range. The combination of BP-ANN and PCA yielded an accuracy of 100%, with r_p of 0.998 and *RMSEP* of 0.18. Hao *et al.* [43] compared the discrimination performance of SIMCA and PLSDA based on diffuse reflectance spectra in the 350–1800 nm range. Both methods could reach a discriminant rate of 100% for all of the four cultivars. These results illustrated that orange varieties could be precisely classified using Vis/NIR spectroscopy technology.

3.2.5. Other Parameters

Deng *et al.* [85] found that reflectance spectrum at 988 nm significantly correlated with SSC ($r = 0.387$**), SSC/acid ratio ($r = 0.440$**) and vitamin C ($r = 0.309$*). Both SSC and vitamin C were positively correlated with second derivative reflectance spectrum at 943 nm, with r of 0.339* and 0.355*, respectively. Cayuela and Weiland [82] measured some other quality attributes of orange including maturity index, firmness, juiciness, and fruit weight. Good results were acquired for most of these parameters, with r_{cv} of 0.66–0.96 and *RPD* of 1.31–4.76.

3.3. Kiwifruit

Kiwifruit is a popular fruit with important medicinal and edible value. As a kind of climacteric variant fruit, the flavor and texture of kiwifruit change over time, so monitoring its internal quality rapidly and nondestructively is of great significance. Aside from indicators commonly used for assessing fruit quality, like sugar content, acidity and dry matter content, the firmness of kiwifruit could greatly affect its consumer acceptance. A number of investigations have been done. A brief review is presented in Table 4.

Table 4. Overview of applications for measuring quality attributes in kiwifruit.

Quality Attribute	Cultivar	Method	Spectral Mode	Spectral Range/nm	Calibration Method	r_p	*RMSEP*	Ref.
	AD	Imaging spectroscopy	Reflectance	650–1100	PLS	0.92	1.18	[2]
	AD	Spectroscopy	Interactance	800–1100	PLS	0.95	0.39	[86]
	HW	Spectroscopy	Transmittance	400–1000	PLS	0.93	0.26	[87]
SSC or sugar content	AC	Spectroscopy	Interactance	520–1100	PLS	0.96	0.80 (*SEP*)	[88]
	Various	Spectroscopy	Interactance	800–1000	PLS	0.97	0.32%	[89]
	YF	Spectroscopy	Interactance	300–1140	PLS	0.96	0.31% (*SEP*)	[90]
	HW	Spectroscopy	Reflectance	800–2500	PLS	Unknown	0.68 (*SEP*)	[91]
	AD	Spectroscopy	Reflectance	408–2492	PLS	0.99	0.49	[92]
Acidity	HW	Spectroscopy	Transmittance	400–1000	PLS	0.94	0.076	[87]
	AD	Spectroscopy	Reflectance	408–2492	PLS	0.95	0.28% (*SEP*)	[92]
	AD	Spectroscopy	Interactance	800–1100	PLS	0.87	7.0	[86]
Firmness	AD	Spectroscopy	Reflectance	408–2492	PLS	0.94	3.32	[92]
	HY	Spectroscopy	Reflectance	833–2500	PLS	0.85	1.89	[19]
	ZH	Spectroscopy	Reflectance	1000–2500	NAP-PLS	0.88	0.88	[27]
	AD	Spectroscopy	Interactance	800–1100	PLS	0.95	0.42%	[86]
Dry matter	Various	Spectroscopy	Interactance	800–1000	PLS	0.97	0.29%	[89]
	YF	Spectroscopy	Interactance	300–1140	PLS	0.98	0.24% (*SEP*)	[90]
	ZH	Spectroscopy	Reflectance	1000–2500	siPLS	0.90	0.53%	[34]

AD: *Actinidia deliciosa*; HW: Hayward; AC: *Actinidia chinensis*; YF: yellow-fleshed kiwifruit; ZH: Zhonghua; HY: Huayou.

3.3.1. SSC

McGlone and Kawano [86] employed interactance spectra to predict the SSC of kiwifruit in the 800–1100 nm wavelength range. A PLS model built with 2nd derivative spectra showed excellent performance, with r of 0.95 and *RMSEP* of 0.39°Brix. The result was much better than those reported by Martinsen and Schaare [2] and Moghimi *et al.* [87], who used reflectance and transmittance spectra, respectively. The superiority of interactance spectra was confirmed by Schaare and Fraser [88]. Lee *et al.* [92] predicted SSC with a broad spectral range (408–2492 nm), and an excellent result was observed (r_p = 0.99; *SEP* = 0.49°Brix). McGlone *et al.* [89] found the 800–1000 nm waveband performed well. They also proved the feasibility of using data obtained from unripe kiwifruit to predict SSC of ripe fruit, with r of 0.96 and *RMSEP* of 0.39%. A similar investigation was done by McGlone *et al.* [90], who proved that the SSC prediction model based on post-storage spectra was better than that based on harvest-time spectra. A possible explanation was that Vis/NIR spectroscopy was better at predicting total carbohydrate concentration, which consists of starch and soluble sugar in about the same amounts at harvest time but mainly of soluble sugar after storage. This was supported by the observation that predicting SSC of post-storage kiwifruits by using the model built with harvest-time spectra could yield more accurate results (*SEP* = ±0.38%). Arazuri *et al.* [91] investigated the influence of temperature. Reflectance spectra in the 800–2500 nm range were obtained at three different sample temperatures (0.5, 10 and 20 °C) and the best performance was achieved using spectra obtained under 0.5 °C, with *SEP* of 0.68.

According to these studies, all three spectral modes provide good accurate estimates of SSC. However, interactance spectra were better for predicting SSC in kiwifruits, and transmittance mode was better than reflectance mode. A possible explanation was that interactance spectra were less susceptible to specular reflections, which was probably the source of larger errors. Spectra in the 800–1000 nm range were very important, while a slightly better result could be achieved by using spectra in the whole Vis/NIR range. The ripening stage of kiwifruit affects greatly the accuracy of SSC prediction and accurate prediction was based on ripe kiwifruit.

3.3.2. Acidity

Moghimi *et al.* [87] predicted the acidity in kiwifruits with transmittance spectra in the 400–1000 nm range. A PLS model based on spectra processed with SNV, median filter and 1st derivative could yield a result with r_p of 0.94 and *RMSEP* of 0.076, close to the values obtained by Lee *et al.* [92], who used spectra in the 408–2492 nm range and predicted acidity with r_p of 0.95 and *SEP* of 0.28%. The feasibility of Vis/NIR spectroscopy for predicting acidity in kiwifruit was thus proved.

3.3.3. Firmness

McGlone and Kawano [86] employed interactance spectra in the 800–1100 nm wavelength range to assess the firmness of kiwifruit. The performance of the PLS model was barely satisfactory, with r of 0.81 and $RMSEP$ of 7.8 N. A better result could be obtained when the samples were sorted in terms of origins and sizes (r = 0.87; $RMSEP$ = 7.0 N). However the model performed poorly against independent data sets, indicating the existence of secondary correlations due to fruit characteristics, which were not directly related to fruit firmness. Liu *et al.* [19] used diffuse reflectance spectra in the 833–2500 nm wavelength range. Both first and second derivative could improve the prediction accuracy, while SNV and MSC could not. The PLS model based on first derivative spectra could yield the optimal result, with r_p of 0.85 and $RMSEP$ of 1.89. A better result was reported by Lee *et al.* [92], who used spectra in the 408–2492 nm range. Flesh firmness was predicted with r_p of 0.94 and SEP of 3.32 N. Lv *et al.* [25] optimized modeling wavelengths and decreased the number of principal components (PCs) with net analyte processing (NAP). Through NAP-PLS, an optimal model was established with five PCs selected in five wavebands (1862–1927, 2164–2198, 1605–1653, 1293–1429 and 1511–1600 nm) and predicted firmness with r_p of 0.88 and $RMSEP$ of 0.88. In all this research, the firmness of kiwifruit could only be coarsely predicted, suggesting that perhaps there is too little pectin in the kiwifruit for NIRS to pick up, so further research should be done to enhance the prediction accuracy.

3.3.4. Dry Matter (DM)

McGlone and Kawano [86] employed a PLS model based on interactance spectra in a narrow waveband (800–1100 nm) to predict the DM of 'Actinidia deliciosa', yielding an r_p of 0.95 and $RMSEP$ of 0.42%. McGlone *et al.* [89] applied interactance spectra processed with S-G smoothing and area normalization sequentially. DM of unripe and ripe fruits could be accurately predicted based on 800–1000 nm spectra. The PLS model based on ripe fruits yielded an r of 0.97 and $RMSEP$ of 0.29%. DM of the ripe kiwifruits could also be predicted with data obtained when the fruit is unripe, with r_p of 0.97 and $RMSEP$ of 0.39%. Lue *et al.* [34] investigated the potential of long-term DM prediction. They obtained spectra and dry matter contents of some unripe and ripe kiwifruits and extracted CWs with synergy interval partial least square (siPLS). A model based on spectra of unripe kiwifruits and DM of ripe fruits was built, with r_p of 0.90 and $RMSEP$ of 0.53%. McGlone *et al.* [90] and Feng *et al.* [93] also proved that DM in kiwifruit could be accurately predicted. In conclusion, these studies proved that dry matter in kiwifruits could be excellently predicted with NIR spectroscopy, and the NIR method gives a good predictive relationship with SCC by finding spectral information that is independent of the DM for kiwifruit. Interactance spectra were most commonly employed.

3.3.5. Other Parameters

Schaare and Fraser [88] applied Vis/NIR spectroscopy in the reflectance, interactance and transmittance mode to measure density and flesh color of kiwifruit. Best performances were all achieved by a PLS model built with interactance spectra in the 520–1100 nm wavelength range. Density and flesh hue angle were predicted with r of 0.86 and 0.91 and SEP of 3.6kg/m^3 and 1.6°, respectively. Tavakolian *et al.* [41] applied reflectance spectra in the 1130–2220 nm range for the classification of kiwifruit varieties with different post-harvest date. PCA was performed and showed that the first three PCs could explain 99% of the variance. SIMCA was applied based on three CWs at 1190, 1450 and 1940 nm. The total classification accuracy was 92.3%.

3.4. Peaches

Peaches are another kind of tasty and nutritious fruit. Like kiwifruit, the firmness of peaches changes greatly over time, thus firmness, along with sugar content and acidity, by which the flavor of peaches are mainly determined, attract the interest of most researchers. A brief review is presented in Table 5.

Table 5. Overview of applications for measuring quality attributes in peach.

Quality Attribute	Cultivar	Method	Spectral Mode	Spectral Range/nm	Calibration Method	r_p	SEP	Ref.
SSC or sugar content	Various	Spectroscopy	Transmittance	800–1050	MLR	0.20–0.91	0.49%–1.63%	[51]
	Unknown	Spectroscopy	Transmittance	1000–2500	Stepwise MLR	0.53	Unknown	[94]
	Mixed	Spectroscopy	Reflectance	325–1075	ICA-LS-SVM	0.95	0.42 (RMSEP)	[36]
	Unknown	Multispectral scattering		632, 650, 670, 900	MLR	0.97	0.69	[3]
	SH	Spectroscopy	Interactance	870, 878, 889, 906	MLR	0.97	0.50	[95]
	Mixed	Spectroscopy	Reflectance	1279–2331	PLS	0.96	0.57	[72]
	DB	Spectroscopy	Reflectance	800–2500	PLS	0.94 (r_{cv})	0.57 (RMSECV)	[96]
Acidity	Mixed	Spectroscopy	Reflectance	928–2331	PLS	0.95	0.13	[96]
	Mixed	Spectroscopy	Reflectance	325–1075	ICA-LS-SVM	0.96	0.047 (RMSEP)	[36]
Firmness	Unknown	Multispectral scattering		670, 780, 850, 900	MLR	0.95	1.56	[3]
	Mixed	Multispectral scattering		680, 880, 905, 940	MLR	0.82	18.55	[97]
	RH	Hyperspectral scattering		500–1000	MLR	0.88	14.2	[98]
	CS	Hyperspectral scattering				0.76	19.1	
	White peach	Spectroscopy	Reflectance	800–2500	PLS	0.89	5.42 (RMSEP)	[99]

SH: Shimizu Hakuto; DB: Dabaitao; RH: Red Haven; CS: Coral Star.

3.4.1. SSC or Sugar Content

Peiris *et al.* [51] and Jiang *et al.* [94] applied transmittance spectroscopy to predict SSC. Neither of the results was satisfactory. Kawano *et al.* [95] got better result using interactance spectra. In 2nd derivative spectra, clear differences among peaches with different Brix values had been observed at the wavelength of 906 nm, which was assigned to sucrose. The best prediction result was yielded by the linear regression model built with spectra at 870, 878, 889 and 906 nm, with r of 0.97 and *SEP* of 0.50°Brix. This result was in the same level as those of Liu *et al.* [72] and Ma *et al.* [96], both of which employed reflectance spectra in the NIR range. The former one reported the first derivative spectra yielded the best result with r_p of 0.96 and *SEP* of 0.54, while the latter one found the original spectra more suitable, with r_{cv} of 0.94 and *RMSECV* of 0.57. Shao *et al.* [36] applied independent component analysis (ICA) and latent variables analysis (LVA) to CWs from reflectance spectra in the wavelength range of 325–1075 nm. PLS and LS-SVM models were built with the CWs selected. LS-SVM always performed better than PLS. The optimal result was achieved by the ICA-LS-SVM model, with r_p of 0.95 and *RMSEP* of 0.42. Liu *et al.* [3] obtained spectral scattering profiles at wavelengths of 632, 650, 670, 780, 850 and 900 nm and fitted them with Lorentzian distribution with three parameters. MLR models were established to relate SSC with Lorentzian parameters based on different number of wavelengths respectively. The optimal performance was achieved by the combination of spectral images at 632, 650, 670 and 900 nm, with r of 0.97 and *SEP* of 0.69°Brix.

Experiments showed that transmittance spectra contain little effective information related with SSC in peaches. We assumed that the incident radiation couldn't permeate the peach core, which could be verified by the fact that when a similar system was utilized to assess SSC in apples, a much more accurate result was obtained [94]. Models built with selected CWs could yield results as good as those based on reflectance or interactance spectroscopy, indicating the feasibility of online detection.

3.4.2. Acidity

Liu *et al.* [72] used FT-NIR reflectance spectroscopy in the 928–2331 nm waveband to measure the valid acidity (pH) in peach. The PLS model could obtain a result with r_p of 0.95 and *SEP* of 0.13. Shao *et al.* [36] got a similar result ($r_p = 0.96$; *RMSEP* = 0.047) from a LS-SVM model built with CWs selected by ICA. The feasibility of spectroscopy was thus verified, but the low acid content in the peaches might cause relative insensitivity for prediction valid of acidity.

3.4.3. Firmness

Lu and Peng [97] predicted peach firmness using spectral scattering profiles at wavelengths of 680, 880, 905 and 940 nm. Soft peaches were found to have broader scattering profiles than firm ones, especially at 680 nm. A Lorentzian distribution function with three parameters was used to fit the scattering profiles with a mean r^2

of above 0.998. MLR was employed to relate Lorentzian parameters with firmness. When models were built with peaches from different orchards, respectively, and the optimal result was acquired with r_p of 0.87 and *SEP* of 14.57 N. When a model was established with samples from two orchards, a lower r_p (0.82) and a higher *SEP* (18.55 N) was obtained. Liu *et al.* [3] got a better result based on the combination of scattering profiles at 670, 780, 850 and 900 nm (r = 0.95; *SEP* = 1.56 N). Moreover, Lu and Peng [98] utilized hyperspectral scattering profiles of 500–1000 nm to assess the firmness of two cultivars: 'Red Haven' and 'Coral Star'. The profiles were fitted using a Lorentzian distribution function with two parameters, with a mean r^2 above 0.99. Then MLR models were established to relate Lorentzian parameters and their contributions at different wavelengths with peach fruit firmness. The highest correlation among all individual wavelengths was found at the wavelength 677 nm, which corresponds to chlorophyll absorption. When two Lorentzian parameters a and b (a represented the peak scattering value while b the full scattering width at one half of the peak value) were used as independent variables, optimal results were achieved for the two cultivars with r of 0.88 and 0.76 respectively.

Fu *et al.* [99] investigated the anisotropy of firmness and spectra, regarding to measuring spots at different latitudes and longitudes. Both spectral absorbance and firmness of peaches were proved to be affected by longitudes and latitudes. The collaboration of spectra from different latitude and longitude, and proper pretreatment methods like scattering correction or derivative could improve the prediction. The best performance was achieved by the holistic model built with spectra processed by MSC (r = 0.89; *RMSEP* = 5.42 N).

These studies above proved that spectral scattering and hyperspectral scattering were feasible for predicting firmness of peach. NIRS does not provide quantitative information on light scattering in peach, which leads to its capability for predicting structurally firmness limited or difficult to justify, and both chlorophyll and water status have an important effect on firmness. More efforts should be made to enhance the prediction accuracy, such as by utilizing hyperspectral scattering.

3.4.4. Variety Discrimination

Wu *et al.* [38] applied reflectance spectra in the 401–1000 nm wavebands for the variety discrimination of three peach cultivars *i.e.*, 'Mengyin', 'Fenghua' and 'Jinhua'. Eight PCs were selected by PCA to build a SDA model, which could yield a discrimination accuracy of 100%, better than that of the PLS model. A similar result was obtained by Li *et al.* [100], who used spectra in the same wavelength range to classify the cultivars 'Milu', 'Hongxianjiu' and 'Dabaitao'. Multiple discriminant analysis (MDA) based on the first eight PCs yielded a recognition rate of 100%. Fu *et al.* [99] compared the discriminant accuracy of discriminant analysis (DA), SIMCA and discriminant PLS (DPLS) using diffuse reflectance spectra in the region of 800–2500 nm. The discriminant accuracy of both DA and SIMCA were above 92%, while that of DPLS was slightly better, reaching 95%. To conclude, Vis/NIR spectroscopy was proved feasible for peach variety discrimination. The full wavelength was useful while spectra in the visible range were essential.

3.4.5. Other Parameters

Zwiggelaar *et al.* [101] combined spectroscopy and machine vision to detect bruises on peaches. After the samples were bruised, they were divided into three groups, one of which was left at room temperature while other two were put in a cold conditions for one week and two weeks, respectively, to inhibit ripening. Reflectance spectra at bruised and un-bruised areas and spectral images at 930 and 970 nm were obtained. Results indicate a success rate of only 65%, perhaps indicating a wrong wavelength choice. More studies should be done to improve the success rate. Takano *et al.* [102] evaluated polyphenols in peach using NIR spectroscopy in the 1100–2500 nm wavelength range. A coarse prediction could be achieved, with r of 0.80 and *SEP* of 14.7 mg/100 g. By multiple regression analysis, spectra at 1720 nm were proved to have the highest correlation with polyphenol content.

To classify peaches according to their maturity, Lleó *et al.* [103] compared two multispectral classification methods based on red images (R) and a combination of R and infrared images (R/IR) respectively. The spectral images were obtained by three CCD cameras (450, 675 and 800 nm). The R/IR method performed better as it eliminated the effect of fruit shape on light reflectance. Reflectance at 680 nm (chlorophyll absorption peak) increased, while firmness decreased during the ripening process.

3.5. Strawberries

With their attractive appearance, luscious taste and rich nutritional value, strawberryies have won the affection of people all over the world. Due to their soft tissues and high moisture content, strawberries are quite perishable, so more effective monitoring and detection methods are required. Being the sugar content, acidity and firmness the most important attributes that could affect quality and price, these draw the attention of most researchers.

3.5.1. SSC

ElMasry *et al.* [52] used hyperspectral imaging in the Vis/NIR region (400–1000 nm) to predict the SSC in strawberry. Some samples were kept in room temperature for several days while others were kept under 5 °C conditions to guarantee wide variations of internal properties. The PLS model built with spectra processed mean-centering and automatic baseline correction performed well, with r_p of 0.85 and *SEP* of 0.184. A MLR model established with CWs selected by β-coefficients of PLS achieved a close result, with r_p of 0.80 and SEP of 0.211. A similar result was obtained by Sanchez et al. [11], who built a modified PLS model with reflectance spectra in the 1600–2400 nm wavelength range ($r = 0.89$; $RPD = 2.15$). Nishizawa et al. [104] reported better results using spectra in the 700–925 nm range ($r = 0.93$; $SEP = 0.9\%$). Pretreatment methods including 2nd derivative and MSC were proved useless for improving the accuracy. The best result was reported by Guo et al. [35], who employed a synergy interval PLS (siPLS) model based on 1st derivative spectra in the 833–2631 nm wavelength range ($r_p = 0.97$; $RMSEP = 0.29$).

Shi et al. [37] attempted to simplify the prediction. CWs were selected using backward interval PLS (BiPLS) combined with simulated annealing algorithm (SAA). Spectra in the 1000–2500 nm wavelength range were divided into 21 subsets and characteristic subsets were determined by BiPLS. Then SAA was applied to select CWs in these informative regions. Finally seven CWs, in the 1135–1322 nm range were selected to build a MLR model. The predictive performance of this BiPLS-SAA-MLR model was better than those of the PLS and BiPLS models, with r_p of 0.94 and $RMSEP$ of 0.43. From these research works, the feasibility of the spectroscopic method for SSC prediction in strawberries were verified. Models built with CWs could yield very good results, indicating the potential for instrument development and online detection.

3.5.2. Acidity

Shao et al. [22] applied reflectance spectroscopy in the 400–1000 nm range to predict the acidity of strawberries. After some defective spectra were eliminated using PCA, a PLS model was built, with r_p of 0.92, SEP of 0.027 and $RMSEP$ of 0.026. After wavelet transform (WT) was applied to compress the spectral data, a PLS model was established with r_p of 0.86, and $RMSEP$ of 0.026. A similar result was observed by ElMasry et al. [52], who applied hyperspectral imaging in the 400–1000 nm range to assess the pH. The PLS model yielded an r_p of 0.87 and SEP of 0.13, while the MLR model achieved better results, with r_p of 0.94 and SEP of 0.091. Sanchez et al. [11] applied reflectance spectra in the 1600–2400 nm wavelength range. The model was built with modified PLS regression. For the prediction of titratable acidity, an acceptable result was obtained, with r of 0.73 and RPD of 1.43, but the predictive performance for pH was not so good ($r = 0.48$), probably indicating an unsuitable wavelength choice. Although the feasibility of the spectroscopy method for predicting acidity in strawberries were confirmed and the ripening stage did not seriously affect pH prediction, further investigations should be done to figure this out.

3.5.3. Firmness

Tallada *et al.* [66] assessed firmness in strawberries using hyperspectral images in the 650–1000 nm range. A stepwise MLR model based on three optimal wavelengths of 685, 865 and 985 nm could give a result with r_p of 0.79 and *SEP* of 0.35 MPa. The performance of a modified PLS model based on reflectance spectra in the 1600–2400 nm range was not satisfactory either, with r of 0.66 and *RPD* of 1.35 [11]. As well as SSC and acidity, firmness prediction is feasible, which is confirmed by the NIRS. Further research should be done to enhance the prediction accuracy.

3.5.4. Variety Discrimination

Niu *et al.* [50] used spectra ranging from 1100 to 2200 nm for the classification of varieties 'Tianbao', 'Fengxiang' and 'Mingxing'. The performance of BP-ANN, LS-SVM and discriminant analysis (DA) was compared and the best result was achieved by the BP-ANN model, with a total discrimination rate of 97.14%. Yan *et al.* [105] tried to simplify the discrimination. Spectra of three varieties were obtained in the 350–2500 nm range. 2nd derivative combined with SNV and moving average was ascertained as the best data pretreatment method. CWs were determined as 548–562 nm by the correlation coefficient and threshold value method. The optimal predictive performance was achieved by the PLS-ANN model, with r_p of 0.97 and *RMSEP* of 0.46, slightly better than those of the PLS and PCR models. However, the discriminant accuracy reported by Sanchez *et al.* [11], who built a PLS-DA model with reflectance spectra in the 1600–2400 nm wavelength range was only 63%, raising some uncertainty as to the feasibility of Vis/NIR spectroscopy for discriminating strawberry varieties.

3.5.5. Other Parameters

Nagata *et al.* [106] investigated the feasibility of NIR hyperspectral imaging in the wavelength range of 650–1000 nm for detecting compression bruises on strawberries. Hyperspectral images of strawberries subjected to different levels of bruising force were obtained during 0–4 days after bruising. Three discriminant methods including linear discriminant analysis (LDA), normalized difference and artificial neural network (ANN) performed equally well. Through stepwise LDA, two optimal wavelengths *i.e.*, 825 and 980 nm were identified and the classification efficiency for bruised and non-bruised pixels could reach 86.5% and 99.7%, respectively. Detected bruises were found to significantly decrease along storage time.

Sanchez *et al.* [11] studied bulk skin or external color values of strawberries including L*, a* and b*. Chroma and hue angle could be calculated with these parameters. The prediction results for L*, a* and chroma were acceptable, with r of

0.77 and *RPD* of 1.56 for all the three parameters, while for b* and hue angle, the result was not good ($r \leqslant 0.44$), which suggested that NIRS prediction for b* and h* was not feasible.

3.6. Grapes

Not only a tasty fruit but also an ingredient for brewing wine, grape is an important economic crop cultivated all over the world. It is the sugar content and acidity in grapes that determine their flavor and indicate the ripeness, thus attracting the attention of researchers. It has been proved that grapes contain some precious nutritional ingredients including anthocyanins and polyphenols, so some work has also been done for measuring these ingredients.

3.6.1. SSC or Sugar Content

Fernandez-Novales *et al.* [107] observed excellent results using transmittance spectra in the 700–1060 nm wavelength range. The PLS model based on spectra processed with Norris first derivative performed excellently, with r_{cv} of 0.99 and *RMSECV* of 0.46. Herrera *et al.* [108] evaluated the feasibility of diffuse transmittance and interactance spectra for SSC prediction. Two spectral regions *i.e.*, 750–1100 and 650–1100 nm, were used to establish PLS models to predict SSC in three grape cultivars: 'Cabernet Sauvignon', 'Chardonnay' and 'Carmenere'. The optimal performance for all cultivars was achieved by models based on 650–1100 nm spectra, with r_p above 0.90 and *RMSEP* lower than 1.2. Transmittance spectra performed slightly better compared to interactance ones. Pretreatment methods had no significant effect, as confirmed by Larrain *et al.* [109], who applied reflectance spectra in the 640–1100 nm range to predict sugar content. The prediction performance of the PLS models they established was excellent, with r_p of 0.93–0.96 and *RMSEP* of 1.01–1.27 for different cultivars.

Wu *et al.* [110] simplified the prediction with BP-ANN. Vis/NIR diffuse reflectance spectra in the 400–1000 nm waveband were obtained. Three principal factors were identified by PLS based on the reliabilities, then the scores of the three selected principal factors were used as the input of a three-layer BP-ANN model. Its prediction accuracy outperformed the PLS model, with r of 0.95 and *RMSEP* of 0.11. Fernandez-Novales *et al.* [29] selected four CWs including 909, 951, 961 and 975 nm by the loading weights of latent variables and used them to build a MLR model for sugar content prediction. A satisfactory result could be obtained, with r of 0.96, *SEP* of 20.0 g/L and *RMSEP* of 20.5 g/L. The CWs they selected were not the same as those identified by Cao *et al.* [30], who applied Vis/NIR reflectance spectra in the 400–1000 nm wavelength range. A genetic algorithm was executed and five wavelengths *i.e.*, 418, 525, 556, 633 and 643 nm were identified as CWs. A PLS model was built for prediction, with r_p of 0.91 and *RMSEP* of 0.93. Omar [69] built a MLR model with spectra of 605, 729, 830, 910 and 950 nm ($r = 0.97$, *RMSE* = 0.18°Brix).

Taking all these studies together, the feasibility of Vis/NIR spectroscopy was verified for SSC prediction. NIRS was sensitive to sugar content changes during the different ripening stages. Spectra in the 400–1000 nm range could yield excellent prediction results and transmittance spectra performed slightly better than interactance spectra.

3.6.2. Acidity

Larrain *et al.* [109] proved that the pH of different grape cultivars could be predicted with r_p of 0.75–0.89 and *RMSEP* of 0.088–0.16 using reflectance spectra ranging from 640 to 1100 nm. Cao *et al.* [30] obtained spectra of samples of three cultivars in the 400–1000 nm wavelength range. The performance of the GA-LS-SVM model was better than that of the PLS model, with r_p of 0.98 and *RMSEP* of 0.13. A very close result was obtained by Omar [69], who found wavebands of 922–923 and 990–995 nm important. Their best result was achieved by a MLR model based on spectra of 605, 923 and 990 nm ($r = 0.87$; *RMSE* = 0.11).

However, Fernandez-Novales *et al.* [107] did a similar prediction for pH and tartaric acid using NIR transmittance spectra in the 700–1060 nm wavelength range and obtained less accurate results ($r_{cv} = 0.52, 0.41$; *RMSECV* = 0.22, 2.02, respectively), indicating some uncertain difficulties involved in the prediction for this attribute in grapes, although there is still a high correlation with the reference pH sensor.

3.6.3. Anthocyanin

Larrain *et al.* [109] evaluated anthocyanin concentration in different grape cultivars using spectra ranging from 640 to 1100 nm. PLS models yielded acceptable results ($r_p = 0.79–0.83$) for most cultivars, except for Pinot Noir ($r_p = 0.63$). This uncertainty was confirmed by Kemps *et al.* [68], who did similar research with reflectance spectra in the 320–1660 nm range. Prediction wasn't feasible except for 'Syrah', in which anthocyanins could be predicted with r_p of 0.8, making the method questionable.

3.6.4. Variety Discrimination

Cao *et al.* [30] applied Vis/NIR reflectance spectroscopy in the 400–1000 nm wavelength range for the discrimination of three grape cultivars, namely 'Manaizi', 'Mulage' and 'Heiti'. Firstly GA was applied to select CWs and four wavelengths *i.e.*, 636, 649, 693 and 732 nm were identified. Then a LS-SVM model was built, which could reach a total discrimination accuracy of 96.6%. For individual cultivars, the accuracy rate was 93.9% for 'Manaizi', 97.6% for 'Mulage' and 100% for 'Heiti'. In another study [40], the combination of BP-ANN and PCA were applied. PCA was first applied for cluster analysis. The spectra of 'Heiti' were found significantly different from the other two cultivars especially in the 520–640 nm wavelength range. Then a BP-ANN model was established by using the first 10 PCs to discriminate the other two cultivars. A total discriminant accuracy of 98.3% could be achieved, slightly better than that yielded by SIMCA, which was 96.6%. A simplified BP-ANN model built with CWs of 452, 493, 542 and 668 nm could yield a very close accuracy rate of 97.4%, indicating the feasibility of online detection.

3.6.5. Other Parameters

Fernandez-Novales *et al.* [107] estimated the maturity of grapes using a maturity index, calculated by dividing reducing sugar content by titratable acid content. This index could be predicted with r_{cv} of 0.77 and *RMSECV* of 10.2. Considering the low accuracy obtained in the prediction of titratable acid, this result was inspiring and indicated a possibility of classifying grapes based on this index. Kemps *et al.* [68] reported that the concentration of polyphenols could not be predicted in any of the cultivars they used.

3.7. Jujube

Having been a kind of popular fruit in China for thousands of years, jujube is increasingly valued worldwide for its rich nutrients. Some studies in which spectroscopy and hyperspectral imaging technology are employed have been carried out for quality assessment of jujubes in recent years.

Zhang *et al.* [64] applied reflectance spectra in the 400–2400 nm range to predict SSC in three jujube cultivars. PCA was used on the spectra processed with smoothing and MSC. Six principal components were selected and employed as input of the BP-ANN model, which could predict SSC with a relative deviation lower than 10%. Wang *et al.* [46] compared reflectance, interactance and transmission spectra in the 310–1100 nm wavelength range for predicting SSC. Interactance spectra were proved to be the best choice, which yielded prediction results with r_p of 0.74–0.91 and *RMSEP* of 2.0–3.2°Brix. The optimal performance was achieved by the model based on 2nd derivative spectra, with r_p of 0.91. The fruit stone of jujubes affects the spectral characteristics of light reflected from or transmitted through the jujube. If each jujube fruit lacks a hard stone, transmission spectral measurements are effective at revealing the SSC. Zhang *et al.* [111] selected CWs for SSC prediction from reflectance spectra in the 350–2500 nm waveband using SPA and stepwise regression analysis (SRA). Wavelengths at 1374 nm and 1718 nm were identified by both methods, suggesting their importance. The best result was acquired by the PLS model based on whole spectra, with r_p of 0.89 and *RMSEP* of 1.09. The LS-SVM model combined with SPA could yielded an acceptable result with r_p of 0.80 and *RMSEP* of 1.40, better than that of the LS-SVM based on whole spectra. He *et al.* [65] employed hyperspectral images in the 900–1700 nm wavelength range for the prediction of SSC in jujubes. Five characteristic wavelengths were identified by PCA. The BP-ANN model yielded an r_p of 0.90 and *RMSEP* of 1.98.

Wang *et al.* [112] tested the applicability of reflectance, interactance and transmittance spectroscopy in the 310–2150 nm wavelength range for detecting internal insect infestations in jujubes. Spectra were divided into three wavebands *i.e.*, 310–1000 nm (VSWNIR), 1000–2150 nm (LWNIR), and 310–2150 nm (Vis/NIR). The highest discriminant rates obtained were 90% for reflectance, 97% for transmittance and 100% for interactance. In the VSWNIR region, transmittance spectra yielded better performance while in the LWNIR region, interactance spectra were the most feasible. This was probably because light in the VSNIR range could transmit through the fruit core more effectively than that in the LWNIR range, leading to increased reflectance and decreased transmittance. Further, Wang *et al.* [47] compared the ability of the three modes of spectra for detecting internal insect infestation with different damage levels. Discriminant functions based on CWs were derived based on stepwise discriminant analysis (SDA). Result showed that reflectance and interactance spectra in the VSWNIR region could discriminate severely infested jujubes from slightly damaged ones the best. Wang *et al.* [113] applied hyperspectral imaging in the 400–720 nm wavelength range to detect external insect infestation. Three CWs *i.e.*, 690, 650 and 500 nm, which corresponded to chlorophyll a, chlorophyll b and carotenoids respectively, were selected by SDA. Over 98.0% of intact jujubes and 94.0% of insect-damaged jujubes could be correctly discriminated, achieving an overall discriminant accuracy of 97.0%. Both the internal and external infestations could be identified with an accuracy of above 97%, indicating the feasibility of Vis/NIR method.

Zhang *et al.* [111] applied NIR reflectance spectroscopy for the detection of subtle bruises on jujubes. Spectra of 350–2500 nm were acquired and processed with MSC. Nine wavelengths *i.e.*, 1869, 2128, 1430, 827, 359, 2477, 1357, 1643 and 762 nm were identified as CWs using SPA. Then four principal components (PCs) were identified from the CWs using PCA and performed as input to establish a LS-SVM model. This MSC-SPA-PCA-LS-SVM model yielded a discriminant accuracy of 100%.

3.8. Bananas

Tarkosova and Copikova [114] applied NIR spectroscopy in the 1100–2500 nm wavelength range to assess sugar content in bananas. Modified PLS models were established for prediction. The total sugar content could be predicted with r_p of 0.99 and *SEP* of 0.80%, for sucrose, glucose and fructose, they could be predicted with r_p of above 0.97 and *SEP* of 0.16%–0.78%.

Jaiswal *et al.* [53] predicted DM, pH, SSC and acid-Brix ratio (ABR) in bananas using transmittance spectra in the 299–1100 nm wavelength range. SSC could be best predicted by a PLS model built with original spectra in the 955–982 nm waveband (r_p = 0.81) and pH could be best predicted by PLS model built with spectra processed with baseline correction in the 1009–1036 nm range (r_p = 0.83). Though a higher r_c could be obtained by MLR, there was a big gap between r_c and r_p, indicating unstability. For DM, the best result was achieved by the MLR model built with original spectra in the 1063–1089 nm wavelength range (r_p = 0.83). As to ABR, the PLS model built with spectra processed by MSC combined with baseline in the 955–982 nm wavelength range yielded the best result (r_p = 0.78).

Davey *et al.* [67] applied Vis/NIR reflectance spectroscopy in the 367–2388 nm wavelength range to measure total carotenoids, α-carotene, β-carotene, c-carotene and lutein in banana. PLS models were built based on 1st S-G derivative spectra. Results showed that total carotenoids and β-carotene could be measured accurately (r_p = 0.98, 0.96; *RPD* = 3.34, 2.74, respectively). For α-carotene and c-carotene, results were acceptable with r_p of 0.91 and 0.90 and *RPD* of 1.68 and 1.96, respectively. However, for lutein the result was not satisfactory, with r_p of 0.75 and *RPD* of 1.16. Considering that 90% of the carotenoids in bananas were α-carotene and β-carotene, it was feasible to measure carotenoids in banana with Vis/NIR spectroscopy.

Subedi and Walsh [115] measured dry matter DM and SSC in banana mesocarps with transmittance spectroscopy in the 500–1050 nm wavelength range. The result obtained for DM was not good, probably due to the thickness of the peel. For SSC, excellent results were obtained from ripening and ripen banana mesocarps (r_{cv} > 0.93; *RMSECV* < 0.80%). However, for green and over-ripe bananas prediction results were not satisfactory, indicating that mesocarp SSC was highly correlated with peel color.

3.9. Mangos

Jha *et al.* [116] applied reflectance spectroscopy to measure SSC and pH in seven mango cultivars. The optimal results were obtained by PLS models based on 2nd derivative spectra in the 1600–1799 nm range, with r_p of 0.76 and 0.70 and *SEP* of 3.23 and 0.72, respectively. Although MLR models yielded higher r_c, the gap between calibration and prediction indicated instability. Their results were inferior to those reported by Schmilovitch *et al.* [117] who applied NIR reflectance spectroscopy to predict firmness, SSC, acidity and storage period of mangos. Spectra were acquired in the 1200–2400 nm wavelength range. Best performances for predicting firmness, SSC, acidity were all achieved by MLR models built with 2nd derivative absorbance spectra, with r_p of 0.91, 0.96, 0.78 and 0.97 and *SEP* of 17.14, 1.223, 0.161 and 37.03, respectively. The result for predicting acidity was not satisfactory, probably due to the low acid content in the samples. Yu *et al.* [71] applied reflectance spectra in the 400–1075 nm wavelength range to predict sugar content and valid acidity in mango fruit. Eighteen PCs were extracted by PLS and employed as inputs of the GA-BPNN to predict sugar content and 17 PCs for valid acidity. The PLS-GA-BP models yielded good predictive results, with r of 0.85 and 0.84 and *SEP* of 0.61 and 0.11, respectively, better than those obtained by the PLS-BPNN models.

Saranwong *et al.* [118] found DM and starch contents increased significantly during ripening, while no obvious differences in individual sugars and fruit density were observed. Interactance spectra of unripe mangos in the 700–1100 nm wavelength range were obtained. For DM, the optimal predictive result was achieved by the MLR model built with 2nd derivative spectra after MSC-treated. Wavelengths of 914, 882, 826 and 954 nm were used for calibration. The r was 0.96 and *SEP* was 0.41%. For starch content, a PLS model based on 2nd derivative spectra in the 850–1000 nm range yielded the best result, with r of 0.93 and *SEP* of 1.71%. Further, DM and starch contents in unripe mangos could be used to predict SSC in ripe ones. The best prediction result was achieved by MLR, with r of 0.92 and *SEP* of 0.55%. The calibration equation was: SSC = 14.755 + 0.812 DM + 0.677 starch.

Jha *et al.* [119] proposed a maturity index (I_m) based on seven mango cultivars:

$$I_m = \eta \frac{\text{SSC} \times \text{DM}}{\text{TA}}$$

where η represents a constant specific for each cultivar. I_m was field tested with less than 10% variation. I_m could be predicted by using a PLS model based on spectra in the 1600–1800 nm wavelength range, with r_p of 0.68 and *SEP* of 0.34.

3.10. Other Fruits

Roger and Bellon-Maurel [9] predicted sugar content in cherry fruit using spectra in the 800–1100 nm wavelength range. A The PLS model based on spectra processed with moving average smoothing yielded a result with r of 0.95 and *RMSEP* of 3.43°Brix. A better result could be obtained employing CWs selected by GA, with r of 0.98 and *RMSEP* of 0.91°Brix. This result was in the same level with that reported by Lu [120], who applied reflectance spectra in the 800–1700 nm wavelength range. The PLS models they built yielded an r_p and *SEP* of 0.95 and 0.71°Brix for 'Hedelfinger', and 0.89° and 0.65°Brix for 'Sam'. They also predicted their firmness, with r_p of 0.80 and 0.65 and *SEP* of 0.79 N and 0.44 N respectively.

Paz *et al.* [12] applied Vis/NIR reflectance spectra to predict SSC and firmness in plum. SSC could be predicted using a modified PLS model based on spectra in the 515–1400 nm range, with r of 0.88 and *SECV* of 0.83°Brix. Firmness could only be predicted using the PLS model built with 515–1650 nm spectra, with r of 0.72 and *SECV* of 2.54 N.

Zhang *et al.* [121] predicted soluble tannin content in persimmon using diffuse reflectance spectra in the 570–1848 nm wavelength range. Different pretreatment methods and calibration methods were compared and the best performance was achieved by the modified PLS model based on 1st derivative spectra processed with de-trending, with r_p and *RMSEP* of 0.82 and 0.18 respectively. More research should be done to enhance the prediction accuracy.

4. Conclusions and Future Research

For their prominent advantages such as simultaneous, precise and rapid analyses compared to traditional methods, spectroscopy, multispectral imaging and hyperspectral imaging have been widely utilized for the measurement of internal and external quality attributes of fruits. One important evaluation criterion for the successful implement of a spectroscopy technique is the accuracy and robustness of the calibration model. According to the overviews above, further studies should be focused on these aspects rather than doing some superficial research or repeating previous studies:

(1)   The optimal spectral acquisition condition, as well as preprocessing and calibration method for each kind of fruit needs to be figured out.

(2)   A large database is crucial, for stable and accurate models should yield satisfactory performance even when applied to fruit from different origins, seasons and climate conditions.

(3)   The model transference between different types of spectrometers hasn't attracted enough attention yet.

(4)   Most of the papers published focused on several major attributes including SSC, acidity and firmness, other important nutrient compositions such as vitamin content, mineral substance and pigments haven't attract enough attention.

(5)   The feasibility of using Vis/NIR spectroscopy to predict some quality attributes has been verified, but the prediction for some other attributes remains uncertain or is definitely less accurate.

Hyperspectral imaging, combining the advantages of imaging technology and spectroscopy technology, could provide abundant information related to fruit quality and thus offers exciting new possibilities. Although hyperspectral imaging with chemometrics frees researchers from laborious measurements and burdensome computations during food quality assessment, hyperspectral imaging has not been applied for online detection, which is restricted by its massive data volume, different prediction results of spectra mode, external characteristics of samples and expensive equipment [4,5,77,78]. Qin and others [122] established a small-scale hyperspectral reflectance imaging for real-time detection of grape canker, but the system only provides a small number of observations from the whole fruit. Multispectral imaging based on selected critical wavelengths derived from hyperspectral imaging has received great attention. Due to their relative little spectral data, low instrument cost and high analytical speed, multispectral imaging systems could be widely used in online detection and practical applications for fruits [78,98,103]. Huang and others [78] selected three effective wavelengths 750, 820 and 960 nm to realize multispectral imaging tests and obtained good prediction results. However, in consideration of the limitations of multispectral imaging, few selected and discrete wavelengths, multispectral imaging has relatively worse performances on the detection of fruit characteristics, such as firmness [98]. However, with the improvement of computer resources, broad prospects are expected. With the help of NIR microscopes and Raman spectroscopy, observation and detection could be achieved at the histological and cellular level [123–125]. A great many of technologies and problems require urgent study and solutions.

Although there is a load of existing problems, the emergence of new technologies and new devices is bringing huge potential to this field. The improved acquisition speed and simplified operation of newly developed spectrographs, multispectral imaging systems and hyperspectral imaging systems combined with the implementation of effective chemometric methods, such as PLS and LS-SVM, have finally make the idea of online detection possible. However, although some efforts have been done to build online detection systems [76,84], real mature and feasible systems are not available in market due to various problems including expensive price, unstable models and complex operation. There is plenty of research left for us to do.

Acknowledgments: This study was supported by 863 National High-Tech Research and Development Plan (2013AA102405), Zhejiang Provincial Public Welfare Technology Research Projects (2014C32103) and the Fundamental Research Funds for the Central Universities (2014FZA6005).

Author Contributions: Y.B. and Y.H. conceived the structure of the paper; H.W., J.P. and C.X. collected the references; H.W. wrote the paper; H.W., J.P., Y.B. and Y.H. revised the paper.

Conflicts of Interest: The authors declare no conflict of interest.

References

1. McClure, W.F. Near-infrared spectroscopy—The giant is running strong. *Anal. Chem.* **1994**, *66*, A43–A53.
2. Martinsen, P.; Schaare, P. Measuring soluble solids distribution in kiwifruit using near-infrared imaging spectroscopy. *Postharvest Biol. Technol.* **1998**, *14*, 271–281.
3. Liu, M.; Fu, P.; Cheng, R. Non-destructive estimation peach ssc and firmness by mutispectral reflectance imaging. *New Zeal. J. Agric. Res.* **2007**, *50*, 601–608.
4. Mendoza, F.; Lu, R.; Ariana, D.; Cen, H.; Bailey, B. Integrated spectral and image analysis of hyperspectral scattering data for prediction of apple fruit firmness and soluble solids content. *Postharvest Biol. Technol.* **2011**, *62*, 149–160.

5. Shan, J.; Peng, Y.; Wang, W.; Li, Y.; Wu, J.; Zhang, L. Simultaneous detection of external and internal quality parameters of apples using hyperspectral technology. *Trans. CSAM* **2011**, *42*, 140–144.

6. Pissard, A.; Pierna, J.A.F.; Baeten, V.; Sinnaeve, G.; Lognay, G.; Mouteau, A.; Dupont, P.; Rondia, A.; Lateur, M. Non-destructive measurement of vitamin c, total polyphenol and sugar content in apples using near-infrared spectroscopy. *J. Sci. Food Agric.* **2013**, *93*, 238–244.

7. Nicolaï, B.M.; Beullens, K.; Bobelyn, E.; Peirs, A.; Saeys, W.; Theron, K.I.; Lammertyn, J. Nondestructive measurement of fruit and vegetable quality by means of nir spectroscopy: A review. *Postharvest Biol. Technol.* **2007**, *46*, 99–118.

8. Sun, T.; Xu, W.; Lin, J.; Liu, M.; He, X. Determination of soluble solids content in navel oranges by vis/nir diffuse transmission spectra combined with cars method. *Spectrosc. Spectr. Anal.* **2012**, *32*, 3229–3233.

9. Roger, J.M.; Bellon-Maurel, V. Using genetic algorithms to select wavelengths in near-infrared spectra: Application to sugar content prediction in cherries. *Appl. Spectrosc.* **2000**, *54*, 1313–1320.

10. Liu, Y.; Zhou, Y. Quantification of the soluble solids content of intact apples by vis-nir transmittance spectroscopy and the ls-svm method. *Spectroscopy* **2013**, *28*, 32–43.

11. Sanchez, M.T.; Jose De la Haba, M.; Benitez-Lopez, M.; Fernandez-Novales, J.; Garrido-Varo, A.; Perez-Marin, D. Non-destructive characterization and quality control of intact strawberries based on nir spectral data. *J. Food Eng.* **2012**, *110*, 102–108.

12. Paz, P.; Sanchez, M.T.; Perez-Marin, D.; Guerrero, J.E.; Garrido-Varo, A. Nondestructive determination of total soluble solid content and firmness in plums using near-infrared reflectance spectroscopy. *J. Agric. Food. Chem.* **2008**, *56*, 2565–2570.

13. Ilari, J.L.; Martens, H.; Isaksson, T. Determination of particle-size in powders by scatter correction in diffuse near-infrared reflectance. *Appl. Spectrosc.* **1988**, *42*, 722–728.

14. Liu, Y.; Sun, X.; Ouyang, A. Nondestructive measurement of soluble solid content of navel orange fruit by visible-nir spectrometric technique with plsr and pca-bpnn. *Lwt-Food Sci. Technol.* **2010**, *43*, 602–607.

15. Shao, Y.; He, Y.; Bao, Y.; Mao, J. Near-infrared spectroscopy for classification of oranges and prediction of the sugar content. *Int. J. Food Prop.* **2009**, *12*, 644–658.

16. Barnes, R.J.; Dhanoa, M.S.; Lister, S.J. Standard normal variate transformation and de-trending of near-infrared diffuse reflectance spectra. *Appl. Spectrosc.* **1989**, *43*, 772–777.

17. Shi, B.; Zhao, L.; Wang, H.; Zhu, D. Signal optimization approaches on the prediction of apples firmness by near infrared spectroscopy. *Sens. Lett.* **2011**, *9*, 1062–1068.

18. Liu, Y.; Sun, X.; Zhou, J.; Zhang, H.; Yang, C. Linear and nonlinear multivariate regressions for determination sugar content of intact gannan navel orange by vis-nir diffuse reflectance spectroscopy. *Math. Comput. Model.* **2010**, *51*, 1438–1443.

19. Liu, H.; Guo, W.; Yue, R. Non-destructive detection of kiwifruit firmness based on near-infrared diffused spectroscopy. *Trans. CSAM* **2011**, *42*, 145–149.

20. Xia, J.; Li, P.; Li, X.; Wang, W.; Ding, X. Effect of different pretreatment method of nondestructive measure vitamin C content of umbilical orange with near-infrared spectroscopy. *Trans. CSAM* **2007**, *38*, 107–111.

21. Xia, J.; Li, X.; Li, P.; Wang, W.; Ding, X. Approach to nondestructive measurement of vitamin c content of orange with near-infrared spectroscopy treated by wavelet transform. *Trans. CSAE* **2007**, *23*, 170–174.

22. Shao, Y.; He, Y. Nondestructive measurement of acidity of strawberry using vis/nir spectroscopy. *Int. J. Food Prop.* **2008**, *11*, 102–111.

23. Sjoblom, J.; Svensson, O.; Josefson, M.; Kullberg, H.; Wold, S. An evaluation of orthogonal signal correction applied to calibration transfer of near infrared spectra. *Chemom. Intell. Lab. Syst.* **1998**, *44*, 229–244.

24. Goicoechea, H.C.; Olivieri, A.C. A comparison of orthogonal signal correction and net analyte preprocessing methods. Theoretical and experimental study. *Chemom. Intell. Lab. Syst.* **2001**, *56*, 73–81.

25. Lv, Q.; Tang, M.J.; Zhao, J.W.; Cai, J.R.; Chen, Q.S. Study of simplification of prediction model for kiwifruit firmness using near infrared spectroscopy. *Spectrosc. Spectr. Anal.* **2009**, *29*, 1768–1771.

26. Araujo, M.C.U.; Saldanha, T.C.B.; Galvao, R.K.H.; Yoneyama, T.; Chame, H.C.; Visani, V. The successive projections algorithm for variable selection in spectroscopic multicomponent analysis. *Chemom. Intell. Lab. Syst.* **2001**, *57*, 65–73.

27. Zhang, S.; Zhang, H.; Zhao, Y.; Zhao, H. Comparison of modeling methods of fresh jujube soluble solids measurement by nir spectroscopy. *Trans. CSAM* **2012**, *43*, 108–112.

28. Chong, I.G.; Jun, C.H. Performance of some variable selection methods when multicollinearity is present. *Chemom. Intell. Lab. Syst.* **2005**, *78*, 103–112.

29. Fernandez-Novales, J.; Lopez, M.I.; Sanchez, M.T.; Morales, J.; Gonzalez-Caballero, V. Shortwave-near infrared spectroscopy for determination of reducing sugar content during grape ripening, winemaking, and aging of white and red wines. *Food Res. Int.* **2009**, *42*, 285–291.

30. Cao, F.; Wu, D.; He, Y. Soluble solids content and ph prediction and varieties discrimination of grapes based on visible–near infrared spectroscopy. *Comput. Electron. Agric.* **2010**, *71*, S15–S18.

31. Li, H.D.; Liang, Y.Z.; Xu, Q.S.; Cao, D.S. Key wavelengths screening using competitive adaptive reweighted sampling method for multivariate calibration. *Anal. Chim. Acta* **2009**, *648*, 77–84.

32. Centner, V.; Massart, D.L.; de Noord, O.E.; de Jong, S.; Vandeginste, B.M.; Sterna, C. Elimination of uninformative variables for multivariate calibration. *Anal. Chem.* **1996**, *68*, 3851–3858.

33. Ouyang, A.; Xie, X.; Zhou, Y.; Liu, Y. Partial least squares regression variable screening studies on apple soluble solids nir spectral detection. *Spectrosc. Spectr. Anal.* **2012**, *32*, 2680–2684.

34. Lue, Q.; Tang, M.; Cai, J.; Lu, H. Long-term prediction of zhonghua kiwifruit dry matter by near infrared spectroscopy. *Scienceasia* **2010**, *36*, 210–215.

35. Guo, Z.; Huang, W.; Chen, L.; Wang, X.; Peng, Y. Nondestructive evaluation of soluble solid content in strawberry by near infrared spectroscopy. In *Piageng 2013: Image Processing and Photonics for Agricultural Engineering*; Tan, H., Ed.; SPIE: Sanya, China, 2013; Volume 8761.

36. Shao, Y.; Bao, Y.; He, Y. Visible/near-infrared spectra for linear and nonlinear calibrations: A case to predict soluble solids contents and PH value in peach. *Food Bioprocess Technol.* **2011**, *4*, 1376–1383.

37. Shi, J.; Zou, X.; Zhao, J.; Mao, H. Selection of wavelength for strawberry nir spectroscopy based on bipls combined with saa. *J. Infrared Millim. Waves* **2011**, *30*, 458–462.

38. Wu, D.; He, Y.; Bao, Y.D. Fast discrimination of juicy peach varieties by Vis/NIR spectroscopy based on bayesian-sda and pca. *Lect. Notes Comput. Sci.* **2006**, *4113*, 931–936.

39. Cen, H.; He, Y.; Huang, M. Combination and comparison of multivariate analysis for the identification of orange varieties using visible and near infrared reflectance spectroscopy. *Eur. Food Res. Technol.* **2007**, *225*, 699–705.

40. Cao, F.W.D.; He, Y.; Bao, Y. Variety discrimination of grapes based on visible-near reflectance infrared spectroscopy. *Acta Opt. Sin.* **2009**, *29*, 537–540.

41. Tavakolian, M.S.S.; Silaghi, F.A.; Fabbri, A.; Molari, G.; Giunchi, A.; Guarnieri, A. Differentiation of post harvest date fruit varieties non-destructively using FT-NIR spectroscopy. *Int. J. Food Sci. Technol.* **2013**, *48*, 1282–1288.

42. Liu, F.; Yusuf, B.L.; Zhong, J.L.; Feng, L.; He, Y.; Wang, L. Variety identification of rice vinegars using visible and near infrared spectroscopy and multivariate calibrations. *Int. J. Food Prop.* **2011**, *14*, 1264–1276.

43. Hao, Y.S.X.; Gao, R.; Pan, Y.; Liu, Y. Application of visible and near infrared spectroscopy to identification of navel orange varieties using simca and pls-da. *Trans. CSAE* **2010**, *26*, 373–377.

44. Baranowski, P.; Mazurek, W.; Wozniak, J.; Majewska, U. Detection of early bruises in apples using hyperspectral data and thermal imaging. *J. Food Eng.* **2012**, *110*, 345–355.

45. Guo, Z.M.; Huang, W.Q.; Chen, L.P.; Zhao, C.J.; Peng, Y.K. Geographical classification of apple based on hyperspectral imaging. In Proceedings of the Sensing for Agriculture and Food Quality and Safety V, Baltimore, MD, USA, 30 April–1 May 2013; SPIE-INT Soc. Optical Engineering: Baltimore, MD, USA, 2013; p. 8.

46. Wang, J.; Nakano, K.; Ohashi, S. Nondestructive evaluation of jujube quality by visible and near-infrared spectroscopy. *Lwt-Food Sci. Technol.* **2011**, *44*, 1119–1125.

47. Wang, J.; Nakano, K.; Ohashi, S. Nondestructive detection of internal insect infestation in jujubes using visible and near-infrared spectroscopy. *Postharvest Biol. Technol.* **2011**, *59*, 272–279.

48. He, Y.; Li, X.L.; Shao, Y.N. Discrimination of varieties of apple using near infrared spectra based on principal component analysis and artificial neural network model. *Spectrosc. SpectR. Anal.* **2006**, *26*, 850–853.

49. He, Y.; Li, X.L.; Shao, Y.N. Fast discrimination of apple varieties using vis/nir spectroscopy. *Int. J. Food Prop.* **2007**, *10*, 9–18.

50. Niu, X.Y.; Shao, L.M.; Zhao, Z.L.; Zhang, X.Y. Nondestructive discrimination of strawberry varieties by nir and bp-ann. *Spectrosc. Spect. Anal.* **2012**, *32*, 2095–2099.

51. Peiris, K.H.S.; Dull, G.G.; Leffler, R.G.; Kays, S.J. Near-infrared spectrometric method for nondestructive determination of soluble solids content of peaches. *J. Am. Soc. Hortic. Sci.* **1998**, *123*, 898–905.

52. ElMasry, G.; Wang, N.; ElSayed, A.; Ngadi, M. Hyperspectral imaging for nondestructive determination of some quality attributes for strawberry. *J. Food Eng.* **2007**, *81*, 98–107.

53. Jaiswal, P.; Jha, S.N.; Bharadwaj, R. Non-destructive prediction of quality of intact banana using spectroscopy. *Sci. Hortic.* **2012**, *135*, 14–22.

54. Park, B.; Abbott, J.A.; Lee, K.J.; Choi, C.H.; Choi, K.H. Near-infrared diffuse reflectance for quantitative and qualitative measurement of soluble solids and firmness of delicious and gala apples. *Trans. ASAE* **2003**, *46*, 1721–1731.

55. Angra, S.K.; Dimri, A.K.; Kapur, P. Nondestructive brix evaluation of apples of different origin using near infrared (nir) filter based reflectance spectroscopy. *Instrum. Sci. Technol.* **2009**, *37*, 241–253.

56. Hadi, A.S.; Ling, R.F. Some cautionary notes on the use of principal components regression. *Am. Stat.* **1998**, *52*, 15–19.

57. Wold, S.; Sjostrom, M.; Eriksson, L. Pls-regression: A basic tool of chemometrics. *Chemom. Intell. Lab. Syst.* **2001**, *58*, 109–130.

58. Liu, Y.; Ouyang, A.; Luo, J.; Chen, X. Near infrared diffuse reflectance spectroscopy for rapid analysis of soluble solids content in navel orange. *Spectrosc. Spectr. Anal.* **2007**, *27*, 2190–2192.

59. Lu, H.; Jiang, H.; Fu, X.; Yu, H.; Xu, H.; Ying, Y. Non-invasive measurements of the internal quality of intact 'gannan' navel orange by vis/nir spectroscopy. *Trans. ASABE* **2008**, *51*, 1009–1014.

60. Bureau, S.; Scibisz, I.; le Bourvellec, C.; Renard, C.M.G.C. Effect of sample preparation on the measurement of sugars, organic acids, and polyphenols in apple fruit by mid-infrared spectroscopy. *J. Agric. Food. Chem.* **2012**, *60*, 3551–3563.

61. Suykens, J.A.K.; Vandewalle, J. Least squares support vector machine classifiers. *Neural Process. Lett.* **1999**, *9*, 293–300.

62. Liu, Y.; Gao, R.; Hao, Y.; Sun, X.; Ouyang, A. Improvement of near-infrared spectral calibration models for brix prediction in 'gannan' navel oranges by a portable near-infrared device. *Food Bioprocess Technol.* **2012**, *5*, 1106–1112.

63. Liu, Y.; Chen, X.; Ouyang, A. Non-destructive measurement of soluble solid content in gannan navel oranges by visible/near-infrared spectroscopy. *Acta Opt. Sin.* **2008**, *28*, 478–481.

64. Zhang, S.; Wang, F.; Zhang, H.; Zhao, C.; Yang, G. Detection of the fresh jujube varieties and ssc by nir spectroscopy. *Trans. CSAM* **2009**, *40*, 139–142.

65. He, J.G.; Luo, Y.; Liu, G.S.; Xu, S.; Si, Z.H.; He, X.G.; Wang, S.L. Prediction of soluble solids content of jujube fruit using hyperspectral reflectance imaging. In *Mechatronics and Intelligent Materials III, Pts 1–3*; Chen, R., Sung, W.P., Kao, J.C.M., Eds.; Trans. Tech. Publications: XiShuangBanNa, China, 2013; Volume 706–708, pp. 201–204.

66. Tallada, J.G.; Nagata, M.; Kobayashi, T. Non-destructive estimation of firmness of strawberries (fragaria x ananassa duch.)   using nir hyperspectral imaging. *Environ. Control Biol.* **2006**, *44*, 245–255.

67. Davey, M.W.; Saeys, W.; Hof, E.; Ramon, H.; Swennen, R.L.; Keulemans, J. Application of visible and near-infrared reflectance spectroscopy (Vis/NIRS) to determine carotenoid contents in banana (musa spp.) fruit pulp. *J. Agric. Food. Chem.* **2009**, *57*, 1742–1751.

68. Kemps, B.; Leon, L.; Best, S.; de Baerdemaeker, J.; de Ketelaere, B. Assessment of the quality parameters in grapes using vis/nir spectroscopy. *Biosyst. Eng.* **2010**, *105*, 507–513.

69. Omar, A.F. Spectroscopic profiling of soluble solids content and acidity of intact grape, lime, and star fruit. *Sens. Rev.* **2013**, *33*, 238–245.

70. Shao, Y.; He, Y. Nondestructive measurement of the internal quality of bayberry juice using vis/nir spectroscopy. *J. Food Eng.* **2007**, *79*, 1015–1019.

71. Yu, J.J.; He, Y.; Bao, Y.D. Nondestructive test on predicting sugar content and valid acidity of mango by spectroscopy technology. *Spectrosc. Spect. Anal.* **2008**, *28*, 2839–2842.

72. Liu, Y.D.; Ying, Y.B.; Chen, Z.M.; Fu, X.P. Appli3 peaches. In *Monitoring Food Safety, Agriculture, and Plant Health*; Bennedsen, B.S., Chen, Y.R., Meyer, G.E., Senecal, A.G., Tu, S.I., Eds.; SPIE: Providence, USA, 2004; Volume 5271, pp. 347–355.

73. González-Caballero, V.; Sánchez, M.T.; López, M.I.; Pérez-Marín, D. First steps towards the development of a non-destructive technique for the quality control of wine grapes during on-vine ripening and on arrival at the winery. *J. Food Eng.* **2010**, *101*, 158–165.

74. Wang, J.; Pan, L.; Li, P.; Han, D. Temperature compensation for calibration model of apple fruit soluble solids contents by near infrared reflectance. *Spectrosc. Spectr. Anal.* **2009**, *29*, 1517–1520.

75. Zou, X.; Zhao, J.; Li, Y. Objective quality assessment of apples using machine vision, nir spectrophotometer, and electronic nose. *Trans. ASABE* **2010**, *53*, 1351–1358.

76. Luo, X.; Takahashi, T.; Kyo, K.; Zhang, S. Wavelength selection in vis/NIR spectra for detection of bruises on apples by ROC analysis. *J. Food Eng.* **2012**, *109*, 457–466.

77. Huang, W.; Zhang, B.; Li, J.; Zhang, C. Early detection of bruises on apples using near-infrared hyperspectral image. In *Piageng 2013: Image Processing and Photonics for Agricultural Engineering*; Tan, H., Ed.; SPIE: Sanya, Chian, 2013; Volume 8761.

78. Huang, W.; Zhao, C.; Wang, Q.; Li, J.; Zhang, C. Development of a multi-spectral imaging system for the detection of bruises on apples. In *Sensing for Agriculture and Food Quality and Safety V*; Kim, M.S., Tu, S.I., Chao, K., Eds.; SPIE: Maryland, USA, 2013; Volume 8721.

79. Zude, M.; Herold, B.; Roger, J.M.; Bellon-Maurel, V.; Landahl, S. Non-destructive tests on the prediction of apple fruit flesh firmness and soluble solids content on tree and in shelf life. *J. Food Eng.* **2006**, *77*, 254–260.

80. Rutkowski, K.P.; Michalczuk, B.; Konopacki, P. Nondestructive determination of 'golden delicious' apple quality and harvest maturity. *J. Fruit Ornam. Plant Res.* **2008**, *16*, 39–52.

81. Kuckenberg, J.; Tartachnyk, I.; Noga, G. Evaluation of fluorescence and remission techniques for monitoring changes in peel chlorophyll and internal fruit characteristics in sunlit and shaded sides of apple fruit during shelf-life. *Postharvest Biol. Technol.* **2008**, *48*, 231–241.

82. Cayuela, J.A.; Weiland, C. Intact orange quality prediction with two portable nir spectrometers. *Postharvest Biol. Technol.* **2010**, *58*, 113–120.

83. Xu, W.; Sun, T.; Wu, W.; Liu, M. Near-infrared spectrum detection result influenced by navel oranges placement position. *Spectrosc. Spectr. Anal.* **2012**, *32*, 3002–3005.

84. Sun, X.; Hao, Y.; Gao, R.; Ouyang, A.; Liu, Y. Research on optimization of model for detecting sugar content of navel orange by online near infrared spectroscopy. *Spectrosc. Spectr. Anal.* **2011**, *31*, 1230–1235.

85. Deng, L.; He, S.; Yi, S.; Zheng, Y.; Xie, R.; Zhang, X.; Mao, S. Study on synchronous correlation between fruit characteristic spectrum and the parameter of internal quality for hamlin sweet orange fruit. *Spectrosc. Spectr. Anal.* **2010**, *30*, 1049–1052.

86. McGlone, V.A.; Kawano, S. Firmness, dry-matter and soluble-solids assessment of postharvest kiwifruit by nir spectroscopy. *Postharvest Biol. Technol.* **1998**, *13*, 131–141.

87. Moghimi, A.; Aghkhani, M.H.; Sazgarnia, A.; Sarmad, M. Vis/nir spectroscopy and chemometrics for the prediction of soluble solids content and acidity (ph) of kiwifruit. *Biosystems Eng.* **2010**, *106*, 295–302.

88. Schaare, P.N.; Fraser, D.G. Comparison of reflectance, interactance and transmission modes of visible-near infrared spectroscopy for measuring internal properties of kiwifruit (actinidia chinensis). *Postharvest Biol. Technol.* **2000**, *20*, 175–184.

89. McGlone, V.A.; Jordan, R.B.; Seelye, R.; Martinsen, P.J. Comparing density and nir methods for measurement of kiwifruit dry matter and soluble solids content. *Postharvest Biol. Technol.* **2002**, *26*, 191–198.

90. McGlone, V.A.; Clark, C.J.; Jordan, R.B. Comparing density and vnir methods for predicting quality parameters of yellow-fleshed kiwifruit (actinidia chinensis). *Postharvest Biol. Technol.* **2007**, *46*, 1–9.

91. Arazuri, S.; Jaren, C.; Arana, J.I. Selection of the temperature in the sugar content determination of kiwi fruit. *Int. J. Infrared Millim. Waves* **2005**, *26*, 607–616.

92. Lee, J.S.; Kim, S.C.; Seong, K.C.; Kim, C.H.; Um, Y.C.; Lee, S.K. Quality prediction of kiwifruit based on near infrared spectroscopy. *Korean J. Hortic. Sci.* **2012**, *30*, 709–717.

93. Feng, J.; McGlone, A.V.; Currie, M.; Clark, C.J.; Jordan, B.R. Assessment of yellow-fleshed kiwifruit (actinidia chinensis 'hort16a') quality in pre- and post-harvest conditions using a portable near-infrared spectrometer. *Hortscience* **2011**, *46*, 57–63.

94. Jiang, M.; Lu, H.; Ying, Y.; Xu, H. Design and validation of software for real-time soluble solids content evaluation of peach by near infrared spectroscopy—Art. No. 638118. In *Optics for Natural Resources, Agriculture, and Foods*; Chen, Y.R., Meyer, G.E., Tu, S.I., Eds.; 2006; Volumr 6381, pp. 38118–38118.

95. Kawano, S.; Watanabe, H.; Iwamoto, M. Determination of sugar content in intact peaches by near-infrared spectroscopy with fiber optics in interactance mode. *J. Jpn. Soc. Hortic. Sci.* **1992**, *61*, 445–451.

96. Ma, G.; Fu, X.; Zhou, Y.; Ying, Y.; Xu, H.; Xie, L.; Lin, T. Nondestructive sugar content determination of peaches by using near infrared spectroscopy technique. *Spectrosc. Spectr. Anal.* **2007**, *27*, 907–910.

97. Lu, R.F.; Peng, Y.K. Assessing peach firmness by multi-spectral scattering. *J. Near Infrared Spectrosc.* **2005**, *13*, 27–35.

98. Lu, R.F.; Peng, Y.K. Hyperspectral scattering for assessing peach fruit firmness. *Biosyst. Eng.* **2006**, *93*, 161–171.

99. Fu, X.; Ying, Y.; Zhou, Y.; Xie, L.; Xu, H. Application of nir spectroscopy for firmness evaluation of peaches. *J. Zhejiang Univ. Sci. B* **2008**, *9*, 552–557.

100. Li, X.L.; Hu, X.Y.; He, Y. New approach of discrimination of varieties of juicy peach by near infrared spectra based on pca and mda model. *J. Infrared Millim Waves* **2006**, *25*, 417–420.

101. Zwiggelaar, R.; Yang, Q.S.; GarciaPardo, E.; Bull, C.R. Use of spectral information and machine vision for bruise detection on peaches and apricots. *J. Agric. Eng. Res.* **1996**, *63*, 323–331.

102. Takano, K.; Senoo, T.; Uno, T.; Sasabe, Y.; Tada, M. Distinction of astringency in peach fruit using near-infrared spectroscopy. *Hortic. Res. (Jpn.)* **2007**, *6*, 137–143.

103. Lleó, L.; Barreiro, P.; Ruiz-Altisent, M.; Herrero, A. Multispectral images of peach related to firmness and maturity at harvest. *J. Food Eng.* **2009**, *93*, 229–235.

104. Nishizawa, T.; Mori, Y.; Fukushima, S.; Natsuga, M.; Maruyama, Y. Non-destructive analysis of soluble sugar components in strawberry fruits using near-infrared spectroscopy. *J. Jpn Soc. Food Sci.* **2009**, *56*, 229–235.

105. Yan, R.; Wang, X.; Qiu, B.; Shi, D.; Kong, P. Discrimination of strawberries varieties based on characteristic spectrum. *Trans. CASM* **2013**, *44*, 182–186.

106. Nagata, M.; Tallada, J.G.; Kobayashi, T. Bruise detection using nir hyperspectral imaging for strawberry (fragaria x ananassa duch.). *Environ. Control Biol.* **2006**, *44*, 133–142.

107. Fernandez-Novales, J.; Lopez, M.I.; Sanchez, M.T.; Garcia-Mesa, J.A.; Gonzalez-Caballero, V. Assessment of quality parameters in grapes during ripening using a miniature fiber-optic near-infrared spectrometer. *Int. J. Food. Sci. Nutr.* **2009**, *60*, 265–277.

108. Herrera, J.; Guesalaga, A.; Agosin, E. Shortwave-near infrared spectroscopy for non-destructive determination of maturity of wine grapes. *Meas. Sci. Technol.* **2003**, *14*, 689–697.

109. Larrain, M.; Guesalaga, A.R.; Agosin, E. A multipurpose portable instrument for determining ripeness in wine grapes using nir spectroscopy. *IEEE Trans. Instrum. Meas.* **2008**, *57*, 294–302.

110. Wu, G.; Huang, L.; He, Y. Research on the sugar content measurement of grape and berries by using vis/nir spectroscopy technique. *Spectrosc. Spectr. Anal.* **2008**, *28*, 2090–2093.

111. Zhang, S.; Zhang, H.; Zhao, Y.; Guo, W.; Zhao, H. A simple identification model for subtle bruises on the fresh jujube based on nir spectroscopy. *Math. Comput. Model.* **2013**, *58*, 545–550.

112. Wang, J.; Nakano, K.; Ohashi, S.; Takizawa, K.; He, J.G. Comparison of different modes of visible and near-infrared spectroscopy for detecting internal insect infestation in jujubes. *J. Food Eng.* **2010**, *101*, 78–84.

113. Wang, J.; Nakano, K.; Ohashi, S.; Kubota, Y.; Takizawa, K.; Sasaki, Y. Detection of external insect infestations in jujube fruit using hyperspectral reflectance imaging. *Biosyst. Eng.* **2011**, *108*, 345–351.

114. Tarkosova, J.; Copikova, J. Determination of carbohydrate content in bananas during ripening and storage by near infrared spectroscopy. *J. Near Infrared Spectrosc.* **2000**, *8*, 21–26.

115. Subedi, P.P.; Walsh, K.B. Assessment of sugar and starch in intact banana and mango fruit by swnir spectroscopy. *Postharvest Biol. Technol.* **2011**, *62*, 238–245.

116. Jha, S.N.; Jaiswal, P.; Narsaiah, K.; Gupta, M.; Bhardwaj, R.; Singh, A.K. Non-destructive prediction of sweetness of intact mango using near infrared spectroscopy. *Sci. Hortic.* **2012**, *138*, 171–175.

117. Schmilovitch, Z.; Mizrach, A.; Hoffman, A.; Egozi, H.; Fuchs, Y. Determination of mango physiological indices by near-infrared spectrometry. *Postharvest Biol. Technol.* **2000**, *19*, 245–252.

118. Saranwong, S.; Sornsrivichai, J.; Kawano, S. Prediction of ripe-stage eating quality of mango fruit from its harvest quality measured nondestructively by near infrared spectroscopy. *Postharvest Biol. Technol.* **2004**, *31*, 137–145.

119. Jha, S.N.; Narsaiah, K.; Jaiswal, P.; Bhardwaj, R.; Gupta, M.; Kumar, R.; Sharma, R. Nondestructive prediction of maturity of mango using near infrared spectroscopy. *J. Food Eng.* **2014**, *124*, 152–157.

120. Lu, R. Predicting firmness and sugar content of sweet cherries using near-infrared diffuse reflectance spectroscopy. *Trans. ASAE* **2001**, *44*, 1265–1271.

121. Zhang, P.; Li, J.; Meng, X.; Zhang, P.; Feng, X.; Wang, B. Research on nondestructive measurement of soluble tannin content of astringent persimmon using visible and near infrared diffuse reflection spectroscopy. *Spectrosc. Spectr. Anal.* **2011**, *31*, 951–954.

122. Qin, J.W.; Thomas, F.B.; Zhao, X.H.; Nikhil, N.; Mark, A.R. Development of a two-band spectral imaging system for real-time citrus canker detection. *J. Food Eng.* **2012**, *108*, 87–93.

123. Schulz, H.; Baranska, M.; Baranski, R. Potential of nir-ft-raman spectroscopy in natural carotenoid analysis. *Biopolymers* **2005**, *77*, 212–221.

124. Baranska, M.; Schutz, W.; Schulz, H. Determination of lycopene and beta-carotene content in tomato fruits and related products: Comparison of ft-raman, atr-ir, and nir spectroscopy. *Anal. Chem.* **2006**, *78*, 8456–8461.

125. Baranska, M.; Roman, M.; Dobrowolski, J.C.; Schulz, H.; Baranski, R. Recent advances in raman analysis of plants: Alkaloids, carotenoids, and polyacetylenes. *Curr. Anal. Chem.* **2013**, *9*, 108–127.

Fruit and Vegetable Quality Assessment via Dielectric Sensing

Dalia El Khaled, Nuria Novas, Jose A. Gazquez, Rosa M. Garcia and
Francisco Manzano-Agugliaro

Abstract: The demand for improved food quality has been accompanied by a technological boost. This fact enhances the possibility of improving the quality of horticultural products, leading towards healthier consumption of fruits and vegetables. A better electrical characterization of the dielectric properties of fruits and vegetables is required for this purpose. Moreover, a focused study of dielectric spectroscopy and advanced dielectric sensing is a highly interesting topic. This review explains the dielectric property basics and classifies the dielectric spectroscopy measurement techniques. It comprehensively and chronologically covers the dielectric experiments explored for fruits and vegetables, along with their appropriate sensing instrumentation, analytical modelling methods and conclusions. An in-depth definition of dielectric spectroscopy and its usefulness in the electric characterization of food materials is presented, along with the various sensor techniques used for dielectric measurements. The collective data are tabulated in a summary of the dielectric findings in horticultural field investigations, which will facilitate more advanced and focused explorations in the future.

Reprinted from *Sensors*. Cite as: El Khaled, D.; Novas, N.; Gazquez, J.A.; Garcia, R.M.; Manzano-Agugliaro, F. Fruit and Vegetable Quality Assessment via Dielectric Sensing. *Sensors* **2015**, *15*, 15363–15397.

1. Introduction

Quality is defined by the Spanish Royal Academy as "the property or set of inherent something, you can judge their values" [1]. In 2006, Choi and his co-workers classified fruit quality into internal and external quality factors [2]. Internal factors include the taste, the texture, the value, the aroma, the nutrition and the lack of biotic and abiotic contaminants, whereas the external factors include the presentation, the appearance, the uniformity, the maturity and the freshness of the fruit. Furthermore, the authors stated that although the internal aspects are not noticeable for consumers, they are highly important next to the external aspects that are considered to be the essential purchase decision.

Due to the large consumption increase in the horticulture field, the fresh fruit and vegetable post-harvest sector is dynamic, and the need for high quality produce is rising [3]. Currently, there is a trend to promote the characteristics of vegetable consumption in the diet. It is suggested that the components of vegetables have the capacity to modulate the complex mechanisms involved in maintaining a healthy physiology and reducing the early onset of age dependant diseases, and the demand for agricultural products such as vegetables and fruits is rising [4]. Increasing consumer demand for high-quality fruit has led to the development of optical, acoustic and mechanical sensors that determine its quality. According to Shewfelt [5], the internal characteristics, which are perceivable by the senses of taste, smell, and touch (mouthfeel), are the ones that will determine the decision to repurchase that product. The other characteristics, such as nutritional value, wholesomeness, and safety, cannot readily be determined by consumers because they require measurement, but, if this information is given to the consumer, it will influence acceptability of the product.

In this context, after highlighting the importance of the fruit and vegetable characterization process from harvest to cold storage, and with all of the rapid technological development, mathematical methods and multiplicity of investigations all over the world, there is a huge need for the development of review methods for electrical characterization in the horticultural field [6]. Improved methods for rapidly sensing quality factors of fruits and vegetables, such as moisture content, maturity defects, and blemishes, would be helpful in the harvesting, sorting and packing operations for these commodities; this rapid technique can save labour costs and provide improvement in the uniformity and quality of the products [7].

Moreover, Pliquett [8] stated that the electrical measurement is a simple innocuous tool for material characterization, which should make the determination of electrical properties a highly effective method to enhance the quality of fruits and vegetables. This being said, this information quickly becomes outdated, such that to obtain sustainable and competitive agriculture, it is necessary to use techniques, systems and tools that provide timely monitoring and measurement with reliable information [9].

2. Dielectric Characterization of Vegetables and Fruits

2.1. Overview

For the past two decades, many researchers have been interested in the study of electrical properties, and numerous experiments were conducted for a large variety of agricultural products (fruits and vegetables). The main factor that affects the dielectric properties of hygroscopic materials is the moisture content; this factor is used as a basis for developing commercial instruments to measure moisture content [10]. As a result of the different mineral substances and organic acids present, along with other components that are susceptible to dissociation, the high electrolytic conductivity of fruits and vegetables is distinguished. Many parameters, such as impedance and permittivity, are quite interesting because dielectric properties are considered to be the most important physical properties associated with radio frequency (RF) and microwave (MW) heating [11].

Due to the interest in the dielectric properties of agricultural materials, which is focused primarily on predicting heating rates that describe these material products when subjected to high frequency heating, the dielectric properties of biological products have become valuable parameters in food engineering and technology [12]. The process of energy absorption through RF or MW energy has been known for a long time and has been widely explored. With the advance of computer modelling tools for RF and MW applications, it is very critical to have data available for the dielectric properties of materials. This issue is noticeable especially in the modern design of heating systems where experiments to meet products have been conducted, and data for the moisture content has been reported for several frequency ranges and temperatures related to the process requirements [13]. Among the factors that are involved in the dielectric properties values, the nature of the material that implies the composition and structure is the most common. Some other factors, such as frequency and temperature, are involved with the maturity stage of the agricultural product. Because MW heating is greatly affected by the presence of water, which is a major absorber of MW energy, the higher the moisture content, the better the heating [14–16]. Moreover, ionic components have significant effects on the dielectric properties [17]. Another factor is the density because the amount of mass per unit of volume (density) has a definite effect on the interaction of the electromagnetic field and the involved mass [18]. Another important factor is the storage time of the agricultural products under measure because ripening processes taking place may affect the dielectric properties [19].

The electrical linear properties of tissues and cell suspensions, because of their variation with frequency, are mostly considered to be unusual. These properties include the dielectric constant ε' and the conductivity, which has been proven to be inversely proportional. To illustrate with an example, a graph is plotted in Figure 1 to show the variations of these parameters with frequency. An interpretation of the parameters behaviour *versus* frequency is to be analyzed in the discussion section to explain the curve patterns. Three distinct major steps accompany the variation of the frequency at low RF and GHz frequencies that are termed as α, β and γ dispersions. Moreover, the dielectric constants reach very high values relative to free space at low frequencies [20]. While the α dispersion remains incomplete for several reasons, the β dispersion is due to the cellular structure of tissues and occurs in the range of 0.1 to 10 MHz. The γ dispersion was noted above 1 GHz for a variety of tissues and protein solutions. In addition to the main dispersion that is due to plasma membranes, the β dispersion possesses additional dispersions on the high frequency side [21].

About the interaction between food and electromagnetic energy at low frequencies, much less is known [22]. At high frequencies, the electric properties of most basic interest are the dielectric properties that affect energy coupling and distribution within the product, which includes the product attenuation constant determining voltage and power penetration depths (Dp) within the product and therefore the temperature at a specific depth [15]. The main advantages of high frequency methods consist of reducing process time, offering more uniform heating patterns and improving product quality in selected applications [23]. By definition, biological materials and their ability to store and dissipate electrical energy are compared to non-ideal capacitors [24]. Energy charging and loss currents related to the material electrical capacitance and resistance are behind these properties, and they are defined as dielectric properties. However, due to the migration of charge carriers, there is a slight difference in the electrical behaviour of a simple resistive-capacitive circuit in conduction and biological materials at high frequency. According to Sarbacher and Edson [25], because the relative magnetic loss of a material is related to the material reluctance and its ability to dissipate magnetic energy, the components of complex permeability, when divided by the permeability of free space, *i.e.*, 1.257×10^{-6} H/m, give the components of the complex relative permeability. Alternately, it is necessary to consider the magnetic coupling effects at high frequencies for fruits and vegetables because their relative magnetic permeability is close to unity (magnetic permeability close to that of free space and zero relative magnetic loss) [26].

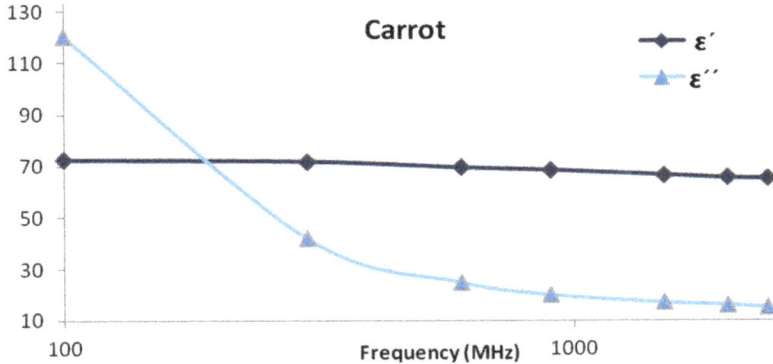

Figure 1. Example of ε' and ε'' variations with frequency on a logarithmic scale.

A material's ability to attenuate or absorb electrical energy coupled by the material from an electromagnetic field is determined by the real component, which is also the major determinant of energy distribution in homogeneous dielectric materials. The attenuation factor determines the ability of the electric component of the field to penetrate the dielectric and is the reciprocal of the material Dp.

2.2. Dielectric Properties

Development of non-destructive and informative sensing techniques to evaluate the properties of living tissues has been a subject of increasing importance for decades [27]. The dielectric properties of food and biological products have become valuable parameters in food engineering and technology [12]. Dielectric spectroscopy is an old experimental tool that has developed dramatically in the last two decades. It currently covers the extraordinary spectral range from 10^{-6} to 10^{12} Hz. Dielectric spectroscopy is a technique used to study the interaction of a material and the applied electric field. It is widely used as a tool for the detection of material aging and fault diagnosis for insulation systems, and hence it has become a popular and powerful research technique [28]. It is based on the phenomena of electrical polarization and electrical conduction in materials [29]. The dielectric property known as complex permittivity is the physical property that describes the interaction between matter and electromagnetic fields, and it is related to the structural and physio-chemical properties, such as water and soluble solids content or water activity, of the material [30]. There are a number of different dielectric polarization mechanisms operating at the molecular or microscopic levels [31]. The analysis of dielectric spectroscopy data gives valuable parameters that characterize the living tissues, such as cell size and shape, the state of the cell membranes and the status of intra- and extracellular media. The dielectric properties of materials that are of interest in most applications can be defined in terms of their relative permittivity. Permittivity is a complex quantity generally used to describe the dielectric properties that influence reflection of electromagnetic waves at interfaces and the attenuation of the wave energy within materials. Based on Maxwell's equation, the complex dielectric function describes the interaction of electromagnetic waves with matter to reflect the underlying molecular mechanisms [32]. The relative complex permittivity ε^* is represented as in Equation (1):

$$\varepsilon^* = \varepsilon' - j\varepsilon'' \tag{1}$$

where ε' and ε'' are commonly called the dielectric constant and loss factor, respectively, and $j = \sqrt{-1}$. The real part, ε', describes the ability of a material to store energy when it is subjected to an electric field and influences the electric field distribution and the phase of waves travelling through the material. The imaginary part, ε'', influences both energy absorption and attenuation and describes the ability to dissipate energy in response to an applied electric field or various polarization mechanisms that commonly result in heat generation [33]. The amount of thermal energy converted in the food is proportional to the value of the loss factor [13].

In the past, dielectric spectroscopy was characterized by its limitations in frequency, as measurements could usually only be carried out within 4–5 decades of frequency; the dielectric constant and the dielectric loss factor data are presented in sets of points in frequency measured at different temperatures. These measurements show that time superposition holds, within the limited frequency range of the measurement, to obtain the following Debye relaxation function, Equation (2):

$$\varepsilon^* = \varepsilon_\infty + \frac{\Delta\varepsilon}{(1 + (i\omega\tau)^\alpha)^\gamma} \tag{2}$$

where $\Delta\varepsilon$ describes the dielectric length, τ is the relaxation time, and α and γ are the quantifications for the symmetric and asymmetric broadening of the relaxation distribution function, respectively. In addition to the molecular dynamic in confining spaces, the scaling of relaxation processes is the major contribution of the broadband dielectric spectroscopy to modern physics [32].

Mechanisms that contribute to the dielectric loss factor include dipole, electronic, ionic and Maxwell-Wagner responses [34], as illustrated by Equation (3):

$$\varepsilon'' = \varepsilon''_d + \varepsilon''_\sigma = \varepsilon''_d + \frac{\sigma}{(\varepsilon_0\omega)} \tag{3}$$

where the subscripts d and σ stand for the contributions due to dipole rotation (d) and ionic conduction (σ), respectively; more specifically, σ is the ionic conductivity in S/m, ω is the angular frequency, and ε_0 is the permittivity of free space or a vacuum (8.854×10^{-12}). For RF (1 to 50 MHz) and MF (915 to 2450 MHz), σ and d are the predominant loss mechanisms [35]. The power in W/m^3 dissipated per unit volume in the dielectric can be expressed via Equation (4):

$$P = E^2\sigma = 55.63 * 10^{-12} f E^2 \varepsilon'' \tag{4}$$

where E represents the rms electric field in V/m.

In this context, we refer to the Kremer statement that although dielectric spectroscopy is old, it is a still-developing experimental technique that has had a strong technological impact and provides variety of novel routes that will open exiting new horizons, such as revealing information about the binding of water in food and other agricultural materials. Thus, further studies should be conducted for the potential applications of dielectric sensing spectroscopy [36].

3. Dielectric Sensing Techniques

In this section, the evolution of dielectric properties sensing techniques is described. Dielectric Spectra can be easily obtained using an automated frequency domain spectrometer (FDS) with high precision over a frequency range of 1 Hz to 10 GHz. The measurement techniques appropriate for any particular application depend on the frequency of interest, the nature, both physical and electrical, of the dielectric material to be measured and the degree of accuracy required. In this context, many different types of instruments can be used, and any measurement instrument providing reliable determinations of the required electrical parameters involving the unknown material in the frequency range of interest can be considered.

In the agricultural field, to understand the dielectric behaviour of agricultural products, it was required to boost the measurements over broad frequency ranges and to develop new techniques for efficient collection of permittivity information [37]. The different measurement techniques of the dielectric properties are summarized in Table 1 along with their characteristics.

Table 1. Dielectric techniques.

	Brief Description	Recommended Materials	Frequency Range	Advantages	Disadvantages
Parallel plate	Material must be placed between two electrodes to form a capacitor	Material with the ability to be formed as a flat smooth sheet <100 MHz	<100 MHz	Inexpensive, high accuracy	Limited frequency range, sheet sample very thin (<10 mm thick)
Lumped circuit	Sample is a part of the insulator in a lumped circuit	All materials with the exception of gazes	<100 MHz	Liquid and solid materials can be measured	Limited frequency range, not suitable for very low loss materials
Coaxial probe	A coaxial line cut off forming a flat plane boundary in contact with food. A vector analyser is needed to measure the reflection	Liquids and semi-solids	200 MHz–20 GHz, even >100 GHz	Easy to use, non-destructive for some materials, sample preparation is not required	Limited accuracy (±5%). Low loss resolution, large samples and solids must show a flat surface
Transmission line	Brick-shaped sample fills the cross section of an enclosed transmission line, causing an impedance change	Liquids and solids	<100 MHz	More accurate and sensitive than the probe method	Less accuracy than resonators, sample preparation is difficult and time-consuming
Cavity resonator	Sample is introduced in a cavity (a high Q resonant structure), which affects the centre frequency and quality factor of the cavity	Solids	1 MHz–100 GHz	Easy sample preparation, adaptable for a wide range of temperatures	Broadband frequency data are not provided and analysis may be complex
Free space	Antennas are used to direct a MW beam at or through the material. A vector network analyser measures the reflection and transmission coefficients of solids	Solids	MW range	Non-destructive, high temperatures can be used	A large flat, thin, parallel-faced sample and special calibration are required
Time domain spectroscopy	Short pulses of THz radiation within a generation and detection scheme that is sensitive to the effect of both the amplitude and phase of the radiation	Homogeneous	10 MHz–10 GHz.	Fast and high accuracy measurement, small sample	Expensive

3.1. Time-Domain Spectroscopy

For some of the studies conducted, time-domain reflectometry and spectroscopy techniques were employed [38]. Time-domain systems generate a fast rise time pulse that is reflected from the sample or transmitted through the sample where the dielectric property information can be extracted by Fourier transform by analysing the waveform. Polymer chemists and physicists have been using time-domain measurements extensively to help in understanding the composition and behaviour of materials [39].

Various techniques over wide ranges of frequency were developed with the presence of suitable equipment for time-domain measurements [40] that have been improved for accuracy [41]. This will enable faster investigations due to the shorter measurement time needed compared with FDS. For measurements below 10 MHz, measuring cells consisting of parallel plate capacitor types are used [42]. However, the residual inductance and capacitance arising from the cell itself and the connecting leads require correction by the measuring cell [43]. At frequencies above 100 MHz, open-ended coaxial probes are suited for measurement with network analysers and time-domain reflectometers. Alternately, for broadband frequency-domain measurements of some products, it was necessary to employ impedance and network analysers with appropriate sample holders and techniques.

3.2. Radio-Frequency

For radio frequencies, the material can be modelled electrically at any given frequency as a series or parallel equivalent lumped element circuit. Therefore, if the RF circuit parameters were measured appropriately, the dielectric properties can be determined from proper equations by relating the way in which the permittivity of the material affects those circuit parameters. However, Nelson stated that the challenge of making accurate dielectric properties or permittivity measurements lies in the design of the material sample holder for those measurements and in an adequate modulation of the circuit for reliable calculation of the permittivity from the electrical measurements [44].

The use of the RF measurements in the early stages of the dielectric measurements of agricultural products was very common. Radio frequency measurements relied basically on instruments such as the Q meter [45], impedance bridges [46], and admittance meters [47]. Various bridges and resonant circuits were used as measurement techniques for permittivity or dielectric properties determination in low-frequency, medium-frequency and high-frequency ranges [48]. However, at very low frequencies, invalid measurement data can occur because of electrode polarization; therefore, attention must be paid to the frequency below which polarization affects measurement. This depends on the nature and conductivity of the material being measured [49]. Among the methods used to eliminate the electrode polarization effect are the four-electrode method and the electromagnetic induction method developed with a pair of toroidal coils [50]. With a Q-meter based on a series resonant circuit, a large number of experiments were conducted, and data were obtained in the 1 to 50 MHz range [51]. For higher frequency ranges, coaxial sample holders modelled as a transmission line section with lumped parameters were developed. For the frequency ranging from 50 to 250 MHz, RX meters were used for measurement, whereas admittance meters are used for frequencies ranging from 200 to 500 MHz. By confining samples in a coaxial sample holder, precision bridges were used for audio frequencies from 250 Hz to 20 KHz [52].

A shielded open-circuit coaxial sample holder was created simply by assembling components of the sample holders used in earlier studies with the RX meter and admittance meter to develop a technique of measurement with a frequency range of 100 KHz to 1 GHz with two impedance analysers, proper calibrations and the invariance of the cross ratio technique [53].

The dielectric properties of grains with high frequency bridge measurement from 1 to 200 MHz were determined using a similar shielded open-circuit coaxial sample by Bussey [54] and by Jones and co-workers [55]. Another coaxial sample holder characterized by full two-port scattering parameters was designed and used to provide dielectric properties of grains over the range from 25 to 350 MHz.

3.3. Microwave

Generally, for frequencies that are higher than the one mentioned earlier (1 GHz and above), several MW measurement techniques are available [56]. Other instruments, such as transmission lines, resonant cavities, free space techniques and waveguide systems, were used [57]. MW dielectric properties measurements can be classified as reflection or transmission measurements using resonant or non-resonant systems, with open or closed structures for the sensing of the properties of material samples [58]. Closed-structure methods include waveguide and coaxial-line transmission measurements and short-circuited waveguide or coaxial-line reflection measurements. Nelson described the classification of the MW techniques as follows; Open structure techniques include free-space transmission measurements and open-ended coaxial-line or open-ended waveguide measurements. Resonant structures can include either closed resonant cavities or open resonant structures operated as two-port devices for transmission measurements or as one-port devices for reflection measurements [44]. A schematic diagram of methods to measure dielectric properties is provided in Figure 2.

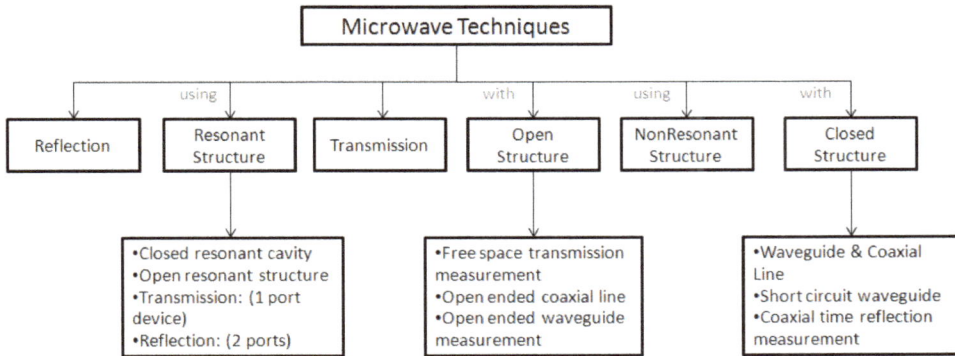

Figure 2. Dielectric property measurement techniques.

With the availability of the modern MW network analyser, the efficiency of the methods used to obtain dielectric properties over a wide range of frequencies has increased; these methods cover both time-domain techniques and frequency-domain techniques [59]. To obtain dielectric properties information over wide ranges of frequencies, several frequency-domain systems have been used for measurements on the samples of interest [60]. Moreover, each of these instruments and systems was designed for use over certain ranges of frequency. Table 2 is a graphical presentation of the different types of instrumentation and their appropriate use according to the frequency and material under test.

Table 2. Material measurement techniques.

	Illustration	Material Under Test	Frequency	Other Comments
Coaxial Probe		Lossy Material (liquids and semi-solids)	Broadband	Non-destructive
Transmission Line		Lossy to low loss material (machineable solids)	Broadband	
Free Space		Best for flat sheets, powder high temperature	Broadband	
Resonant Cavity		Low loss materials, small samples	Single Frequency	Accurate
Parallel Plate		Flat sheets	Low Frequencies	Thin
Inductance measurement		Toroidal structures required		Accurate, simple measurement

3.4. Dielectric Sample Holder

An important aspect of the measurement technique is the dielectric sample holder. A suitable method for many materials was provided by the short-circuited technique of Roberts and Hippel [61]. Applying this method, the sample holder is simply a short section of coaxial-line or rectangular or circular termination at the end of the line against which the sample rests. Because the slotted line or slotted section to which the sample holder is connected can be mounted in a vertical orientation, this holder is convenient. In other words, the top surface of the sample can be maintained perpendicular to the axis of wave propagation, which satisfies the measurement requirement.

For fruit and vegetable samples, a rectangular waveguide K-band system was used for measurement [62]. The standing wave ratio (SWR) can be determined by the Roberts and von Hippel [61] method by using the shift of the standing wave node and change in node widths related to the SWR's. This allows for calculating the sample length and waveguide dimensions ε' and ε'' with suitable computer programs [63]. However, a network analyser or other instrumentation can provide similar determinations by measurement of the complex reflection coefficient of the empty and filled sample holder [44]. With the development of special calibration methods to eliminate errors caused by unknown reflections in the coaxial line [64], the automatic measurement of dielectric properties has been facilitated by computer control of impedance analysers and network analysers [65]. Moreover, the accuracy and reliability of free space measurements of the MW dielectric properties of agricultural products has always been a priority for developing researchers.

Recently, the technique that is mostly used for measuring dielectric properties of fruits and vegetables and is showing success for convenient broadband permittivity is the open-ended coaxial line. However, some errors might arise with this technique, such as for significant density variations or in case of the presence of air gaps or air bubbles between the end of the coaxial probe and the sample [66]. Construction of a cylindrical cavity might be advantageous for fruit and vegetable applications, especially a cavity with provision for alternate dielectric properties measurements and MW heating of the sample for temperature control [67].

To avoid the disturbances resulting from multiple reflections within the sample and between the sample and antennas, a minimal 10 dB attenuation through the sample layer should be maintained for free space permittivity. For extensive studies, it was advised to use the network analysers and impendence analysers, but, for the common limited cases studies, a suitable sample holder can be constructed in available MW laboratories. For cases where the data are required at only one MW frequency or for a limited number of frequencies, a resonant cavity technique is a logical choice [68]; a resonant cavity can be used to measure other permittivity-related characteristics, such as moisture content, mass volume and mixture proportions [69].

Figure 3 is a graphical presentation of the system measurement suite according to the type of material under test and the frequency range adopted for measurement.

3.5. Novel Perspective of Dielectric Techniques

In his paper, "dielectric spectroscopy today yesterday and tomorrow", Kremer developed various novel Perspectives of Dielectric Techniques that are expected to broaden, in his view, the scope of dielectric spectroscopy considerably. First, the principle of hole-burning spectroscopy has been extended to dielectrics and has enabled the answering of questions that cannot be tackled with conventional dielectric spectroscopy [70]. For instance, non-resonant dielectric hole-burning spectroscopy has made it possible to determine experimentally if relaxation processes are homogeneously or heterogeneously broadened. Second, the extraordinary high sensitivity of dielectric spectroscopy enables (frequency-dependent) thermal expansion spectroscopy with high precision [71,72]. Third, dielectric spectroscopy is one of the few spectroscopic techniques that measures directly a molecular fluctuation and has an increasing sensitivity with decreasing thickness, which is a good advantage (this requires refined techniques to deposit electrodes). In this context, a novel approach that allows for avoiding the direct evaporation of metal electrodes has been published [73]. Kremer [32] described an exciting perspective in his view, which is the employment of AFM to carry out dielectric measurement on the very local scale of approximately 50 nm.

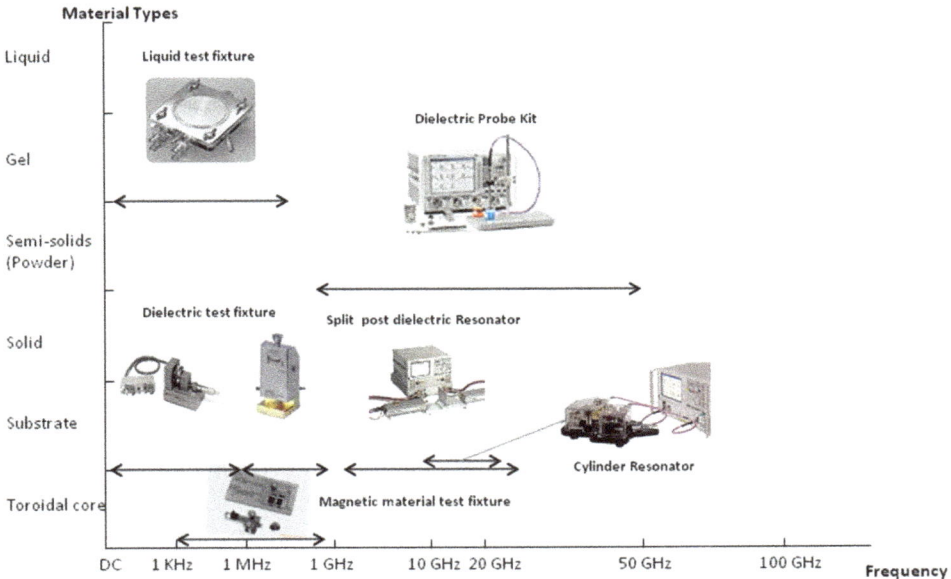

Figure 3. The material measurement fixtures.

111

4. Dielectric Application Data

4.1. Apple

Guo and his co-workers [19] measured the dielectric properties of "Fuji", "Pink Lady" and "Red Rome", three cultivars of fresh apple *Malus domestica* Borkh, to study their sense of quality. The measurements, executed over 10 weeks in storage at 4 °C, aim to detect whether the dielectric properties are useful for determination of quality factors such as SSC, firmness, moisture content and pH. The experiments were conducted for 51 frequencies on a logarithmic scale from 10 MHz to 1800 MHz for external surface and internal tissue measurements, initially and at 2-week intervals during the 10-week storage period.

During the 10 weeks of storage, the dielectric properties did not change much and instead remained relatively constant. This fact agrees with earlier reported literature regarding the permittivity differences for apple surface and interior tissue measurements [74]. Another point to state is that there are some considerable differences among the three cultivars. The same as for watermelon, the dielectric relaxation phenomenon is also exhibited on the surface measurements that are most likely related to the exocarp structure of the fruit. Taking into consideration the difference in scale, it is noted that the dielectric constant measurements are similar for both the external and internal measurements, but, for the loss factor, the internal tissue measurements show a greater variation than those obtained for the external tissues, especially at lower frequencies. From the obtained results, it is noted that all of the dielectric properties of the three cultivars have the same behaviour with respect to frequency. Additionally, the decline in the dielectric constant is a notable characteristic with increasing frequency. The same phenomena of ionic conduction and dipolar losses at low and high frequencies, respectively, are detected for internal apple measurements as well. As for the surface exterior measurements and considering the frequency range, an overriding dielectric relaxation behaviour that might be due to bound water relaxation is revealed.

To determine the state of fruit maturity through the potential use of dielectric spectroscopy, a study was conducted in 2010 to find relations with apple physiological compounds, such as sugar content and malic acid. A non-destructive control method was presented for the prediction of climacteric fruits maturity. Measurements were taken at a frequency ranging from 500 MHz to 20 GHz. A new good correlation was found between apples' Thiault Index and a new defined Dielectric Maturity Index. This Dielectric Maturity Index was related to the loss factor at two punctual frequencies (0.5 GHz and dipolar relaxation frequencies) [75].

At the frequency range of 500 MHz to 20 GHz, another dielectric measurement was applied to apples with different sugar contents during ripening. The objective of the study was to determine the optimal time for eating the fruit. Various good new correlations between the newly defined maturity index and the Thiault index were found. Prospective data of some chemical components of apples were presented in the study [76].

To distinguish between diseased and normal fruits, a new theoretical basis for non-destructive inspection and a research method based on electrical properties was developed. The study investigates the change of the law of dielectric properties on the Fuji apple superficial scale. For a frequency ranging from 100 Hz to 3.98 MHz, the dielectric properties of fruits embracing the impedance, reactance, conductance, capacitance and loss coefficient were measured using the Inductance Capacitance Resistance analyser (LCR). The results show that fruit impedance and reactance decrease as frequency increases; the capacitance and dielectric loss coefficient changes are irregular. Only the conductance increases in spiral form for the same storage times. A significant positive correlation exists between impedance and reactance. Moreover, values of capacitance differ between diseased and normal fruits form 100 Hz to 3.98 MHz, which suggests that the quality of apples can be reflected by capacitance to some extent [77].

Using a network analyser and open-ended coaxial line probe, dielectric properties measurements were effected on the external surface, internal tissue and juice of "Fuji" apples during the last two months of tree-ripening. The objective of the study was to determine the relation of permittivity with apple quality by measuring the firmness, SSC, pH, moisture content and electrical conductivity. The observations reveal that the permittivity and electrical conductivity did not show a specific pattern during the ripening period; the moisture content and SSC remained constant, but the firmness and pH increased with maturity. Thus, the study did not find a correlation between permittivity, firmness, moisture content or pH, and the linear relationship between surface permittivity and SSC was poor (<0.2). Calculating the linear regression at 4.5 GHz, the best correlation between the loss tangent of the tissue and SSC and between the dielectric constant of juice and SSC was found to be 0.61 and 0.67, respectively [78].

In the same year, Guo and his co-workers also investigated the dielectric constant and loss factors of apples with skin (skin on), without skin (skin off) and the flesh juice of "Fuji" cultivar for a frequency range from 10 to 4.5 GHz. The firmness, moisture content of the flesh, SSC, pH and direct current conductivity of the juice were under measurement. The results show that with storage time, the pH increases and the firmness decreases, but no pattern was detected for the moisture content, SSC, conductivity or permittivity. With increasing frequency, the depth of penetration in skin-on and skin-off apples and apple juice decreases. Because the correlation between apple permittivity, internal quality and limited Dp was weak, it was difficult to sense the apple internal qualities from the permittivity of skin-on apples, skin-off apples and apple juice. The study concludes that information provided by dielectric properties can be useful in developing thermal treatments for postharvest insect control based on RF, MW energy and non-destructive methods for detecting internal fruit quality as well. Additionally, designing the treatment bed thickness in RF and MW systems and estimating the heating rates of apples samples can be conducted with dielectric properties data. The limited penetrations at frequencies higher than 3 GHz may suggest some unpractical apple quality sensing under the experimental conditions. The study points also to some weak correlations between permittivity and quality indices [79].

Physical information about apples that is closely related to chemical information has been investigated by monitoring apples during aging via electrical impedance spectroscopy as a fast and non-invasive method. The study proposes two different analytical techniques for assessing the apple properties changes. The first method is based on a single measurement at a low frequency range (approximately 100 Hz), while the other is based on an Argand plot. The major changes observed in electric impedance spectra were attributed to changes in the apples' relative moisture content. Moreover, an equivalent circuit scheme helped to model the apples' behaviour and derive the apoplastic and simplistic resistances and relaxation times [80].

A two-stage study was conducted in 2013 based on the necessity of finding effective methods to select key features from all other dielectric features to reduce the cost of dielectric signal application in non-destructive detection of fruits and crops. For this, a compact discriminative dielectric feature subset was found at first, and then, based on the first step, a non-destructive apple internal quality estimation system is evaluated. Measurements are executed with nine frequency points ranging from 158 Hz to 3.98 MHz using an LCR tester. Apple samples are graded according to their freshness by weight loss rate (WLR). The apple with WLR equal to 0%, 5%, 10% and 15% are assigned grade one, two, three and four, respectively, whereas grade five apples are those with brown stains. The most discriminative apple internal quality estimation is the compact set of dielectric features. Three classifiers are evaluated at the internal quality estimation stage, the sparse representation classification (SRC), artificial neural network (ANN) and support vector machine (SVM). The results show that fast clustering-based feature subset selection (FAST) selects only four dielectric features with 80% classification rate, while sparse principal component analysis (SPCA) accuracy is mediocre and the performance of the greedy selector is significantly outstanding with classification rates of 91.22% and 95.95%, respectively. The results show that dielectric features are highly relevant to the apple internal quality that can be estimated with a compact set of dielectric features [81].

Another experiment studied the effect of high-voltage pulsed electric field treatment on the dielectric properties of apples using the theory of electromagnetic fields. The results show that when the pulse voltage and the pulse frequency increase, both the equivalent capacitance and equivalent impedance of apples decrease. In

114

fact, the dielectric properties are affected by the high-voltage pulsed electric field parameters [82].

The soluble solid content in "Fuji" apples was predicted by applying the BP network model and support vector regression (SVR) over a frequency range from 10 MHz to 4.5 GHz during 21 weeks of storage. Effects of the full frequency (FF), principal component analysis (PCA) and successive projection algorithm (SPA) prediction models were compared and evaluated. The results show that the PCS-SVR results were better than SPA-BP; the predicted correlation coefficient of PCA-SVR was 0.883, and the root mean square error (RMSE) was 0.552. Moreover, the effect of the PCA-BP was worse than that of PCA-SVR. The RMSE of the established model by SPA was generally smaller than by other methods, and the predicted correlation coefficient of the models established by PCA was generally higher. Based on the frequency spectrum of dielectric parameters, some useful technologies in developing non-destructive sensors for fruits' soluble acid content were offered as well [83].

Based on the concept that dielectric properties are highly relevant to the internal quality of many fruits and crops but their application is restricted because of their changing frequency, a study was conducted by Li and his co-workers. Its objective was to find an effective dielectric feature selection method to reduce the cost of the dielectric signal application in non-destructive detection of fruits and crops. For this, "Fuji" apples were graded into five levels according to their dielectric features using several methods, such as SPCA, FAST and a ranker method with the attribute evaluator of information gain. The experimental results show that some key dielectric features are sufficient for high classification accuracy [84].

Recently, a study investigated the feasibility of using dielectric spectroscopy as a non-destructive technique to determine the soluble solid contents of fruits. Three varieties of apples were chosen ("Fuji", "Red Rome" and "Pink Lady"), and their dielectric constants and loss factors were obtained at 51 discrete frequencies from 10 to 1800 MHz using an open-ended coaxial line probe and an impedance analyser. Methods, PCA and the uninformative variables elimination method (UVE-PLS) were applied for the extraction of the dielectric spectra. To establish models that predict the SSC of apples, other methods such as the generalized regression neural network (GRNN), SVM and extreme learning machine (ELM), have served, with calibrated root mean square error and predicted root mean square error of 0.840 and 0.822, respectively. The study also reveals that non-destructive determination of SSC of apples can be done through a combination of dielectric spectra, artificial neural networks and chemometric methods [85].

It is important to mention the existence of other non-destructive methods that have been used for the detection of apples firmness and SSC, such as the biospeckle based on the analysis of laser light variation scattered from the apple sample [86]. Monitoring of quality and ripening of fruits and vegetables, analysis of seeds and assessment of mobility parameters are all applications of the biospeckle technique in the agricultural area [87]. Overall, one can conclude that ionic conduction occurs at low frequencies and dipole relaxation at higher frequencies. A combination of effects including Maxwell-Wagner, bound water, ion related phenomena and molecular cluster are the reason behind the overriding dielectric relaxation of loss factors. For non-destructive surface measurements, the correlations between permittivity and quality indices are not high enough for sensing internal quality. On the other hand, a potential for on-line quality sensing applications may be offered at the tissue and juice levels. All the studies conducted admit that further analysis through radio frequency electromagnetic fields is necessary for a satisfactory quality assessing.

4.2. Avocado

In 2003, Nelson [37] conducted measurements of the dielectric properties of the avocado *Persea americana* Miller var. Americana obtained over the temperature range of 5 °C to 95 °C, in the frequency range from 10 MHz to 1.8 GHz. The results show that the dielectric constant dependence with the temperature disappears at some frequency in the range shown; above that frequency, the temperature coefficient for the dielectric constant is negative, and, below that frequency, the temperature coefficient for the dielectric constant is positive. As explained by Nelson, dipole relaxation accounting for most of the energy loss occurs at above that frequency, while ionic conduction is the dominant loss mechanism below that frequency. This critical edge frequency is approximately 100 MHz for the avocado. It is also noted that for lower frequencies, the dielectric constant of the avocado increases in a regular fashion with the temperature increase. The same pattern is observed for the dielectric loss factor, which increases with temperatures at lower frequencies regularly.

Another experiment on avocado products was conducted in 2008 to measure the dielectric properties of food having the potential to be processed using a continuous flow MW heating system at 915 MHz and in the temperature range of 10 to 90 °C. The results show that with the temperature increase, the dielectric constant decreases and the dielectric loss factor increases [88].

Although only poor correlations were observed between moisture content, density and soluble solid content, the research on avocado provided new information that could be useful in understanding the dielectric behaviour for sensing quality.

4.3. Carrot

In 2003, Nelson [37] measured the frequency and temperature dependence of the dielectric constant and the dielectric loss factor of the carrot (*Daucus carota* subsp.sativus (Hoffm) Arcang) over the temperature range from 5 °C to 95 °C. The critical edge frequency obtained is approximately 100 MHz for the carrot. It is also noted that for lower frequencies, the dielectric constant of the carrot increases in a regular fashion with the temperature increase. A remarkable peak at 65 °C is noticeable for the carrot tissues before decreasing slightly with the temperature increase to 95 °C. The same pattern is observed for the dielectric loss factor, which increases with temperatures at lower frequencies regularly. As for the frequency increase, both the dielectric constant and loss factor have a decreasing pattern [66].

Recently, an experiment was conducted to find the glass transition temperature from frozen and fried mixtures of carrot fibre plus fructose using dielectric analysis in the range of 200 Hz to 1 MHz. The temperature of the peak derivative assumed to be the transition temperature was found to be frequency-dependent. The moisture was in the range of 2% to 4% when the fructose carrot fibre mixture was spray dried and was free flowing. When the fibre content was higher, no significant difference in stickiness and moisture content was detected. Differential scanning calorimetry of 40%, 50%, 60% and 70% carrot fibre fructose showed transition temperatures of 107 °C, 114.5 °C, 122.9 °C and 130 °C, respectively. The values imply that wall build-up might be avoided in a larger scale dryer [89].

Research concludes that there is point in the frequency range between 10 MHz and 100 MHz where the dielectric constant dependence with the temperature was minimal. The behaviour of the permittivity with temperature can be interpreted by the dominance of ionic conduction and dipole relaxation for frequencies below and above that point respectively.

4.4. Coconut Water

Kundu and Gupta applied the same experiment on coconut water. However, because the sample is liquid, the permittivity sensor probe needs to be inserted totally within the liquid to yield accurate measurements. The results show that similarly to other fruits (brinjal, tomato, guava and apple), the real component of permittivity decays in an inverse manner with frequency. The energy storage capability of coconut water was reduced with frequency, and the energy loss appeared to be, at minimum, in the frequency range of approximately 2 GHz [90].

4.5. Eggplant (Brinjal)

Under the same conditions of the experiment conducted on tomatoes, Kundu and Gupta measured the permittivity of eggplant (brinjal). The real component of the permittivity decays almost exponentially similarly to tomatoes (to be discussed in section 4.13 of this article) with frequency increase. Over the measured frequency, the dielectric constant is higher at 16 °C than at 25 °C. The relative permittivity of brinjal falls from 36 at 200 MHz to 26 at 8.5 GHz, implying that the energy storage capability decreases with frequency. The value of permittivity is reasonably low due to the presence of less moisture and more air in the brinjal body. The observations of the minimum field energy loss and loss tangent curve indicate that energy storage capability and energy loss within the vegetable vary with ambient temperature [90].

4.6. Grape

Using a parallel plate electrode system, the dielectric parameters of red globe grapes were studied as a function of the room temperature and storage time for a frequency range of 50 Hz to 1 MHz. The results show that with the frequency increase, both electric conductance and equivalent parallel capacitance increase, whereas equivalent impedance decreases. At 25 KHz, the loss coefficient reaches its minimum and the quality factor reaches its maximum. With increasing storage time, both the equivalent impedance and quality factor decrease for the same frequency, whereas the equivalent parallel capacitance, the loss coefficient and the electric conductance increase. The study concludes that there is a strong correlation between the electrical characteristic parameters and electric field frequency [91].

4.7. Guava

Guava fruit has been measured under the same conditions as tomatoes by Kundu and Gupta [90]. The fruit shows a similar pattern of dielectric constant decrease with frequency increase. The first observation of the study shows that relative permittivity goes up with a temperature increase from 16 °C to 25 °C in GHz frequency scale, implying that the energy storage capability of guava is reduced with frequency. The energy conversion to heat is at its minimum in the frequency band of approximately 1 GHz. The loss within guava, as noted by the loss tangent curves, slightly decreases with the temperature increase from 16 to 25, which goes against Nelson's reports in 2003 and 2005 [90].

4.8. Mango

Sosa-Morales and his co-workers measured the dielectric properties of mangos to understand the interaction between the fruit and electromagnetic energy. Measurements were performed using an open-ended coaxial line probe and an impedance analyser for a frequency ranging from 1 to 1800 MHz, a temperature from 20 °C to 60 °C and 16 days of storage. The results show that the dielectric constant was decreasing with increasing frequency, and the loss factor was decreasing even more so. With increasing temperature, the dielectric constant decreases while the loss factor increases. With storage time, both the dielectric constant and loss factor were decreasing, which is attributed to the reduction in moisture content and increase in pH. The penetration depth (Dp) at which the power drops to 36.8% of its value at the surface of the material was calculated. Obtained results guide that Dp decreases with both temperature and frequency increase, implying that energy penetrates deeper into mangos than MW [92]. This should be expected from the equation developed by Metaxas and Meredith [34] that reflects the relative variation of Dp in function of temperature and frequency.

In 2013, a study was conducted by Yoiyod and Kaririskh [93] on the ripeness monitoring of mangos through a cost-effective remote sensing system. The dielectric properties of the peel and pulp of mango fruit were measured at different maturity stages. A frequency ranging between 6 and 18 GHz is considered to be the most suitable operating frequency. The results show that a significant difference in the reflection coefficient exists between ripe and unripe mangos [93].

4.9. Melon

In 2007, Nelson and his co-workers [94] conducted a measurement of honeydew melon dielectric properties to study if a useful correlation exists between the dielectric properties and melon sweetness, as measured by soluble solid content. The measurements show that the mean values of dielectric properties from six probe measurements on internal watermelon tissue reveal a linear relationship for the loss factor between 10 and 500 MHz, which can be interpreted by the dominant influence of ionic conduction in that range [95]. It is essential to mention that a significant difference exists between the measurements executed on internal tissue and those on the surface. Due to the influence of bound water and the complex combination of Maxwell-Wagner and ion-related phenomena, broad dielectric relaxation occurs at the surface measurements, explaining this result difference. Despite the success of relating the dielectric properties of solids in complex plane plots, Nelson [94] stated that the search for correlations for predicting melon sweetness from the dielectric properties was not successful so far.

In 2011, Nelson and Trabelsi [96] re-examined the earlier dielectric spectroscopy measurements of honeydew melons and watermelons from 10 MHz to 20 GHz. The objective of the study was to find useful correlations with SSC (sweetness) for non-destructive sensing of melon quality. The study could not reveal any new information regarding the melon quality through dielectric properties. The internal edible tissues useful for quality sensing were attenuated by RF signals. However, better coefficients of variations were obtained at higher frequencies.

4.10. Orange

Nelson [37] measured the dielectric properties for the navel orange (*Citrus aurantium* sbsp. Bergamia). The frequency at which the temperature coefficient reverses its sign, known as the frequency of zero temperature dependence, is noted to be at approximately 50 MHz [36]. A phenomena occurs at lower frequencies where large ionic conduction masks the biological cell constituents and the dielectric relaxation forms of bound water [97]. This explains why the reversal of the sign of the temperature coefficient is evident only for the dielectric constant and not for the case of the dielectric loss factor. However, for high frequencies, a shift in the relaxation frequency for liquid water to higher frequencies occurs with temperature increase, resulting in the reversal of the temperature coefficient for the loss factor at the high end of the frequency range. The relaxation frequency of liquid water is 10.7 GHz at 50 °C; thus, the associated dispersion is evident above 1 GHz. The polarization at lower frequencies in cellular structures is mainly due to the increased ionic diffusion and ionic conductivity processes, which cause the high values of the dielectric constant at low frequencies [98]. As evidenced by the data, when the temperature increases, this dispersion might also shift to higher frequencies. Nelson also concluded that the ionic conductivity phenomena which increases with frequency falls in keeping the convention of the Kramers-Kronig relations. This critical edge frequency is approximately 40 MHz for the navel orange. It is also noted that for lower frequencies, the dielectric constant of the orange increases in a regular fashion with temperature increases. The same pattern is observed for the dielectric loss factor, which increases with the temperature at lower frequencies regularly.

4.11. Peach

In 1995, Nelson [7] measured the dielectric properties of tree-ripened peaches of different maturities. Among the three categories of peaches included in the study, the "Dixired" were the first to mature, followed by "Redheaven" and "Windblo." An attempt to narrow the maturity range was taken via non-destructive measurements that have been associated with maturity in experimental work [99], taking into account both the blush and the ground sides of the peach fruit to accommodate any differences that might be noted on the same fruit (5% accuracy as specified by Helwett-Packard).

The results of their experiments prove that the dielectric constant of the peaches at different maturities are diverging at the lower end of the frequency range; therefore, it seems interesting to explore the behaviour of the constants for other fruits and vegetables at a lower range of frequencies (below that point). Generally, standard deviations for the 15 permittivity measurements on tissue samples of the five fruits were less than 2% for the dielectric constant and less than 10% of the loss factor. It is also noticeable that maturity is accompanied by a slight increase in the dielectric constant at 200 MHz; however, the dielectric loss factor shows only a slight dependence on the stage of maturity. At a higher frequency, e.g., 10 GHz, the dielectric loss factor tends to increase with maturity, whereas the dielectric constant has no consistent trend across the samples. In sensing the stage of maturity, Nelson and his co-workers suggested from the data obtained that values of ε' at lower frequencies and values of ε'' at higher frequencies should be useful. Therefore, a permittivity maturity index was developed, as shown in Equation (5):

$$Mp = \frac{(\varepsilon')_{0.2} + (\varepsilon'')_{10}}{100} \tag{5}$$

A second maturity index based on permittivity, utilizing both real and imaginary components, was formulated as Equation (6):

$$Mltr = \frac{(\tan\delta)_{10}}{(\tan\delta)_{0.2}} \tag{6}$$

This new ratio takes into account the fact that ε' shows more variation with maturity at 0.2 GHz and ε'' shows more maturity dependence at 10 GHz. Thus, their product should be useful. To cover both real and imaginary components of the permittivity, the ratios $\varepsilon'/\varepsilon''$ at 0.2 GHz and $\varepsilon'/\varepsilon''$ at 10 GHz would be relevant with the use of the loss tangent tan δ (defined as the ratio of $\varepsilon'/\varepsilon''$). Based on the experiments, the dielectric constant and loss factor, which are the real and imaginary components of the complex permittivity, respectively, are showing a significant variation with frequency over the range from 200 MHz to 20 GHz. Because the values appear to be diverging with respect to maturity indices, more effort should be required in this area for more reliable maturity indices in the non-destructive sensing of maturity for peaches. This being said, Nelson's experiment could reach two maturity indices development but could not reveal any new information behind the data variations of ε' and ε''.

In 2007, an experiment based on electrical properties measurements was conducted to investigate the electrical and physiological properties of peaches. The study was seeking a better understanding of the electrical properties of post-harvest fruits and a deeper exploration of new quality sensing methods. With peach aging, the loss tangent was observed to decrease and the relative dielectric constant to vary with the cosine law, roughly. At the peak of respiration, the maximum relative dielectric constant appeared [100].

Also in 2007, the internal quality of just-harvested early- and mid-season peaches was studied using MW spectroscopy. Establishing the feasibility of MW measurements for the firmness and sugar estimation of the peaches was the objective of the study. Using a contact coaxial probe from 1 to 20 GHz, the return loss (LR), a MW parameter, was measured simultaneously with reference parameters such as firmness, acidity, sugar content and optical reflectance. Because the MW response changes significantly with the treated samples, the results show that the dielectric response in the peach fruits is affected highly by moisture content and temperature. In fact, when fruits are submerged in water for 1 h (moist) and the temperature is cold (1 °C), the LR increases 50%–100% compared with fruits at ambient temperatures

(20 °C). Return loss values obtained at different measured frequencies (1, 7, 9.9 and 19 GHz) show high correlations for the same fruit. However, low correlation is observed with fruit firmness, which is not consistent enough to be applied for fruit inspection and sorting purposes. The study demonstrates a low repeatability of the LR and concludes that the most significant independent variable for estimating peach firmness is the reflectance at 680 nm. Moreover, the variance that is the coefficient of determination of the firmness models reaches approximately 50% to 60%, with lower values for sugar estimation models [101].

In 2010, peaches with different maturities were selected to explore their dielectric properties and determine their internal quality. Experiments were held at 25 °C using coaxial open-ended probe technologies over a frequency range from 10 to 4.5 GHz. Measurements of the moisture content of peach pulp, soluble solid contents (SSC), pH value and electrical conductivity show that the dielectric constants of the pulp and juice decreased with increasing frequency, while the loss factors change with "V" type. The major loss mechanisms at lower and higher frequencies were the ionic conduction and the dipolar polarization. Moreover, the relationship between permittivity and soluble content, pH value and moisture content was non-linear [102].

In order to understand the interaction between the electromagnetic fields and to design treatment beds in industrial applications, an experiment investing the peaches dielectric properties was conducted recently. The study was based on the fact that MW and radio frequency methods hold potential for postharvest thermal disinfestations to replace chemical fumigation. Using an impedance analyser, the dielectric properties of the peaches were determined between 10 and 1800 MHz over a temperature ranging from 20 to 60 °C. The results show that the dielectric constant varies between 60 and 75, accounting for an 8% to 10% change due to the temperature effect. When temperature increased from 20 to 60 °C, the loss factor decreased linearly with frequency on the log scale by approximately 110%, and the loss factor ratio was 1.66 at 20 °C. The penetrating depth decreased with increasing frequency [103].

4.12. Potato

Dielectric properties of potatoes were measured with a system consisting mainly of an impedance analyser, high temperature coaxial cable and a metal sample holder. When increasing the temperature from −20 °C to 0 °C, both the dielectric constant and loss factor rapidly increase. From 0 °C to 100 °C, the dielectric constant decreases linearly, while the loss factor decreases first and then linearly increases [104]. With a frequency increase, both the dielectric constant and loss factor have a declining pattern [66]. Results show that the thermal properties found are critical input parameters for a microwave heat transfer model. Thus, authors suggested implementing this model to be used by food scientists in developing novel products that minimize non-uniform heating during cooking in domestic microwave oven.

4.13. Tomato

De los Reyes [30] measured the dielectric properties of tomatoes at three locations: the centre of the pericarp, the locular cavity and the skin. However, the measurements at the skin were not considered to be successful because of the non-permeability of the cereus skin to the mass transfer or the bad connection between the probe and the tomato skin. The mean values of six replicates of tomato samples were measured by the authors from 200 MHz to 20 GHz at 21 °C. It is remarkable that large drops in the Dp may damage the fruits due to elevated power absorption.

Over a frequency ranging from 30 MHz to 3000 MHz, a study was executed in 2013 to determine the tomatoes dielectric properties. Open-ended coaxial probe technique was used. The effects of NaCl and $CaCl_2$ were investigated on three tomato tissues separately: the pericarp (including the skin), the locular tissue (includes the seeds) and the placental tissue. The results show that with increasing frequency and salt addition, the loss factors of the three tomato tissues decrease. However, with increasing temperature, the loss factors decrease initially but then increase at 915 MHz. The ionic conductivity is the main reason behind the differences in the loss factors of the three tomato tissues [105].

Recently, Kundu and Gupta measured the dielectric properties of tomato fruits for a frequency range from 0.5 GHz to 5.5 GHz at two different temperatures, 16 °C and 25 °C, using the Agilent 85070E Dielectric Probe Kit and Network analyser. The objective is to note the switch in dielectric constant and dielectric loss factor. Although the permittivity curve of tomatoes is higher at 16 °C than at 25 °C, results show that the dielectric constant decays exponentially with frequency increase at both temperatures. The relative permittivity of tomatoes falls from 63 at 200 MHz to 40 at 8.5 GHz. This implies that the energy storage capability of tomatoes is reduced with frequency, but a high permittivity is always expected due to high moisture content. Moreover, the loss tangent is minimal at approximately 1 GHz to 2 GHz, and the dielectric loss factor increases. This implies that the energy conversion is also minimal in the frequency range of 1 GHz to 2 GHz [100].

5. Discussion

Sensing food quality through dielectric spectroscopy reveals some very impressing results and opens the horizon for better food consumption in a society where quality is the key for food industries success. Thus, the focus was to build a solid literature review of the various species of fruits and vegetables that have been measured for their dielectric properties. The topic might seem straightforward but experiments varied much in their hypotheses and circumstances. As the paper shows, for the same type of fruit "apple", authors might come out with different results according to the nature of experiments conducted. For example, apples in each experiment were tested for different quality indices (firmness, ripening, soluble solid content, sugar content, moisture content) and the factor under scope was considered by changing different environmental parameters (temperature, frequency, storage period). Moreover, the quality index varies from one fruit to another which multiplies the possibilities of experiments and makes it an unlimited field. This explains the variation in the experiments that were never unified for only one aim. For the same type of fruit, quality estimation could be measured from different angles seeking improvement.

Regarding the frequency range, it is not consistent within the different apple experiments and for the other fruits as well. Some were conducted at low frequency range; however recent ones have been applied at higher frequency ranges. Recently, high-frequency sensing measurement has been approachable with the availability of advanced technology instruments. It is important to note that a higher frequency requires higher budget cost which might be the obstacle for many of the researches. Throughout the paper, it was shown that higher frequency lead to more focused results. The various frequency ranges executed on the different species of vegetables and fruits are presented graphically in Figure 4.

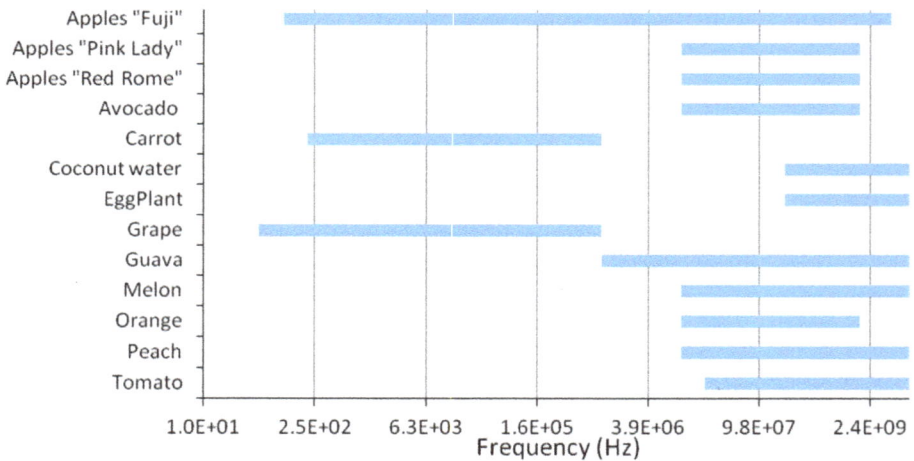

Figure 4. Frequency Ranges applied for fruit and vegetable measurements.

After comparing the results of several experiments conducted on fruits and vegetables, only the dielectric constant measurements of apple fruit by Kundu and Gupta were similar and repeated in most of the studies and verified Nelson's work in 2005, 2007 and 2008. In most of the vegetables rich in water, the real component of the permittivity shows reasonably high results due to the moisture content with the exception of brinjal, where relative permittivity was lower due to lower moisture. Thus, the results demonstrate the fact that higher permittivity is expected in fruits and vegetables with higher moisture content [100]. The loss tangent curves show that the loss is at a minimum at approximately 1 GHz to 2 GHz for most of the measured samples, which implies that dielectric loss is minimum at that frequency band and that the RF field provides an optimum angular velocity for the polar particles in those samples. In other words, the friction becomes nominal to reduce the heat generation within vegetables and fruits. For the apple, tomato, eggplant and coconut water, it is remarked that the relative permittivity decreases and the loss tangent increases with increases in temperature (from 16 to 25). Only the guava fruit is an exception. Both observations of Nelson, in 2003 and 2005, are in concordance with Kundu and Gupta. It may be that maturity levels of guavas were not exactly the same, which resulted in different temperature dependence for the dielectric loss and loss curve for guava.

For apples [19], the dielectric measurements applied on different parts of the fruits did not show similar results; the same was true for the surface and internal measurements that were considerably different. Lower dielectric constants were observed at the surface of apples, and much lower dielectric loss factors at the surface than at the internal tissues with increasing frequency, regularly pointing to the dominance of ionic conduction at lower frequencies and dipolar relaxations at high frequencies. Additionally, the dielectric properties remained relatively constant throughout the storage duration of 10 weeks. To sense apples' quality factors, wider frequency range analysis must be necessary through RF electromagnetic fields. Moreover, the fact that extracellular resistance is lower than intracellular resistance implies a disproportionate relation between resistance and permittivity constants. This statement is verified by Keller and Licastro, who proved that the high resistivity was associated with lower water content and that a high dielectric constant is associated with high water content [106]. In practice, this would mean that the surfaces of the apples having high resistivity and lower permittivity is poor in water content, whereas the internal part of the apple showing lower intracellular resistance and higher permittivity values is rich in water. One can conclude that permittivity and water content are positively correlated; then, because resistivity and water content are negatively correlated, resistivity and permittivity are thus negatively correlated. Table 3 shows a summary of the different dielectric parameter variations with various experimental factors; a conclusion on the permittivity pattern is accompanied.

Table 3. Parameter Variation overview (Δ & ∇ for parameter increases and decreases, respectively).

Fruit/ Vegetable	Frequency			Temperature			Storage Time			Conclusions
	ε'	ε''	Others	ε'	ε''	Others	ε'	ε''	Others	
Apple	∇		Impedance ∇ Reactance ∇				No pattern	No pattern	Conductance Δ Firmness Δ pH Δ	Linear decrease with frequency
Avocado				At low freq. Δ At high freq. ∇	At low freq. Δ					Inflection point at Critical edge freq. 100 MHz
Carrot	∇	∇								Inflection point at Critical edge freq. 100 MHz
Coconut	∇									Linear decrease with frequency
Eggplant	∇									Linear decrease with frequency
Grape			Conductance Δ Capacitance Δ Impedance ∇					Δ	Equ. Capacitance ∇ Equ. Parallel capacitance Δ conductance Δ	Linear increase with storage time
Guava	∇		Energy storage capability ∇	Relative permittivity Δ	Relative permittivity Δ					Linear decrease with frequency
Mango	∇	∇	Dp ∇	∇	Δ	Dp ∇	∇	∇		Linear decrease with frequency
Melon										Frequency linear relationship between 10 and 500 MHz
Orange				At low freq.Δ	At low freq.Δ	Dispersion shift to higher freq.				Temperature linear increase below 50 MHz
Peach	∇ & Std. < 2%	V type & Std. < 10%	Dp Δ High correlation of:LR & freq. and LR & fruit firmness		linear ∇					Frequency and temperature linear decrease
Potato	∇	∇		From −20 °C to 0 °C Δ From 0 °C to 100 °C ∇	From −20 °C to 0 °C Δ From 0 °C to 100 °C Δ then ∇					Frequency linear decrease, varying temperature pattern
Tomato	Exponential ∇	∇			∇ then Δ at 915 MHz					Exponential relationship with frequency

The experiments shown earlier contribute in framing the permittivity constant variations and determining its major influencing factors. The composition of the fruit, especially the water content, has a positive correlation with the dielectric constant and dielectric loss factor. Indeed, the water is the major absorber of energy in foods; thus, the higher the moisture content, the better is the heating [107]. Therefore, dehydrated fruits show a decrease in their permittivity level compared to other fruits rich in water.

Regarding ionic components, it was observed that salts result in augmenting the dielectric loss factor in mashed potatoes, while the dielectric constant was not affected [17]. Because the physical structure considerably affects the dielectric properties, density is another important factor observed to change with measurements, e.g., the lower the density, the lower is the permittivity. In fact, Guo suggested some simple relations to estimate the chickpea flour dielectric properties from the density and vice versa where it was concluded that the dielectric properties increased with the increase in density and moisture content (from 1.2698 to 1.321 g/cm^3 and 1.9 to 20.9%, respectively) [108].

In regard to temperature, this factor has a strong yet complex influence over the dielectric properties but depends initially on the food composition and moisture content [13]. It can be stated generally that at low frequencies, due to the ionic conductance, the loss factor increases with the temperature increase [17]. In contrast, at high frequencies, due to free water dispersion, the dielectric constant decreases with increasing temperature [109]. These theories have been verified through the experiments. As was discussed for the avocado fruit, the temperature coefficient is positive below 100 MHz, where both the dielectric constant and dielectric loss factor were increasing with temperature; in contrast, the temperature coefficient was negative for frequencies higher than 100 MHz. The pattern was also observed for the navel orange measurements, where the loss factor was increasing with temperature at lower frequencies in a regular fashion, and a massive explanation of the ionic diffusion was a subject of interpretation. Similarly, the peak temperature of carrots reached 65 °C.

In regard to frequency, most lossy materials, *i.e.*, materials that absorb and loose energy from RF or MW heating, have dielectric properties that vary considerably with the frequency variation. The imposed electric field and its orientation influence the polarization of molecules, resulting in the dependence of dielectric properties and frequency [48]. At MW frequencies, both σ and d (of free water) play a major role, whereas only σ is dominant at lower frequencies (<200 MHz). This phenomenon was observed for the avocado fruit, for which Nelson attributed the energy loss at high frequency to the dipole relaxation and the ionic conduction at low frequencies [37]. Figure 5 shows the loss factor and dielectric constant behaviour of carrots [66] and peaches [110]. The same pattern is concluded after the navel orange measurement conduction as well. While the dielectric constant is always decreasing with the frequency increase, the loss factor patterns prove to have either a declining curve (carrots) or a V-type curve with a point of inflection (peaches). This critical frequency point identifies each product and characterizes its behaviour. Thus, fruits and vegetables cannot be distinguished according to the dielectric constant that has similar behaviour, but the loss factor can be characterizing because it differs from one product to another.

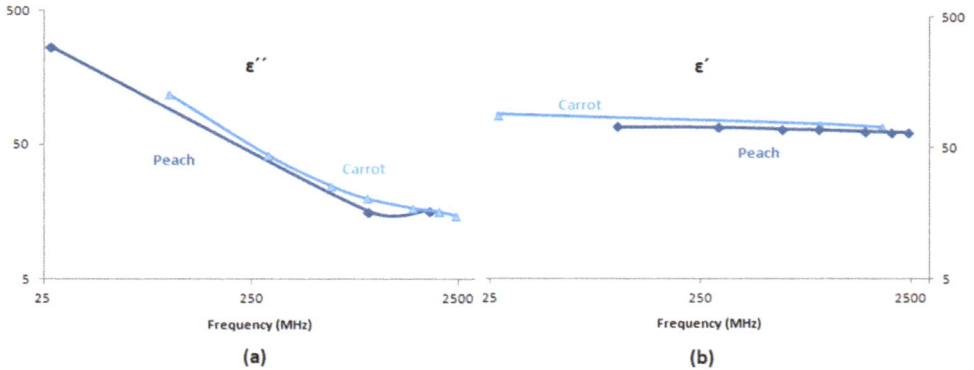

Figure 5. (a) Loss factor behaviour *versus* frequency for carrots and peaches on a logarithmic scale; (b) Dielectric Constant behaviour *versus* frequency for carrots and peaches on a logarithmic scale.

Concerning the peel of the fruit, an experiment tackling the disinfestation of fruits adopted the radio frequency heating as a fast and volumetric method to overcome the problems of conventional heating. The results of the measurements conducted via open-ended coaxial probe methods show that the RF heating behaviour of the fruits is greatly influenced by dissimilarity in peel and pulp dielectric properties. For apples, peeled oranges and grapefruits, core heating was prominent, whereas it was subsurface or peripheral heating for whole oranges, grapefruits and avocados (fruits with thicker peels). This is expected due to the difference in dielectric and physical properties of the pulp and peel [110].

Last but not least is the storage time factor, which directly affects a large number of investigated fruits. Because fruit and vegetable ripening occurs during storage time, storage time effect studies are one of the main goals of the dielectric properties experiments conducted so far. Storage time, being considered as a major quality factor, was significant in the apple study; dielectric constants and loss factors of the apples remained constant during the 14 weeks of storage, which revealed the importance of further research at wider frequency ranges. The case of apples is quite different than that of mangos, whose dielectric properties decrease with storage time.

The phenomena of decreased permittivity in consideration of storage time can be attributed to the increase in pH and decrease in moisture content. The effect of the storage time on the dielectric properties differs from that of the electrical conductivity, which increases with temperature during the storage time [107].

The dielectric studies reached various good conclusions regarding the development of maturity indices. However, further investigation at higher ranges of frequency are always needed for better accuracy, especially in the areas where dielectric properties data do not show significant variations upon which the author can build an effective hypothesis. Moreover, dielectric measurements are distinguished by not being destructive, at least for the experiments shown in this review. The destructivity of such method could be reached at very high frequencies, causing damages to the fruit membranes. In addition, for fruits or vegetables where the membrane is thick, damage is less likely than for fruits with very thin membranes, or by performing external or internal measurements. The importance of these measurements is not only for the detection of the typical dielectric properties but also to provide information on the dependence on variables, such as the alternating field applied, the moisture content and the temperature of the products [95]. In many of the cases, measurements for the same type of agricultural product were duplicated over time with more improved measurements, using several frequency domain measurement systems [60] starting with the RF measurements, the MW measurements, time-domain reflectometry and open-ended coaxial line probes. Investigations of dielectric heating treatments of materials at different frequency ranges may show a higher level of mortality. Some new measurements have been taken to establish a database for a few different fruits and vegetables [37], but comparisons of tissue from fruits or vegetables of different maturities have not yet been tested.

All these indices contribute positively to the hypothesis of valid dielectric characterization of fruits and vegetables. Each of the horticultural products has its appropriate comportment for ε' and ε''. It was sufficient for some products to be characterized through one constant behaviour while others need both and some others are not yet determined. It is a three dimensional system of frequency temperature and dielectric parameters.

To summarize, measuring permittivity and loss parameters of different agricultural products can be useful in employing RF dielectric heating. According to the dielectric materials to be measured, the choice of measurement techniques, equipment and sample holder design are performed; additionally, the frequency range of interest is an important factor as well [44]. In addition to dielectric spectroscopy, other sensors for quality detection are being used, such as E-noses sensors that use metal oxide conductors. Theses sensors have been applied to a wide range of food and beverages [111]. Once the measurement systems and techniques have been developed and their reliability has been verified, the recently developed instrumentations and techniques make the assembly of dielectric properties information or other techniques much more efficient [95].

6. Conclusions

Sensing food quality through dielectric spectroscopy reveals some very impressing results and opens the horizon for better food consumption in a society where quality is the key for food industries success. The paper aims to enlighten the importance of this sensing technique, describing the availability of its instruments, and presenting a state of art of the topic. So, this review summarizes the potential of dielectric properties to investigate functional relationships with temperature, frequency, MW, soluble solid content, moisture content and other processing parameters with sensing the fruit maturity and firmness, knowing that materials undergo physiological changes that will affect the electric property measurements. The application section summarizes from the recent practical research conducted in 13 fruits or vegetables: apple, avocado, carrot, coconut, eggplant, grape, guava, mango, melon, orange, peach, potato, and tomato. All of the obtained conclusions lead to a strong correlation of the dielectric properties with the quality fruit factors at various levels. They aim for a more reliable fruit characterization through the development of rigid permittivity-based indices or solid resistance-based indices in non-destructive environments. The multiple experiences were executed using a variety of instruments and techniques, which implies that a huge market is available for such systems. However, the difficulties faced in many MW applications should be directed towards better designs that are adaptable to the structural changes that might occur during heating or other physical interactions. This review collects a variety of experiments that have been tested, but further research experiments should be conducted for better oriented conclusions and more precise hypotheses.

Acknowledgments: The Erasmus Mundus PHOENIX project of the European Commission has funded the research of Dalia El Khaled at the University of Almeria (Spain) as a Doctoral Fellow (2014-16). This work was supported in part by the Ministry of Economic and Competitiveness of Spain, under Project TEC201460132P, in part by the European Union FEDER Program, in part by Innovation, Science and Enterprise, Andalusian Regional Government through the Electronics, Communications and Telemedicine TIC019 Research Group and want to thank BITAL (Research Center on Agricultural and Food Biotechnology—University of Almeria) for its support.

Conflicts of Interest: The authors declare no conflict of interest.

Acronyms

Nomenclature	Definition
α	Quantification of the symmetric broadening of the relaxation distribution
τ	Relaxation time
σ	Ionic conductivity
γ	Quantification of the asymmetric broadening of the relaxation distribution
ω	Angular frequency
Δ	Increase
∇	Decrease
d	Dipole rotation
E	Rms electric field
ε^*	Complex relative permittivity
ε'	Dielectric constant
ε''	Loss factor
$\Delta\varepsilon$	Dielectric length
α dispersion	Alpha dispersion
β dispersion	Beta dispersion
λ dispersion	Gamma dispersion
f	Frequency
L_R	Return loss
M_{ltr}	Permittivity maturity index
M_p	Permittivity maturity index
$\tan\delta$	Loss tangent
BP	BP network model
Dp	Penetration depth
ELM	Extreme Learning Machine
FAST	Fast clustering based feature subset selection
FDS	Frequency domain spectrometer
FF	Full Frequency
GRNN	Generalized Regression Neural Network
LCR	Inductance Capacitance Resistance analyser
MW	Microwave
P	Power dissipated
PCA	Principal Component Analysis
PCA-BP	Principal Component Analysis using the BP model
PCA-SVR	Principal Component Analysis using the SVR model
PCA-SVR	Principal Component Analysis using the SVR model
RF	Radio Frequency
RMSE	Root Mean Square Error
SPA	Successive Projection Algorithm
SPA-BP	Successive Projection Algorithm using the BP model
SPCA	Sparse Principal Component Analysis
SRC	Sparse representation classification
SSC	Soluble Solid Content
SVM	Support Vector Machine
SVR	Support Vector Regression
SWR	Standing Wave Ratio
UVE-PLS	Uninformative Variation Eliminations
WLR	Weight Loss Rate

References

1. González-Araiza, J.R. Impedancia Bio-Electrica Como Técnica No-Destructiva para Medir la Firmeza de la Fresa (Fragaria x Ananassa Duch) y su Relación Con Técnicas Convencionales. Ph.D. Thesis, Universidad Politécnica de Valencia, Valencia, Spain, 2014. (In Spanish).

2. Choi, K.H.; Lee, K.J.; Kim, G. Nondestructive quality evaluation technology for fruits and vegetables using near—Infrared spectroscopy. In Proceedings of the International Seminar on Enhancing Export Competitiveness of Asian Fruits, Bangkok, Thailand, 18–19 May 2006.

3. Moreda, G.P.; Ortiz-Cañavate, J.; García-Ramos, F.J.; Ruiz-Altisent, M. Non-destructive technologies for fruit and vegetable size determination—A review. *J. Food Eng.* **2009**, *92*, 119–136.

4. Cortés-Olmos, C.; Leiva-Brondo, M.; Roselló, J.; Raigón, M.D.; Cebolla-Cornejo, J. The role of traditional varieties of tomato as sources of functional compounds. *J. Sci. Food Agric.* **2014**.

5. Shewfelt, R.L. Quality of minimally processed fruits and vegetables. *J. Food Qual.* **1987**, *10*.

6. Butz, P.; Hauffman, C.; Taushard, B. Recent Developments in Noninvasive Techniques for Fresh Fruit and Vegetable Internal Quality Analysis. *J. Food Sci.* **2005**, *70*, 131–141.

7. Nelson, S.O.; Forbus, W.R.; Lawrence, K.C. Assessment of MW permittivity for sensing peach maturity. *Trans. ASAE* **1995**, *38*, 579–585.

8. Pliquett, U. Bioimpedance: A Review for Food Processing. *Food Eng. Rev.* **2010**, *2*, 74–94.

9. Ortiz Meléndez, A.D.J. Caracterización de nitratos presentes en lechuga (*Lactuca sativa* L.) mediante el uso de bioimpedancia eléctrica. Ph.D. Thesis, University Autonoma de Queretaro, Santiago de Querétaro, Mexico, 2014. (In Spanish).

10. Chen, P.; Sun, Z. A review of non-destructive methods for quality evaluation and sorting of agricultural products. *J. Agric. Eng. Res.* **1991**, *49*, 85–98.

11. Żywica, R.; Pierzynowska-Korniak, G.; Wójcik, J. Application of food products electrical model parameters for evaluation of apple purée dilution. *J. Food Eng.* **2005**, *67*, 413–418.

12. Içier, F.; Baysal, T. Dielectric properties of food materials-2: Measurement Techniques. *Crit. Rev. Food Sci. Nutr.* **2004**, *44*, 473–478.

13. Tang, J. Dielectric properties of foods. In *The MW Processing of Foods*; Schubert, H., Regier, M., Eds.; Woodhead Publishing Limited: Cambridge, UK, 2005; pp. 22–38.

14. Von Hippel, A.R. *Dielectric Properties and Waves*; John Wiley & Son: New York, NY, USA, 1954.

15. Mudgett, R.E. Electrical properties of foods. In *Engineering Properties of Food*, 2nd ed.; Tabulates Electrical and Dielectrical Properties of Food Materials; Rao, M.A., Rizvi, S.S.H., Eds.; Marcel Dekker: New York, NY, USA, 1986; pp. 329–339.

16. Nelson, S.O.; Kraszewski, A.W. Dielectric properties of materials and measurement techniques. *Dry. Technol.* **1990**, *8*, 1123–1142.

17. Guan, D.; Cheng, M.; Wang, Y.; Tang, J. Dielectric properties of mashed potatoes relevant to MW and radio-frequency pasteurization and sterilization processes. *J. Food Sci.* **2004**, *69*, FEP30–FEP37.

18. Nelson, S.O. MW dielectric properties of fresh onions. *Trans. ASAE* **1992**, *35*, 963–966.
19. Guo, W.; Nelson, S.O.; Trabelsi, S.; Kays, S.J. 10–1800-MHz dielectric properties of fresh apples during storage. *J. Food Eng.* **2007**, *83*, 562–569.
20. Schwan, H.P. Electrical properties of tissues and cell suspensions: Mechanisms and models. Engineering in Medicine and Biology Society. Engineering Advances: New Opportunities for Biomedical Engineers. In Proceedings of the 16th Annual International Conference of the IEEE Engineering in Medicine and Biology Society, Baltimore, MD, USA, 3–6 November 1994; pp. 70–71.
21. Asami, K.; Yonezawa, T.; Wakamatsu, H.; Koyanagi, N. Dielectric spectroscopy of biological cells. *Bioelectrochem. Bioenerg.* **1996**, *40*, 141–145.
22. Grant, E.; Halstead, B.J. Dielectric parameters relevant to MW dielectric heating. *Chem. Soc. Rev.* **1998**, *27*, 213–224.
23. Ramaswamy, H.S.; Voort, F.V.D.; Ghazala, S. An analysis of TDT and Arrhenius methods for handling process and kinetic data. *J. Food Sci.* **1989**, *54*, 1322–1326.
24. Castro-Giráldez, M.; Botella, P.; Toldrá, F.; Fito, P. Low-frequency dielectric spectrum to determine pork meat quality. *Innov. Food Sci. Emerg. Technol.* **2010**, *11*, 376–386.
25. Sarbacher, R.I.; Edson, W.A. *Hyper and Ultrahigh Frequency Engineering*; Wiley: New York, NY, USA, 1943; Volume 19442, p. 174.
26. Rotman, W. Plasma simulation by artificial dielectrics and parallel-plate media. *IRE Trans. Antennas Propag.* **1962**, *10*, 82–95.
27. Sanchez, B.; Bandarenka, A.; Vandersteen, G.; Schoukens, J.; Bragos, J. Novel approach of processing electrical bioimpedance data using differential impedance analysis. *Med. Eng. Phys.* **2013**, *35*, 1349–1357.
28. Zaengl, W.S. Dielectric spectroscopy in time and frequency domain for HV power equipment. I. Theoretical considerations. *IEEE Electr. Insul. Mag.* **2003**, *19*, 5–19.
29. Castellon, J.; Nguyen, H.N.; Agnel, S.; Toureille, A.; Frechette, M.; Savoie, S.; Schmidt, L.E. Electrical properties analysis of micro and nano composite epoxy resin materials. *IEEE Trans. Dielectr. Electr. Insul.* **2011**, *18*, 651–658.
30. De los Reyes, R.; Heredia, A.; Fito, P.; de los Reyes, E.; Andrés, A. Dielectric spectroscopy of osmotic solutions and osmotically dehydrated tomato products. *J. Food Eng.* **2007**, *80*, 1218–1225.
31. Liu, T.; Fothergill, J.; Dodd, S.; Nilsson, U. Dielectric spectroscopy measurements on very low loss cross-linked polyethylene power cables. *J. Phys. Conf. Ser.* **2009**, *183*.
32. Kremer, F. Dielectric spectroscopy—Yesterday, today and tomorrow. *J. Non-Cryst. Solids* **2002**, *305*, 1–9.
33. Ikediala, J.N.; Tang, J.; Drake, S.R.; Neven, L.G. Dielectric properties of apple cultivars and codling moth larvae. *Trans. ASAE* **2000**, *43*, 1175–1184.
34. Metaxas, A.C.; Meredith, R.J. *Industrial MW Heating*; IEE Power Engineering Series; Peter Peregrinus Ltd.: London, UK, 1993.
35. Ryynänen, S. The electromagnetic properties of food materials: A review of basic principles. *J. Food Eng.* **1995**, *26*, 409–429.

36. Nelson, S.O. Dielectric spectroscopy in agriculture. *J. Non-Cryst. Solids* **2005**, *351*, 2940–2944.

37. Nelson, S.O. Frequency and Temperature permittivities of fresh fruits and vegetables from 0.01 to 1.8 GHz. *Trans. ASAE* **2003**, *39*, 281–289.

38. Bellamy, W.L.; Nelson, S.O.; Leffler, R.G. Development of a time-domain reflectrometry system for dielectric properties measurement. *Trans. ASAE* **1985**, *28*, 1313–1318.

39. Einfeldt, J.; Meißner, D.; Kwasniewski, A. Polymerdynamics of cellulose and other polysaccharides in solid state-secondary dielectric relaxation processes. *Prog. Polym. Sci.* **2001**, *26*, 1419–1472.

40. Kwok, B.P.; Nelson, S.O.; Bahar, E. Time-domain measurements for determination of dielectric properties of agricultural materials. *IEEE Trans. Instrum. Meas.* **1979**, *28*, 109–112.

41. Feldman, Y.D.; Zuev, Y.F.; Polygalov, E.A.; Fedotov, V.D. Time domain dielectric spectroscopy. A new effective tool for physical chemistry investigation. *Colloid Polym. Sci.* **1992**, *270*, 768–780.

42. Asami, K. Characterization of heterogeneous systems by dielectric spectroscopy. *Prog. Polym. Sci.* **2002**, *27*, 1617–1659.

43. Schwan, H.P. Determination of biological impedances. In *Physical Techniques in Biological Research*; Academic Press: New York, NY, USA, 1963; Volume 4, pp. 323–407.

44. Nelson, S.O. Agricultural applications of dielectric measurements. *IEEE Trans. Dielectr. Electr. Insul.* **2006**, *13*, 688–702.

45. Nelson, S.O.; Soderholm, L.H.; Yung, F.D. Determining the dielectric properties of grain. *Agric. Eng.* **1953**, *34*, 608–610.

46. Jorgensen, J.L.; Edison, A.R.; Nelson, S.O.; Stetson, L.E. Bridge method for dielectric measurements of grain and seed in the 50-to 250-MHz range. *Trans. ASAE* **1970**, *13*, 18–20.

47. Stetson, L.E.; Nelson, S.O. A method for determining dielectric properties of grain and seed in the 200- to 500-MHz Range. *Trans. ASAE* **1970**, *13*, 491–495.

48. Venkatesh, M.S.; Raghavan, G.S.V. An overview of dielectric properties measuring techniques. *Can. Biosyst. Eng.* **2005**, *47*, 15–30.

49. Kuang, W.; Nelson, S.O. Low-frequency dielectric properties of biological tissues: A review with some new insights. *Trans. ASAE* **1998**, *41*, 173–184.

50. Wakamatsu, H. A dielectric spectrometer for liquid using the electromagnetic induction method. *Hewlett-Packard J.* **1997**, *48*, 37–44.

51. Nelson, S.O. Improved sample holder for Q-meter dielectric measurements. *Trans. ASAE* **1979**, *22*, 950–954.

52. Corcoran, P.T.; Nelson, S.O.; Stetson, L.E.; Schlaphoff, C.W. Determining Dielectric properties of grain and seed in the audio frequency range. *Trans. ASAE* **1970**, *13*, 348–351.

53. Lawrence, K.C.; Nelson, S.O.; Kraszewski, A.W. Automatic system for dielectric properties measurements from 100 kHz to 1 GHz. *Trans. ASAE* **1989**, *32*, 304–308.

54. Bussey, H.E. Dielectric measurements in a shielded open circuitline. *IEEE Trans. Instrum. Meas.* **1980**, *29*, 120–124.

55. Jones, R.N.; Bussey, H.E.; Little, W.E.; Metzker, R.F. *Electrical Characteristics of Corn, Wheat, and Soya in the 1–200 MHz Range*; US Department of Commerce, National Bureau of Standards: Washington, DC, USA, 1979.

56. Mätzler, C. Applications of the interaction of microwaves with the natural snow cover. *Remote Sens.Rev.* **1987**, *2*, 259–387.

57. Nelson, S.O. Electrical properties of agricultural products—A critical review. *Trans. ASAE* **1976**, *16*, 384–400.

58. Kraszewski, A. MW aquametry—A review. *J. MW Power* **1980**, *15*, 209–220.

59. Afsar, M.N.; Birch, J.R.; Clarke, R.N.; Chantry, G.W. The measurement of the properties of materials. *Proc. IEEE* **1986**, *74*, 183–199.

60. Nelson, S.O.; Charity, L.F. Frequency dependence of energy absorption by insects and grain in electric fields. *Trans. ASAE* **1972**, *15*, 1099–1102.

61. Roberts, S.; von Hippel, A. A new method for measuring dielectric constant and loss in the range of centimeter waves. *J. Appl. Phys.* **1946**, *17*, 610–616.

62. Nelson, S.O. Dielectric properties of some fresh fruits and vegetables at frequencies of 2.45 to 22 GHz. *Trans. ASAE* **1983**, *26*, 613–616.

63. Nelson, S.O.; Stetson, L.E.; Schlaphoff, C.W. A general computer program for precise calculation of dielectric properties from short-circuited waveguide measurements. *IEEE Trans. Instrum. Meas.* **1974**, *23*, 455–460.

64. Kraszewski, A.; Stuchly, S.S.; Stuchly, M.A. ANA calibration method for measurement of dielectric properties. *IEEE Trans. Instrum. Meas.* **1983**, *32*, 385–386.

65. Waters, D.G.; Brodwin, M.E. Automatic material characterization at MW frequencies. *IEEE Trans. Instrum. Meas.* **1988**, *37*, 280–284.

66. Tran, V.N.; Stuchly, S.S.; Kraszewski, A. Dielectric properties of selected vegetables and fruits 0.1–10 GHz. *J. MW Power* **1984**, *19*, 251–258.

67. Bosisio, R.; Akyel, C.; Labelle, R.C.; Wang, W. Computeraided permittivity measurements in strong RF Fields (Part II). *IEEE Trans. Instrum. Meas.* **1986**, *35*, 606–611.

68. Rzepecka, M.A. A cavity perturbation method for routine permittivity measurements. *J. MW Power* **1973**, *8*, 3–11.

69. Kraszewski, W.; Nelson, S.O. Resonant-cavity perturbation measurement for mass determination of the perturbing object. In Proceedings of the IEEE Instrumentations and Measurements, Technology Conference, Hamamatsu, Japan, 10–12 May 1994; pp. 1261–1264.

70. Böhmer, R.; Diezemann, G. Principles and applications of pulsed dielectric spectroscopy and nonresonant dielectric hole burning. In *Broadband Dielectric Spectroscopy*; Kermer, F., Schönhals, A., Eds.; Springer: Berlin/Heidelberg, Germany, 2003; pp. 523–569.

71. Bauer, C.; Böhmer, R.; Moreno-Flores, S.; Richert, R.; Sillescu, H.; Neher, D. Capacitive scanning dilatometry and frequency-dependent thermal expansion of polymer films. *Phys. Rev. E* **2006**, *61*.

72. Fukao, K.; Miyamoto, Y. Glass transition temperature and dynamics of α-process in thin polymer films. *Europhys. Lett.* **1999**, *46*.

73. Hartmann, L.; Kratzmuller, T.; Braun, H.G.; Kremer, F. Molecular dynamics of grafted PBLG in the swollen and in the dried state. *Macromol. Rapid Commun.* **2000**, *21*, 814–819.

74. Seaman, R.; Seals, J. Fruit pulp and skin dielectric properties for 150 MHz TO 6400 MHz. *J. MW Power Electromagn. Energy* **1991**, *26*, 72–81.

75. Fito, P.; Castro-Giráldez, M.; Fito, P.J.; Chenoll, C. Development of a dielectric spectroscopy technique for the determination of apple (Granny Smith) maturity. *Innov. Food Sci. Emerg. Technol.* **2010**, *11*, 749–754.

76. Castro-Giráldez, M.; Fito, P.J.; Chenoll, C.; Fito, P. Development of a dielectric spectroscopy technique for determining key chemical components of apple maturity. *J. Agric. Food Chem.* **2010**, *58*, 3761–3766.

77. Ma, H.; Feng, M.; Zheng, J. Properties of Fuji apples superficial scald in the 100 Hz to 3.98 MHz range. *Trans. Chin. Soc. Agric. Mach.* **2010**, *41*, 105–109.

78. Guo, W.; Zhu, X.; Nelson, S.O.; Yue, R.; Liu, H.; Liu, Y. Maturity effects on dielectric properties of apples from 10 to 4500 MHz. *Food Sci. Technol.* **2011**, *44*, 224–230.

79. Guo, W.; Zhu, X.; Yue, R.; Liu, H.; Liu, Y. Dielectric properties of fuji apples from 10 to 4500 MHz during storage. *J. Food process. Preserv.* **2011**, *35*, 884–890.

80. Yovcheva, T.; Vozáry, E.; Bodurov, I.; Viraneva, A.P.; Marudova, M.; Exner, G. Investigation of apples aging by electric impedance spectroscopy. *Bulg. Chem. Commun.* **2013**, *45*, 68–72.

81. Cai, C.; Li, Y.; Ma, H.; Li, X. Nondestructive classification of internal quality of apple based on dielectric feature selection. *Trans. Chin. Soc. Agric. Eng.* **2013**, *29*, 279–287.

82. Ma, F.; Guo, Y. Influence experiment and mechanism analysis of high-voltage pulsed on dielectric properties of fruits and vegetables. *Trans. Chin. Soc. Agric. Mach.* **2013**, *44*, 177–180.

83. Guo, W.; Shang, L.; Wang, M.; Zhu, X. Soluble solids content detection of postharvest apples based on frequency spectrum of dielectric parameters. *Trans. Chin. Soc. Agric. Mach.* **2013**, *44*, 132–137.

84. Li, Y.; Cai, C.; Ma, H. Nondestructive Fuji apple internal quality classification with Dielectric feature selection. *ICIC Express Lett.* **2014**, *8*, 145–149.

85. Guo, W.; Shang, L.; Zhu, X.; Nelson, S.O. Nondestructive Detection of Soluble Solids Content of Apples from Dielectric Spectra with ANN and Chemometric Mehtods. *Food Bioprocess Technol.* **2015**, *8*, 1126–1138.

86. Zdunek, A.; Cybulska, J. Relation of Biospeckle Activity with Quality Attributes of Apples. *Sensors* **2011**, *11*, 6317–6327.

87. Adamiak, A.; Zdunek, A.; Rutkowski, K. Application of the Biospeckle Method for monitoring Bull's eye Rot Development and Quality Changes of Apples Subjected to Various Storage Methods-Preliminary Studies. *Sensors* **2012**, *12*, 3215–3227.

88. Coronel, P.; Simunovic, J.; Sandeep, K.P.; Kumar, P. Dielectric properties of pumpable food materials at 915 MHz. *Int. J. Food Prop.* **2008**, *11*, 508–518.

89. Cheuyglinstase, K.; Morison, K.R. Carrot Fiber as a Carrier in Spray Drying of Fructose. In *Water Properties in Food, Health, Pharmaceutical and Biological Systems: ISOPOW 10*; Centre for Formulation Engineering, Chemical Engineering, University of Birmingham: Birmingham, UK, 2010; p. 191.

90. Kundu, A.; Gupta, B. Broadband dielectric properties measurements of some vegetables and fruits using open ended coaxial probe technique. In Proceedings of the International Conference on Control, Instrumentation, Energy and Communication, Calcutta, India, 31 January–2 February 2014; pp. 480–484.

91. Bian, H.; Tu, P. Influence of fall height and storage time on dielectric properties of red globe grape. In Proceedings of the International Conference on Energy and Power Engineering, Hong Kong, China, 26–27 April 2014; pp. 209–213.

92. Sosa-Morales, M.E.; Tiwari, G.; Wang, S.; Tang, J.; Garcia, H.S.; Lopez-Malo, A. Dielectric heating as a potential post-harvest treatment of disinfesting mangoes, Part I: Relation between dielectric properties and ripening. *Biosyst. Eng.* **2009**, *103*, 297–303.

93. Yoiyod, P.; karirikh, M. Analysis of a Cost-effective Remote Sensing System for Fruit Monitoring. In Proceedings of the IEEE Antennas and Propagation Society International Symposium (APSURSI), Orlando, FL, USA, 7–13 July 2013.

94. Nelson, S.O.; Guo, W.; Trabelsi, S.; Kays, S.J. Dielectric spectroscopy of watermelons for quality sensing. *Meas. Sci. Technol.* **2007**, *18*, 1887–1892.

95. Nelson, S.O. Dielectric properties of agricultural products and some applications. *Res. Agric. Eng.* **2008**, *54*, 104–112.

96. Nelson, S.O.; Trabelsi, S. Examination of dielectric spectroscopy data for correlations with melón quality. In Proceedings of the American Society of Agricultural and Biological Engineers Annual International Meeting, Louisville, Kentucky, 7–10 August 2011; pp. 5123–5128.

97. Hasted, J.B. *Aqueous Dielectrics*; Chapman and Hall, Ed.; John Wiley & Sons: New York, NY, USA, 1973.

98. Foster, K.R.; Schwan, H.P. Dielectric properties of tissues and biological materials: A critical review. *Crit. Rev. Biomed. Eng.* **1988**, *17*, 25–104.

99. Forbus, W.R.; Dull, G.G. Delayed light emission as indicator of peach maturity. *J. Food Sci.* **1990**, *55*, 1581–1584.

100. Guo, W.; Zhu, X.; Guo, K.; Wang, Z. Electrical properties of peaches and its application in sensing freshness. *Trans. Chin. Soc. Agric. Mach.* **2007**, *38*, 112–115.

101. Llero, L.; Ruiz-Altisent, M.; Hernandez, N.; Gutierrez, P. Application of MW return loss for sensing internal quality of peaches. *Biosyst. Eng.* **2007**, *96*, 525–539.

102. Guo, W.; Chen, K. Relationship between dielectric properties from 10 to 4500 MHz and internal quality of peaches. *Trans. Chin. Soc. Agric. Mach.* **2010**, *41*, 134–138.

103. Ling, B.; Tiwari, G.; Wang, S. Pest Control by MW and radio frequency energy: Dielectric properties of stone fruit. *Agron. Sustain. Dev.* **2014**, *35*, 233–240.

104. Chen, J.; Pitchai, K.; Birla, S.; Gonzalez, R.; Jones, D.; Subbiah, J. Temperature-dependant dielectric and thermal properties of whey protein gel and mashed potato. *Trans. Am. Soc. Agric. Biol. Eng.* **2013**, *56*, 1457–1467.

105. Peng, J.; Tang, J.; Jiao, Y.; Bohnet, S.G.; Barret, D.M. Dielectric properties of tomatoes assisting in the development of MW pasteurization and sterilization processes. *Food Sci. Technol.* **2013**, *54*, 367–376.

106. Keller, G.V.; Licastro, P.H. Dielectric constant and electrical resistivity of natural-state cores. U.S. Geological Survey bulletin, 1052-H. *Exp. Theor. Geophys.* **1959**, *IV*, 257–285.

107. Sosa-Morales, M.E.; Valerio-Junco, L.; López-Malo, A.; García, H.S. Dielectric properties of foods: Reported data in the 21st century and their potential applications. *LWT Food Sci. Technol.* **2010**, *43*, 1169–1179.

108. Guo, W.; Tiwari, G.; Tang, J.; Wang, S. Frequency, moisture and temperature-dependent dielectric properties of chickpea flour. *Biosyst. Eng.* **2008**, *101*, 217–224.

109. Wang, Y.; Wig, T.; Tang, J.; Hallberg, L. Dielectric properties of food relevant to RF and MW pasteurization and sterilization. *J. Food Eng.* **2003**, *57*, 257–268.

110. Birla, S.L.; Wang, S.; Tang, J.; Tiwari, G. Characterization of radio frequency heating of fresh fruits influenced by dielectric properties. *J. Food Eng.* **2008**, *89*, 390–398.

111. Berna, A. Metal Oxide Sensors for Electronic Noses and Their Application to Food Analysis. *Sensors* **2010**, *10*, 3882–3910.

Distributed Wireless Monitoring System for Ullage and Temperature in Wine Barrels

Wenqi Zhang, George K. Skouroumounis, Tanya M. Monro and
Dennis K. Taylor

Abstract: This paper presents a multipurpose and low cost sensor for the simultaneous monitoring of temperature and ullage of wine in barrels in two of the most important stages of winemaking, that being fermentation and maturation. The distributed sensor subsystem is imbedded within the bung of the barrel and runs on battery for a period of at least 12 months and costs around $27 AUD for all parts. In addition, software was designed which allows for the remote transmission and easy visual interpretation of the data for the winemaker. Early warning signals can be sent when the temperature or ullage deviates from a winemakers expectations so remedial action can be taken, such as when topping is required or the movement of the barrels to a cooler cellar location. Such knowledge of a wine's properties or storage conditions allows for a more precise control of the final wine quality.

Reprinted from *Sensors*. Cite as: Zhang, W.; Skouroumounis, G.K.; Monro, T.M.; Taylor, D.K. Distributed Wireless Monitoring System for Ullage and Temperature in Wine Barrels. *Sensors* **2015**, *15*, 19495–19506.

1. Introduction

Over the past decade, the development and integration of wireless sensor networks (WSN) within the agriculture and food industry, along with a greater understanding of the theory and potential application of such devices, has seen considerable growth [1–9]. Indeed, there is now a multitude of devices designed to provide information on precision agriculture, environmental monitoring, machine and process control, facility automation, food packaging, food inspection and quality control [3,7].

Surprisingly, the use of WSN in vineyards and wineries is still quite rare. In terms of vineyard monitoring, several reports have appeared recently detailing the use of WSN to monitor not only a vineyard's microclimate but also the risk of vine damage due to frost, pests and disease [10–12]. There have also been several reports utilising WSN to monitor the quality of wine in terms of cellaring [13], and temperature control at the various stages of vinification [14,15]. Controlling the temperature during primary fermentations of wine is extremely important in terms of the development of the "bouquet" or "aroma" of a wine. For example, white wines are usually fermented at around 15 °C with higher temperatures (e.g., 20 °C) potentially resulting in the loss of volatile "aroma" due to the sweeping away of these volatiles by the carbon dioxide gas generated during fermentation [16]. In addition, the control of the temperature of wines undergoing malolactic fermentation or maturation in barrels is vital to ensure the control of potentially unwanted bacterial growth or oxidative damage [17]. Currently, the temperature of wines in barrels is simply measured with a thermometer, which means the barrels need to be opened and potentially exposed to oxygen, whilst the temperature is controlled by simply moving the barrels from one cold room to another.

Another key parameter that is monitored for wines in barrels is the extent of ullage. Ullage is defined as the empty space that lies between the wine and the closure, *i.e.*, the space between a bung in a barrel and the surface of the wine, or the space between a cork or screw cap in a bottle and the surface of the wine. Given that this air contains around 20.95% oxygen, minimising ullage is very important to avoid chemical oxidative damage or bacterial damage of the wine [18]. For example, acetobacter (acetic acid bacteria) which converts ethanol to acetic acid is facilitated by the presence of oxygen thereby increasing the volatile acidity of the wine [19]. To our knowledge ,there is yet to be any report on the development of a WSN to monitor ullage.

In this paper, we demonstrate a WSN platform that can be embedded inside wine barrels bung and used to monitor the temperature and ullage of the wine in each individual barrel during both the fermentation and maturation stages of winemaking. Similar pioneer works of potential for the deployment of WSN in wineries have been reported in recent years [20,21]. However, the size, cost and energy consumption are still not ideal for large scale winery use. Our design is particularly focused on these aspects. For instance, Di Gennaro reported that the basic components of the WineDuino node (excluding the actual sensors) cost more than 90 euros each [21]. The aim was to design a device that was of low cost, low-energy consumption and could be used in large barrel rooms containing thousands of barrels with the information being sent to remote computer workstations or a winemaker's iPhone for consideration. This early warning system thus allows a winemaker to make immediate modifications to winemaking processes to ensure the highest quality wines are being produced. The "smart-bung" and WSN platform developed also allows for the addition of other sensor modules in the future in order to extend its functionality and allow for even closer monitoring of all important analytes and parameters of wine during the vinification and maturation stages of winemaking.

2. The Overall Systems Architecture

The distributed wine monitoring platform consisted of a small central node and a subsystem where sensor modules may be attached, a schematic of which is depicted in Figure 1. The subsystem was an energy-saving small signal chip computer with an on-board FM radio system and general purpose IO interface that could be used to connect to the sensor modules. The subsystem and the sensor modules were to be powered by a long-life battery and consume no energy when in an idle state (most of the time), whilst the subsystem and the sensor modules attached would only be powered on when in data requisition mode. This ensures that the subsystem would run on a single battery for an extended period of time (12 months) without recharging of the battery. The subsystem was small enough that it could be embedded into a barrel bung. The central node was to be built upon a single computer board running a custom build standard Linux operation system which can support running a wide range of programs for gathering and processing the data received from the subsystems. The central node also had the capability of receiving and transmitting both FM radio and WIFI signals. It should be noted that the FM radio modules would consume much less power when compared to WIFI technology which would be used in the central nodes to communicate with the remote computer center.

The subsystems acquire data, push it to the central node and power off immediately at a fixed interval. The central nodes receive the data from the subsystems, identify the source barrels of the data, conduct preliminary data processing and upload the data to the server in the computer room. The server logs and analyses the data, puts them on the web interface and sends out alerts when necessary.

Specifically, a FriendlyARM Mini210s single computer board computer was used as the basis of the central node whilst an ATMega328 chip based Arduino-like single chip microcontroller (Moteino) was used as the basis of the subsystem. An additional Moteino chip was connected to the Mini210s via a USB port serving as a FM radio transceiver. Currently, the Moteino software supports 65,280 chips in its networks, but there is no hardware limitation. The ullage and temperature modules were built using a SHARP GP2Y0A41SK0F (measures 4–30 cm) infrared distance sensor and Dallas DS18B20 1-wire temperature sensor (measures from $-55\,°C$ to $+125\,°C$ with $\pm0.5\,°C$ from $-10\,°C$ to $+85\,°C$), respectively. Figure 2 depicts the distance sensor assembled in a barrel bung (top left) and an unprotected temperature sensor (bottom left). The temperature sensor was protected with a heat conducing shell and hanging from the bung into the wine. The entire assembled subsystem is depicted in the right photo of Figure 2 and had a total cost of $27 AUD retail price, which could be further scaled down if the system was produced at scale.

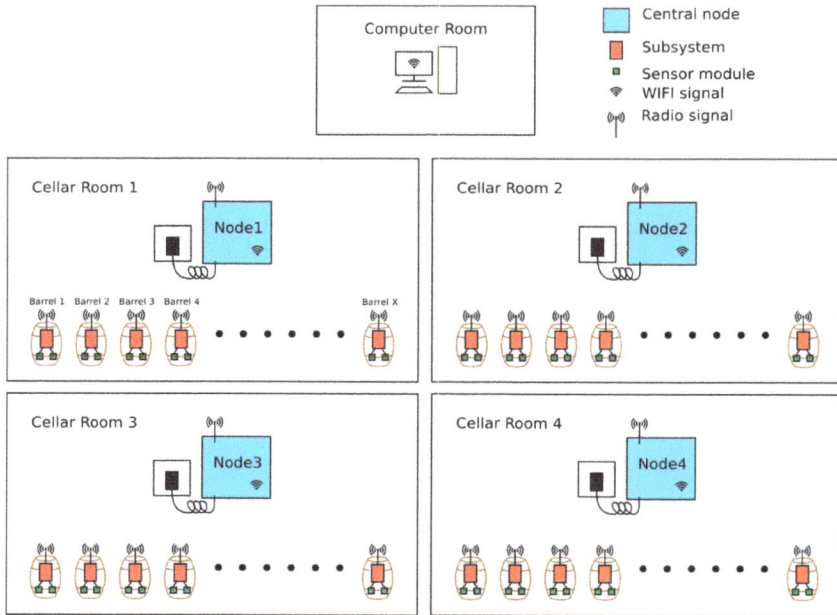

Figure 1. The schematics of the distributed wine monitor system.

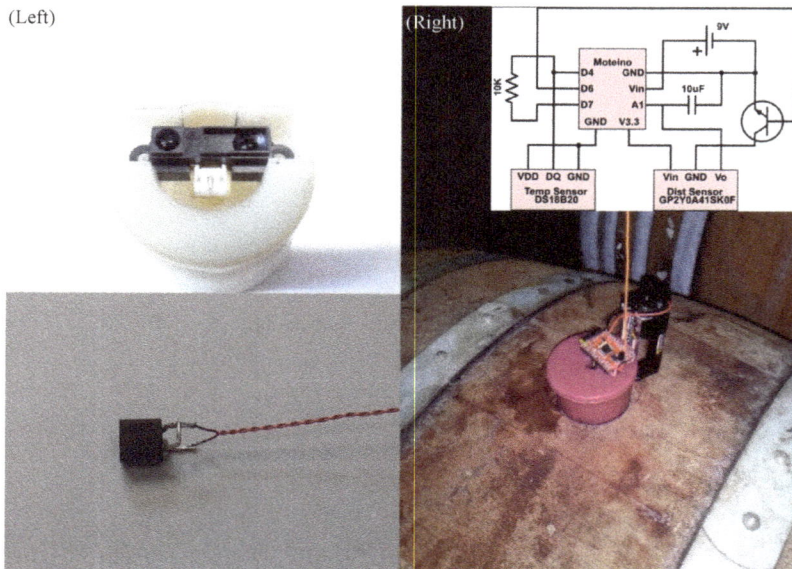

Figure 2. (**Left**) The distance sensor inside a bung and a naked temperature sensor; (**Right**) An assembled subsystem on a wine barrel and its schematic as the inset.

In the work by Sainz *et al.*, bluetooth was used as the wireless network interface and an analogue temperature sensor TM35DZ for temperature measurement. Such modules suffer from several drawbacks including that the usual transmission range for bluetooth devices is only a few meters which makes the monitoring system unsuitable for practical use in large wineries where the barrels are constantly being moved around, whilst the accuracy of the analogue temperature sensor would not be as good if a digital temperature was used [14]. Consequently, it appeared to us that our system, which employs a Moteino microcontrollor with an integrated low power FM radio as the sensor platform and main network interface along with a digital temperature probe, will offer additional advantages over what has thus far been reported. Such a radio system would have a coverage of 300–400 m radius at the BAUD rate of 55 kbps (or 1.5 mile range at 1200 bps), more than enough to cover most winery cellar areas. In addition, whilst Boquete *et al.* employed high-precision digital temperature sensors (DS1631 ± 0.5 °C from 0 to 70 °C) and an XBee wireless interface with a range of 20–30 m [15], it is now recognised that the price of a XBee wireless module alone is more than the price of an entire microcontrollor plus wireless module, thus our WNS would come in at a much lower cost.

The assembly of the subsystem was rather straightforward. As shown in the top left photo in Figure 2, a slot is cut in the bottom of a bung where the SHARP infrared distance sensor is to be installed. The edge of the distance sensor is coplanar to the bottom of the bung to allow easy measurement of the position of the distance sensor. A small hole is drilled through the bung from bottom to top for any wires including the temperature sensor. The Moteino chip is installed on the top-side of the bung and connected to a battery. The distance sensor and the temperature sensor were soldered onto the Moteino chip according to the schematic shown in Figure 2. A 10 K pull-up resistor was used together with temperature sensor. The 10 µF capacitor between pin A1 and GND was used for reducing voltage fluctuation from the distance sensor. The NPN transistor was used as a switch allowing the program to switch the power to the distance sensor to save energy. The Moteino chip allows the user to use any power supply from 3.3 V to 12 V. We used both packs of 4 AA batteries (6 V) and 9 V batteries to power the subsystem in this work. The central node can be installed anywhere convenient inside the wine cellar next to a power source. Each central node hosts a unique radio network. Each subsystem was pre-programmed with a unique ID and can be assigned to any radio network, which allows each wine barrel to be easily identified by the winemaker. Our design approach greatly reduces the cost of the entire WSN. Table 1 shows the cost of each of the components used.

Table 1. The cost of the elements used in this work.

Central Node	
FriendlyARM Mini210S with WIFI	$225.89
Moteino R4 with RFM69WH	$19.95
Total: $245.84 AUD	
Subsystem	
Moteino R4 with RFM69WH	$19.95
SHARP GP2Y0A41SK0F	$6.20
Dallas DS18B20	$0.60
Heat Conducting Shell	$0.40
10K Resistor	$0.01
10 uF Capacitor	$0.01
2N2222 NPN Transistor	$0.03
Total: $27.20 AUD	

Specific software was designed which would run on both the central node and computer server. It was written in C/C++ utilizing Qt [22] and Wt [23] packages to ensure portability. The software on the central node collects and processes the sensor data whilst the software on the central computer node serves as a webpage server as shown in Figure 3. The web page consists of two important sections. On the left, the operator can select one of many parameters to be monitored by the "smart bungs". For example, temperature, ullage, pH *etc*. To the right of this menu is a pane, which displays each individual barrel of wine and graphically or numerically displays what the current status of that parameter is. This setup allows users to access the sensor data using any remote web browser with early warning messages being sent out via emails to the winemakers if pre-set parameters are not being adhered to. For example, if the winemaker wished the wine to be stored in barrels at 15 °C for six months and the temperature deviated by more than 2 °C (pre-set in the software by the winemaker), then an alert would be sent wirelessly to warn of such an instance so remedial action could be taken. The ullage readings were not calibrated to physical units, however, it could be given that the ullage values are inversely proportional to the voltages read from the IR distance sensors. To calibrate the inversed voltage to physical units, one may use the relationship: $L = a * 1/V + b$, where L is the distance, and V is the voltage, to fit the voltages manually measured for different wine levels to obtain the calibration coefficients a and b, and input them into the software. The code on Moteino was developed using the Arduino toolkit. Moteino is fully compatible with the Arduino Uno platform.

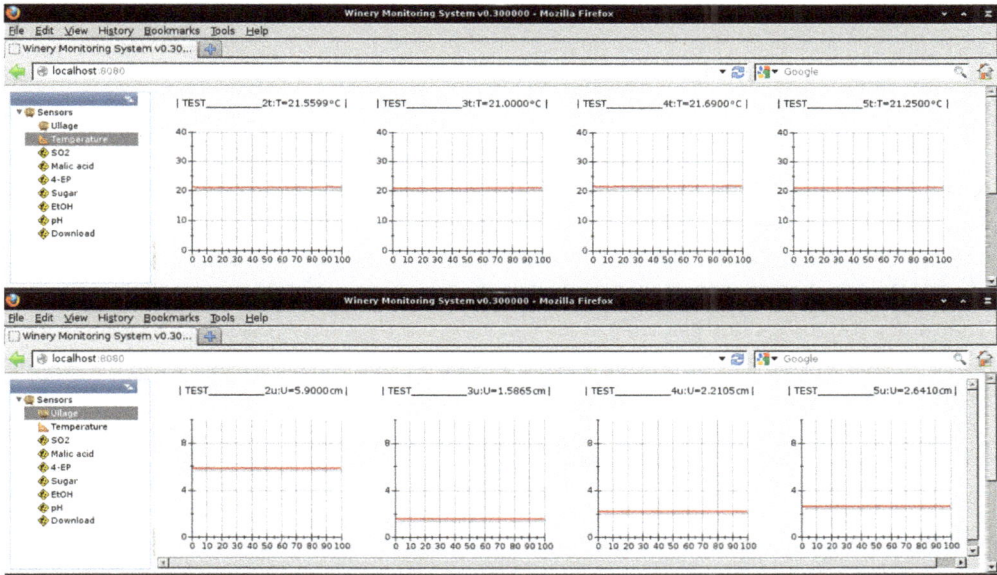

Figure 3. The temperature sensor page of the web interface shows eight virtual barrels.

3. Results and Discussion

After building the bungs with associated temperature and ullage sensors incorporated and designing the software, we then tested the system under several real world winemaking conditions. The first was to monitor the temperature and ullage of still wine on the cellar floor over time (maturation in barrels) whilst the second test was to monitor the same parameters during fermentation. In both test trials, four barrels of wine were monitored simultaneously. The sampling rate in our tests was set to once every 8 s for both temperature and ullage.

3.1. Trial I—Temperature and Ullage of Still Wine on the Cellar Floor

The first trial was performed on four barrels of red wine stored on the cellar floor for a period or nearly three months, where the temperature was not actively controlled, Figure 4. The temperature sensors were placed 30 cm below the bung holes using insulated magnet wires. The four test barrels were set next to each other and were subject to daily variances in temperature in the cellar. It should be noted that day zero was at the beginning of the Australian autumn when the daily temperatures are higher than at the end of autumn when the trial was completed.

It can be clearly seen that the readings from the individual temperature sensors within each bung are remarkably identical except for some minor variations due to small differences in the heat capacity of each barrel. Given that the winemaker wished the wines to be kept in the range of 10–15 °C during maturation, which was observed by our WSN, it provided confidence to the winemaker that the cellar hands did not need to manually measure the temperature of each barrel. On occasions, we opened one of the barrels and manually measured the temperature with a thermometer and found the same values for the temperature of the wine. Furthermore, the barrels did not need to be opened which aids in avoiding potential contamination, and oxidative spoilage. During this first trial, one of the subsystems (bottom left in Figure 4) went offline for about two weeks due to a power glitch. It was recovered by manually rebooting it and could be avoided in the future by installing a "watchdog" timer program, which will automatically reboot the system whenever there is a problem detected.

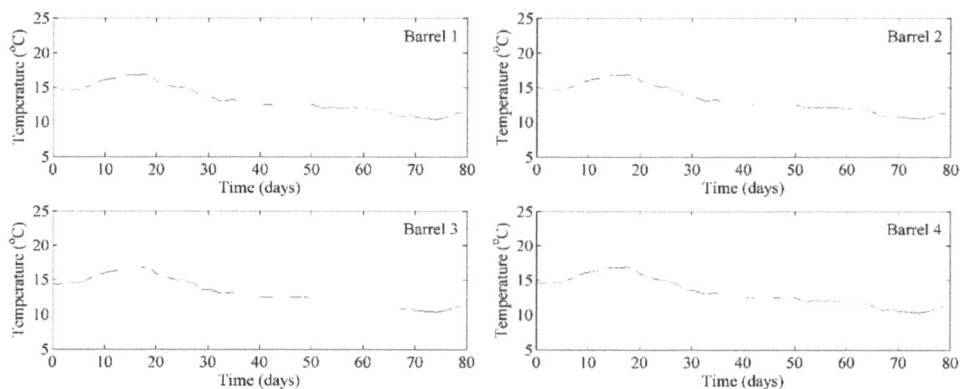

Figure 4. Continual monitoring of temperature of four barrels of wine in storage.

At the same time, we also measured the ullage in the four barrels on the cellar floor. The sensors were turned on at day 12 as we found that after fermentation and transfer of the wine into barrels, there was a large amount of foaming that required a number of days to dissipate. The settling down of the wines surface can still be seen during the day 12 and day 25 time-points in the plots of Figure 5. Importantly, all barrels showed a decrease in ullage height at around day 26 which corresponded to the barrels being opened and topped up with additional wine. After this time point, the ullage again began to increase slowly due to slow evaporational losses, as expected. Two of the barrels showed some slight fluctuations in daily ullage levels after topping (right column of Figure 5), which we believe is due to the disturbance of floating films resulting in fluctuations in the ullage depth readings. We inspected one of these barrels and found that there was indeed a thin film floating on the surface of the wine, Figure 6. Thus, given that the sensor we chose for ullage is based on light reflection, consistency of the reflective surface would be extremely important. Any disturbance by floating films would be expected to cause some variation seen in the ullage measurements as found here, although the overall trends found in our trials are what a winemaker would expect to observe. If the wines were to be stored in barrels for extended periods of time (up to two years), as many are, then the winemaker could simply set a minimum ullage distance required before the barrels needed to be topped; when reached, the sensors would trigger an alert that this action needs to be taken.

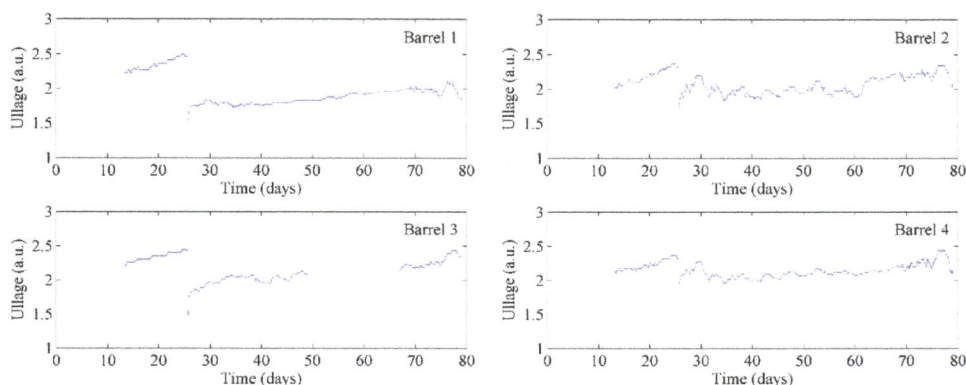

Figure 5. Continual monitoring of ullage of four barrels of wine in storage.

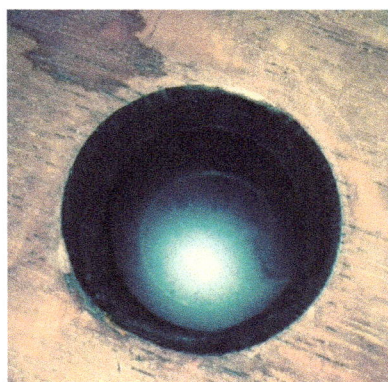

Figure 6. A thin film formed on top of the wine surface that influenced the ullage sensor reading.

3.2. Trial II—Temperature and Ullage during Fermentation

Given that our wirelessly distributed system successfully gathered the temperature and ullage information from wines undergoing maturation in barrels

147

on the cellar floor and sent the information to a remote computer server room which can be accessed by winemakers from a simple webpage, we next examined its performance at monitoring temperature and ullage during a real wine fermentation. Again, fours barrels were monitored with the data from the temperature sensors collated in Figure 7. The winemaker instructed the cellar hands to place the juice in barrels in a 15 °C cold room and perform inoculation. We affixed our remote sensors and began monitoring the barrels. Pleasingly, our sensors detected that the juice was around the initial temperature of the cold room, *i.e.*, 15 °C. Naturally, each ferment in each individual barrel will progress through fermentation at slightly different speeds and consequently subtle temperature differences are expected to be observed, as seen in Figure 7. As the ferments began over the first few days, the observed temperature was found to rise several degrees to between 17 °C and 18 °C. Given that the winemaker wished the ferments to be conducted near 15 °C and wished to avoid excessive fermentations of 20+ °C, they made the decision to move the barrels to the 10 °C cold room on the third night. As can be seen, our remote detectors picked this up and the temperatures within the barrels began to decrease by several degrees by day 5. The ferments were then at their full exponential growth phase and thus the temperature began to rise again up to 18 °C as they pushed through to completion over the following five days or so. All the ferments finished around day 12 as the winemaker expected and the temperatures began to drop back to the outside ambient cellar room temperature of 10 °C. They were then moved out of the cold room onto the normal cellar floor for further processing on day 18 which again resulted in a gradual warming of the wine.

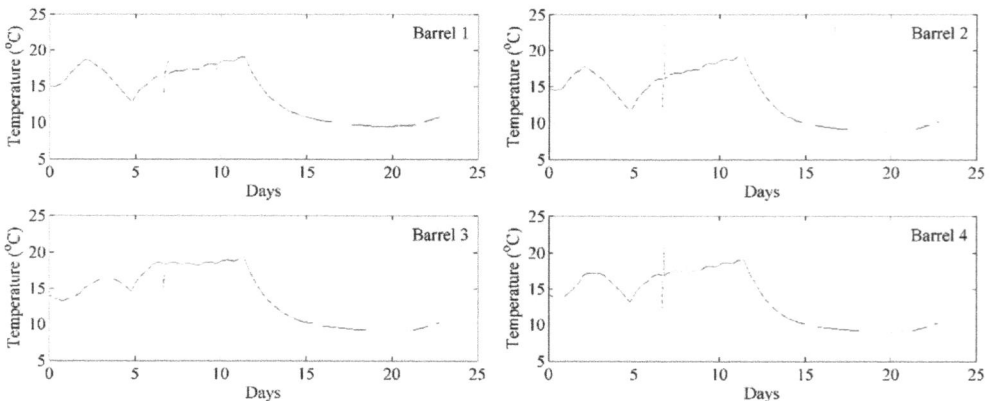

Figure 7. Continual monitoring of temperature of four barrels of wine during fermentation.

148

(a)

(b)

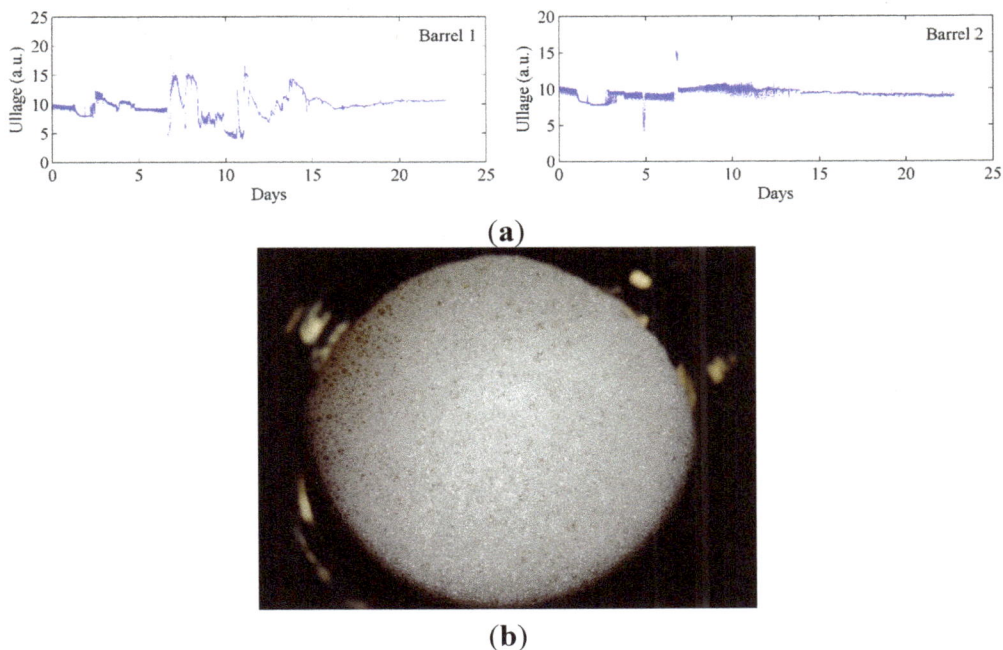

Figure 8. (**a**) Continual monitoring of ullage of two barrels of wine during fermentation; (**b**) the overflowing wine resulting from the excessive bubbling due to carbon dioxide evolution.

At the same time, we also measured the ullage in the barrels during fermentation, the results of which are displayed for two barrels, Figure 8a. Fermentation results in excessive bubbling due to carbon dioxide evolution and in some cases results in the wine overflowing the barrels (see picture in Figure 8b). The bungs come with a hole imbedded in them to allow for such overflows. Given that our ullage sensor is based on infrared light reflection between the surface of the liquid and the bung itself, extensive fluctuation in ullage height was expected as the wines level rises and falls over time. Furthermore, this fluctuation would be exacerbated with changes in the fermentation temperature. Indeed, extensive fluctuations were observed when monitoring the ullage height during fermentation, Figure 8a. After days 12–15, when the fermentations were beginning to be completed and the foams dissipate, the noise in the ullage curves significantly reduces. Importantly, the time-point of 12–15 days indicating completion of fermentation by ullage also corresponds to that observed by our temperature sensors highlighted above. Whilst it was intriguing to measure ullage under real fermentation conditions, the measurement of ullage is of most importance during long-term storage and maturation of wine in barrels.

149

4. Conclusions

In this work, we have designed a simple low-cost energy saving wireless distributed sensor network that allows for the simultaneous monitoring of a wine's temperature and ullage both in the fermentation and maturation stages of wine making in barrels. The distributed sensor subsystem runs on battery for a period of at least 12 months and costs around $27 AUD for all the parts. In addition, software was designed which allows for the remote transmission and easy visual interpretation of the data for the winemaker, displaying "virtual barrels" that can be monitored remotely. Early warning signals can be sent when the temperature or ullage deviates from a winemaker's expectations so remedial action can be taken, such as topping up or the movement of the barrels to a cooler cellar. Such knowledge of a wine's properties or storage conditions will allow for a more precise control of a final wine's quality. Moreover, the WSN has been designed so that additional modules of analysis (e.g., pH, sulfur levels, ethanol content *etc.*) can be simply added in the future to the "smart bungs" upon development. Finally, we anticipate that in the future these "smart bungs" could be manufactured via the use of 3D printing which may further reduce the cost and allow installation in thousands of barrels in each winery.

Acknowledgments: This project was supported by the Australian Research Council through a Linkage Project grant (#LP120100483). We would also like to thank Louisa Rose and Alana Seabrook (The Yalumba Wine Company), Kate Lattey (Pernod Richard Winemakers) and Iain Jones (Treasury Wine Estates) for their financial support along with access to wine samples and personnel. Tanya Monro acknowledges the support of an ARC Georgina Sweet Laureate Fellowship.

Author Contributions: Tanya Monro and Dennis Taylor are joint group leaders of the project and were responsible for the project management, planning of experiments, data analysis, as well as the preparation of this manuscript. Wenqi Zhang carried out the majority of the design work including writing the software, conducted the trials, performed the data analysis and assisted with the manuscript. George Skouroumounis assisted with experimental work and sections of the manuscript.

Conflicts of Interest: The authors declare no conflict of interest.

References

1. Akyildiz, I.F.; Su, W.; Sankarasubramaniam, Y.; Cayirci, E. A Survey on Sensor Networks. *Commun. Mag.* **2002**, *40*, 102–114.
2. Rajaravivarma, V.; Yang, Y.; Yang, T. An Overview of Wireless Sensor Network and Applications. In Proceedings of the 35th Southeastern Symposium, Morgantown, WV, USA, 18 March 2003; pp. 432–436.
3. Ruiz-Garcia, L.; Lunadei, L.; Barreiro, P.; Robla, J.I. A Review of Wireless Sensor Technologies and Applications in Agriculture and Food Industry: State of the Art and Current Trends. *Sensors* **2009**, *9*, 4728–4750.

4. Sahota, H.; Kumar, R.; Kamal, A. A wireless sensor network for precision agriculture and its performance. *Wirel. Commun. Mob. Comput.* **2011**, *11*, 1628–1645.

5. Zhang, S.; Zhang, H. A Review of Wireless Sensor Networks and Its Applications. In Proceeding of the IEEE International Conference on Automation and Logistics, Zhengzhou, China, 15–17 August 2012; pp. 386–389.

6. Gupta, D.K. A Review on Wireless Sensor Networks. *Netw. Complex Syst.* **2013**, *3*, 18–23.

7. Wang, N.; Zhang, N.; Wang, M. Wireless sensors in agriculture and food industry—Recent development and future perspective. *Comput. Electron. Agric.* **2006**, *50*, 1–14.

8. Garcia, M.; Bri, D.; Sendra, S.; Lloret, J. Practical deployments of wireless sensor networks: A survey. *Int. J. Adv. Netw. Serv.* **2010**, *3*, 163–178.

9. Ammari, H.M.; Gomes, N.; Jacques, M.; Axim, B.M.; Yoon, D. A Survey of Sensor Network Applications and Architectural Components. *Ad Hoc Sens. Wirel. Netw.* **2015**, *25*, 1–44.

10. Lloret, J.; Bosch, I.; Sendra, S.; Serrano, A. A wireless sensor network for vineyard monitoring that uses image processing. *Sensors* **2011**, *11*, 6165–6196.

11. Catania, P.; Vallone, M.; Re, G.L.; Ortolani, M. A wireless sensor network for vineyard management in Sicily (Italy). *Agric. Eng. Int. CIGR J.* **2013**, *15*, 139–146.

12. Valente, J.; Sanz, D.; Barrientos, A.; Cerro, J.; Rossi, C. An Air-Ground Wireless Sensor Network for Crop Monitoring. *Sensors* **2011**, *11*, 6088–6108.

13. Anastasi, G.; Farruggia, O.; Re, G.L.; Ortolani, M. Monitoring High-Quality Wine Production using Wireless Sensor Networks. In Proceedings of the 42nd Hawaii International Conference on System Sciences, Waikoloa, HI, USA, 5–8 January 2009; pp. 1–7.

14. Saintz, B.; Antolin, J.; López-Coronado, M.; Castro, C. A Novel Low-Cost Sensor Prototype for Monitoring Temperature during Wine Fermentation in Tanks. *Sensors* **2013**, *13*, 2848–2861.

15. Boquete, L.; Cambralla, R.; Rodríguez-Ascariz, J.M.; Miguel-Jiménez, J.M.; Cantos-Frontela, J.J.; Dongil, J. Portable system for temperature monitoring in all phases of wine production. *ISA Trans.* **2010**, *49*, 270–276.

16. Fischer, U. Wine aroma. In *Flavours and Fragrances*; Springer Berlin-Heidelberg: Berlin, Germany, 2007; pp. 241–267.

17. Ribereau-Gayon, P.; Dubourdieu, D.; Doneche, B.; Lonvaud, A. *Handbook of Enology*, 2nd ed.; Wiley: Hoboken, NJ, USA, 2000.

18. Boulton, R.B.; Singleton, V.L.; Bisson, L.F.; Kunkee, R.E. *Principles and Practices of Winemaking*; Chapman and Hall: New York, NY, USA, 1995.

19. Bartowsky, E.J.; Henschke, P.A. Acetic acid bacteria spoilage of bottled red wine-a review. *Int. J. Food Microbiol.* **2008**, *125*, 60–70.

20. Di Gennaro, S.F.; Matese, A.; Primicerio, J.; Genesio, L.; Sabatini, F.; Di Blasi, S.; Vaccari, F.P. Wireless real-time monitoring of malolactic fermentation in wine barrels: The Wireless Sensor Bung system. *J. Grape Wine Res.* **2013**, *19*, 20–24.

21. Di Gennaro, S.F.; Matese, A.; Mancin, M.; Primicerio, J.; Palliotti, A. An Open-Source and Low-Cost Monitoring System for Precision Enology. *Sensors* **2014**, *14*, 23388–23397.
22. Qt (software). Wikipedia, the free encyclopedia. Available online: http://en.wikipedia.org/wiki/Qt_(software) (accessed on 5 August 2015).
23. Wt (web toolkit). Wikipedia, the free encyclopedia. Available online: http://en.wikipedia.org/wiki/ Wt_(web_toolkit) (accessed on 5 August 2015).

Unmasking of Olive Oil Adulteration Via a Multi-Sensor Platform

Marco Santonico, Simone Grasso, Francesco Genova, Alessandro Zompanti, Francesca Romana Parente and Giorgio Pennazza

Abstract: Methods for the chemical and sensorial evaluation of olive oil are frequently changed and tuned to oppose the increasingly sophisticated frauds. Although a plethora of promising alternatives has been developed, chromatographic techniques remain the more reliable yet, even at the expense of their related execution time and costs. In perspective of a continuous increment in the number of the analyses as a result of the global market, more rapid and effective methods to guarantee the safety of the olive oil trade are required. In this study, a novel artificial sensorial system, based on gas and liquid analysis, has been employed to deal with olive oil genuineness and authenticity issues. Despite these sensors having been widely used in the field of food science, the innovative electronic interface of the device is able to provide a higher reproducibility and sensitivity of the analysis. The multi-parametric platform demonstrated the capability to evaluate the organoleptic properties of extra-virgin olive oils as well as to highlight the presence of adulterants at blending concentrations usually not detectable through other methods.

Reprinted from *Sensors*. Cite as: Santonico, M.; Grasso, S.; Genova, F.; Zompanti, A.; Parente, F.R.; Pennazza, G. Unmasking of Olive Oil Adulteration Via a Multi-Sensor Platform. *Sensors* **2015**, *15*, 21660–21672.

1. Introduction

Olive oil is the most popular vegetable oil produced and consumed in Mediterranean countries. According to international standards [1], olive oils have to be obtained exclusively from the fruit of the olive tree (*Olea europaea*) using cold pressing techniques and in conditions that do not alter the organoleptic properties of the oil at all. Current European Union regulation [2] and the International Olive Committee (IOC) require olive oils to be graded in function of sensory assessment and three fundamental chemical parameters: free acidity, peroxide value, and UV absorbance [2]. By comparing oils scores with threshold values, these are classified as extra virgin olive oil (EVOO), virgin olive oil, and other low-quality olive oil typologies. Olive oil is a very complex matrix [3,4]. The main compounds are triacylglycerols and fatty acids contributing to 94%–96% of their total weight. However, triacylglycerols and fatty acid contents show a broad variability in olive oils chemical composition and this is largely dependent on both cultivar and geographical origin [5]. Recently, the authentication of products labeled as olive oil has become a fundamental issue for either commercial or health aspects [6,7]. In fact, the high price of olive oil and its increased popularity as a potential health food have made it an ideal target for frauds [8]. Common olive oil adulterations include accidental contaminations during production stages, deliberate mislabeling of less expensive oil categories and, more often, the admixtures of expensive olive oils with low quality oils. Although advances in knowledge and technology have undoubtedly led to greater success over frauds, even more complex forms of adulteration have been developed to invalidate the usefulness of official methods, thus leaving the authenticity verification still an unsolved matter [9]. Actually, no rapid and universal method exists that is officially recognized for all the authenticity issues [10]. Liquid and gas chromatographic techniques represent the elective methods for the authentication and characterization of individual olive oil compounds [11–15]. Nevertheless, these analytic verifications require valuable instrumentation and highly-qualified staff. All of these features together make authentication a time consuming and expensive process which is not applicable as routine analysis. In this context the BIONOTE (BIOsensor-based multisensorial system for mimicking Nose, Tongue and Eyes), a recently developed sensor platform [16], has been employed. The system, which embeds gas and liquid sensors having a common biologically-derived sensing interface, allows the simultaneous analysis of the vapor and liquid phase of the samples. As a consequence, the integrated multi-sensorial platform led different sensors to catch more comprehensive information which, in turn, requires a further elaboration through multivariate data analysis techniques. At the end of the analytical procedure, similarities and differences between the samples are highlighted. In this multi-parametric study, the correct discrimination of twelve EVOOs made up of dissimilar olive cultivars and having different geographical origin has been achieved. Furthermore, the high sensitivity and reproducibility of the analysis, which were guaranteed by the innovative electronic interface of the system, permitted the detection of fraudulent admixing of extraneous vegetable oils (pomace, soybean, sunflower seeds, and peanut oils) up to concentrations lower than 5%. These promising results altogether present BIONOTE as a rapid and economic tool for high-throughput screening analysis.

2. Materials & Methods

2.1. Oil Samples

Twelve EVOO samples, indicated in the paper as EVOO #1, #2, #3, and so on, were obtained from twelve different Italian orchards. Several characteristics of the oils are reported (Table 1). The commercial EVOO as well as the pomace, soybean, sunflower seeds, and peanut oils were bought at a local market.

Table 1. General EVOOs specifications.

Oil Sample	Geographical Origin	Year of Production	Oil Variety
EVOO #1	Laterba	2013/2014	Picoline
EVOO #2	Castellaneta	2013/2014	Leccino
EVOO #3	Laterba	2013/2014	Picoline (organic)
EVOO #4	Laterba	2013/2014	Arbequina (organic)
EVOO #5	Grottaglie and Crispiano	2013/2014	Picoline (50%), Nociara (35%), Leccino (15%)
EVOO #6	Crispiano	2013/2014	Leccino
EVOO #7	Grottaglie	2013/2014	Ogliarola
EVOO #8	Grottaglie	2013/2014	Picoline
EVOO #9	Grottaglie	2012/2013	Cellina di Nardò
EVOO #10	Laterba	2013/2014	Leccino
EVOO #11	Crispiano	2012/2013	Cellina di Nardò
EVOO #12	Crispiano	2012/2013	Cima di Melfi

2.2. Gas Analysis

Quartz Micro Balances (QMBs) with six functionalized piezoelectric sensors were used as transducers for the gas sensor array as already described [16]. In order to perform homogeneous gas measurements the following experimental set-up was used. A volume of 2 mL for each olive oil sample was placed in a 50 mL glass flask and kept for 10 min at room temperature to obtain an adequate headspace. Dehumidified reference air was pumped into the sensors chamber at a flow rate of 3 L/min for 10 min to desorb any volatile trace from sensors surface before every measure. Oil samples were analyzed five times, setting a sampling interval of 90 s.

2.3. Liquid Analysis

Electronic interface and sensors employed in the liquid analyses were the same described in Santonico *et al.* Cyclic voltammetry in the range from −1 to 1 V was performed using a triangular function at 10 mHz and a sampling interval of 1 second. Olive oil samples for liquid sensor analysis were prepared following the procedure reported below. Briefly, a volume of 1 mL of oil was poured into a tube with 3 mL of methanol 70% (*v/v*) and mixed vigorously for 1 min. The vial containing the oil-alcohol emulsion was centrifuged for 5 min at 1000 RCF and 4 °C to separate the two phases efficiently. Finally, the methanol phase was collected and stocked in ice up until the analysis.

2.4. Chemical Quality Control Analyses

Polyphenol content, free acidity, peroxide value, ΔK, and refractive index of olive oil samples have been assessed following the standard chemical testing methods [15]. Briefly, polyphenol content was evaluated by means of Folin-Ciocalteu method, according to the procedure reported by Singleton and Rossi [17]. Free acidity content [18] was evaluated, dissolving the samples in a mixture of equal parts by volume of ethyl ether (95%) and ethyl alcohol, thus titrating with an ethanolic solution of potassium hydroxide, using phenolphthalein as indicator. Results were reported as grams of oleic acid per 100 g of oil. To determine the peroxide value [19], oil samples were dissolved in chloroform and glacial acetic acid, then a solution of potassium iodide was added, leaving the mixture incubating for five minutes in the dark, and finally a titration of the generated iodine with a standard sodium thiosulphate solution, using starch solution as indicator, was performed. The peroxide value was expressed in terms of milliequivalents of active oxygen per kilogram able to oxidize potassium iodide under the operating conditions. The quality of the olive oils employed in this study was also assessed measuring the absorption bands between 200 and 300 nm [20]. Samples were dissolved in iso-octane to obtain 1% (w/v) solutions and the specific absorbance at 232 and 270 nm with reference to pure solvent was determined. These absorptions were expressed as specific extinctions, conventionally indicated by K. Finally, a ΔK value was calculated relating the maximum recorded absorbance at 270 nm against the absorption of surrounding spectral region (±4 nm). The refractometric index of olive oils was determined using the Abbé refractometer, paying attention to correct the recorded value on a temperature basis. Three independent parameter's determinations were carried out for each test sample. All the reagents used in this study were of certified analytical quality.

2.5. Data Analysis

Multivariate data analysis: Principal Component Analysis (PCA) and Partial Least Square Discriminant Analysis (PLS-DA), was performed using PLS-Toolbox (Eigenvector Research Inc., Manson, WA, USA) in the Matlab Environment (The MathWorks, Natick, MA, USA). PLS-DA models have been calculated in order to detect EVO adulteration and investigate BIONOTE relevance to the chemical parameters.

3. Results

3.1. Olive Oil BIONOTE Characterization

Twelve Italian EVOOs having different geographical origin and olive variety compositions have been characterized through the BIONOTE system, performing five measuring cycles each. Gas analysis was performed on EVOOs without any modification of the samples. Volatile compounds released in the system headspace at room temperature were characterized through their interaction with the functionalized sensors, resulting in a reproducible pattern response (Figure 1). Olive oil as such is not applicable for electrochemical analysis due to the absence of conductivity and the high viscosity of the media. Therefore, oil samples underwent liquid extraction with methanol and the deriving alcoholic fractions were analyzed by the liquid sensor (Figure 1). Cyclic voltammetry in the range from -1 to 1 V was performed using a triangular function at 10 mHz and a sampling interval of 1 second. By means of this setup, an array of 100 virtual sensor responses has been obtained from one physical sensor for each voltammetric measuring cycle. Finally, a data fusion of the information deriving from the last three measuring cycles of gas and liquid sensors was accomplished. The obtained data set has been evaluated by Principal Component Analysis (PCA) and the ability of the system to sharply discriminate the twelve EVOOs was demonstrated. The score plot of the first two Principal Components (PCs), accounting for 76.94% of the explained variance, is reported (Figure 2). Ten of the twelve oil samples clustered in three separate regions along the Principal Component 2 (PC2). EVOOs #1, #6, and #12 formed a group in the bottom part of the plane. EVOOs #5, #8, #10, and #11 distributed in a second area at the interception of the two PCs. EVOOs #2, #4, and #9 clustered in the upper portion of the plane (Figure 2). Nevertheless, within the groups almost every oil sample can be discriminated from the others along the Principal Component 1 (PC1). EVOOs #3 and #7 were distinguished from the rest of the analyzed samples by positioning at the upper end and at the left edge of the plane, respectively (Figure 2). Additionally, a Partial Least Square Discriminant Analysis (PLS-DA) model using the leave one out criterion has been calculated showing a correct classification rate of 100% for the twelve different EVOOs (five independent repetitions each).

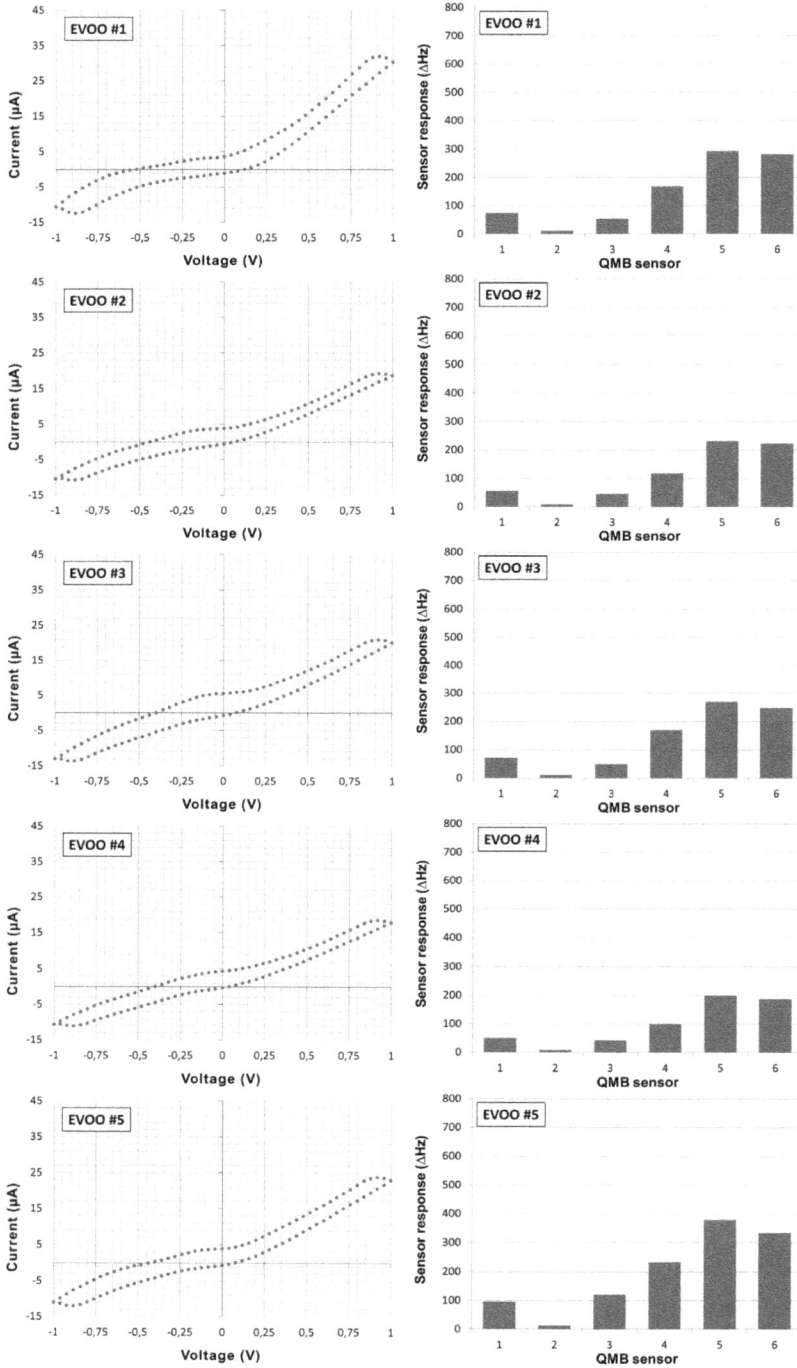

Figure 1. *Cont.*

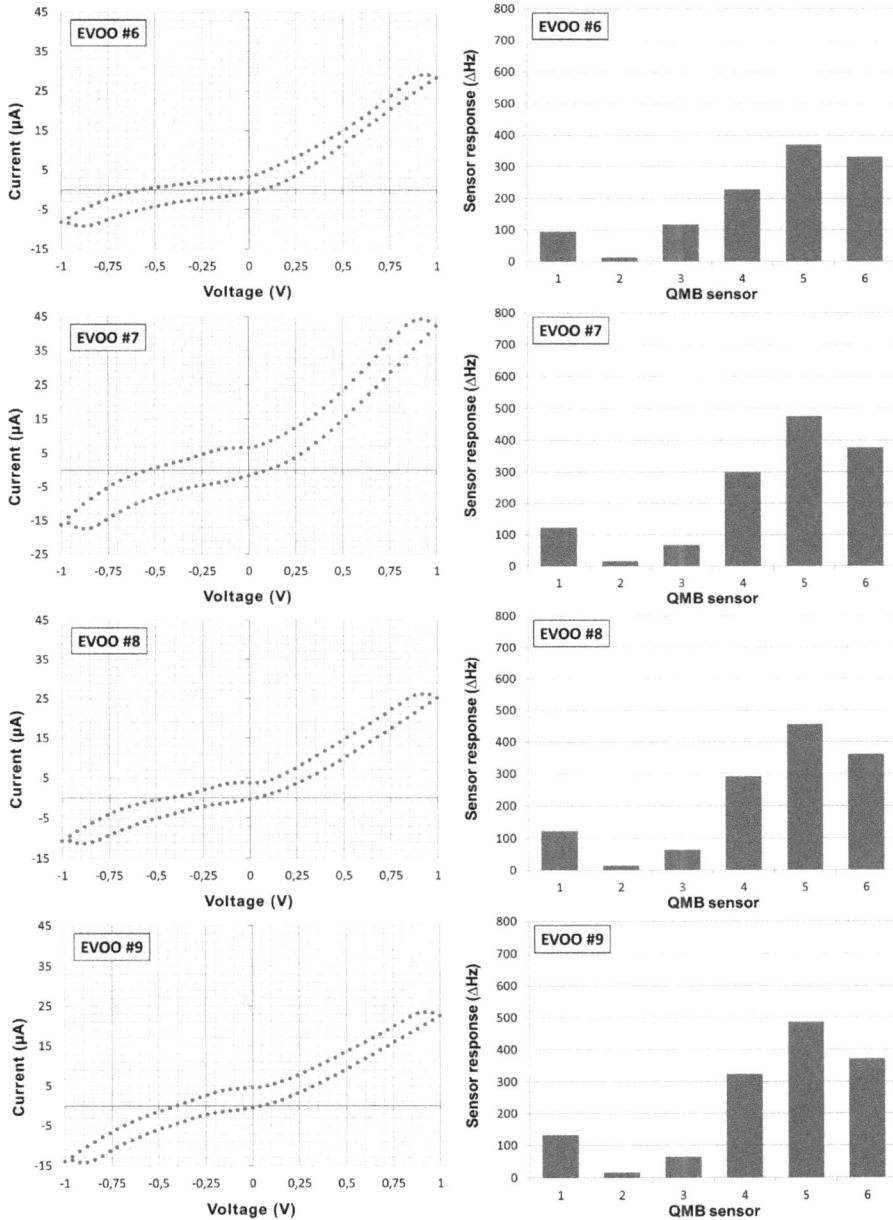

Figure 1. *Cont.*

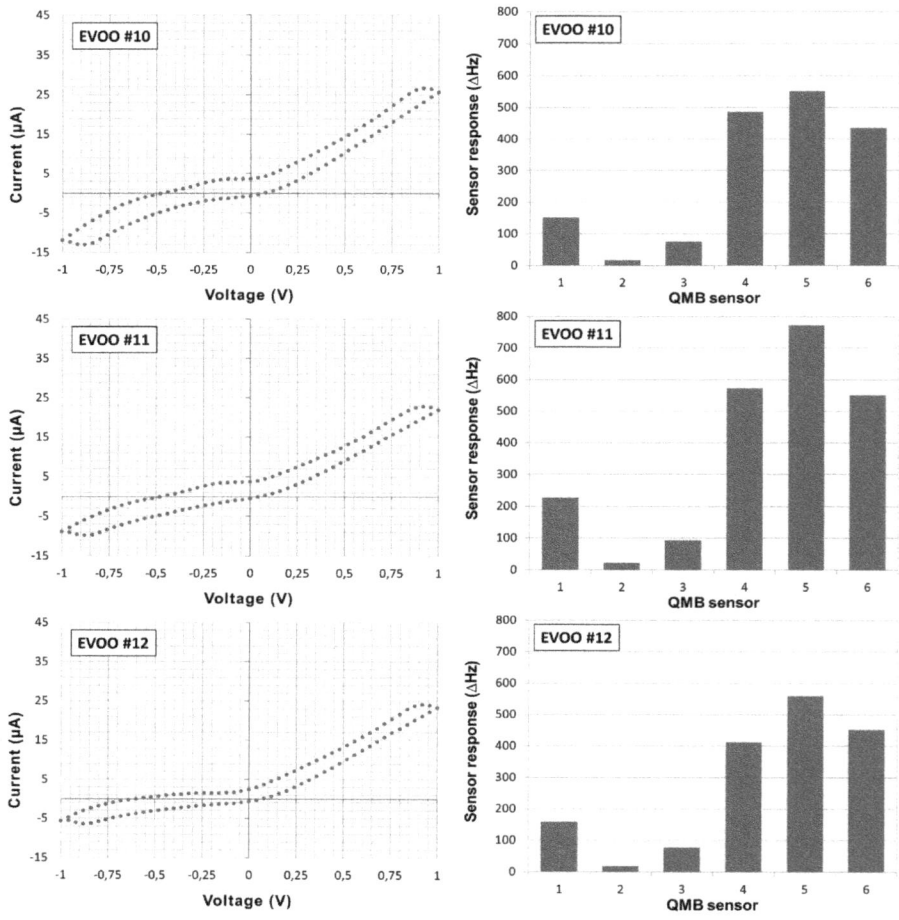

Figure 1. BIONOTE characterization of different EVOO samples. Liquid (**left panels**) and gas (**right panels**) fingerprints.

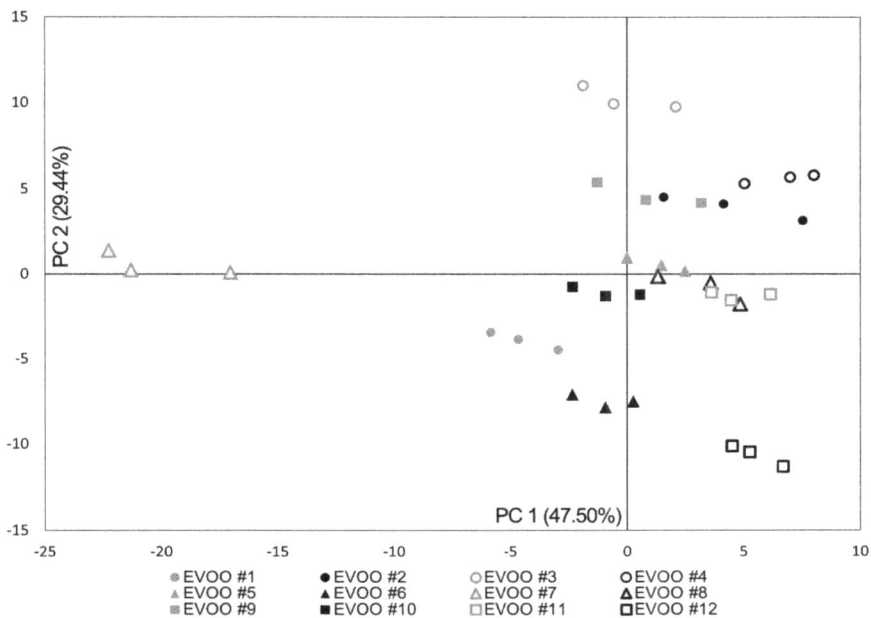

Figure 2. Score Plot of the first two principal components deriving from the data fusion of the BIONOTE liquid and gas sensors responses.

3.2. Olive Oil Chemical Characterization

To assess the quality of the EVOOs, common chemical analyses were also performed. All the EVOOs got parameters satisfying the imposed normative limits, even though some slight differences between the samples were found (Table 2), thus supporting BIONOTE discrimination evidence. Free acidity and ΔK values were significantly lower than normative standard ones being, however, slightly different among each other. The refractive index of the twelve oil samples was almost the same, while the peroxide parameter showed the greatest variability. The obtained results confirmed the excellent quality of the oil samples, highlighting the absence (in terms of usual parameters) of significant differences between the EVOOs themselves.

Table 2. EVOO purity and quality characteristics according to the International Olive Council [1].

Oil Sample	Free Acidity (mg/100 g Oleic Acid)	Peroxide Value (mEq O$_2$/Kg)	ΔK	Refractive Index
EVOO #1	3.4 ± 0.1	15.0 ± 0.4	0.0020	1.469
EVOO #2	3.4 ± 0.1	12.2 ± 0.1	0.0045	1.468
EVOO #3	4.9 ± 0.2	6.0 ± 0.1	0.0065	1.468
EVOO #4	2.0 ± 0.1	6.9 ± 0.1	0.0015	1.467
EVOO #5	7.3 ± 0.1	8.7 ± 0.1	0.0015	1.468
EVOO #6	6.0 ± 0.1	9.5 ± 0.3	0.0005	1.467
EVOO #7	5.3 ± 0.1	7.2 ± 0.4	0.0030	1.467
EVOO #8	4.3 ± 0.2	18.1 ± 0.2	0.0045	1.468
EVOO #9	2.8 ± 0.1	9.9 ± 0.2	0.0035	1.468
EVOO #10	3.9 ± 0.1	13.5 ± 0.4	0.0015	1.467
EVOO #11	6.1 ± 0.2	9.4 ± 0.5	0.0030	1.467
EVOO #12	3.1 ± 0.2	9.9 ± 0.3	0.0160	1.467

3.3. Olive Oil Adulteration

A commercial EVOO was bought at local market and mixed with four vegetable oils (pomace, soybean, sunflower seeds, and peanut oils) at different blending concentrations (1.25%, 5%, 10%, and 25% (v/v)). The prepared EVOO's admixtures were characterized through the BIONOTE system, performing five measuring cycles each. Sophisticated EVOO samples were treated as already described (see Materials & Methods section) before being analyzed through either the liquid or the gas sensors. A comprehensive array containing the overall sensors' responses was built for each EVOO sophistication independently and the collected data were further analyzed using multivariate data analysis techniques. The calculated PLS-DA models highlighted the ability of the system to distinguish an authentic EVOO from an adulterated one in all the tested cases, showing also a rather high degree of efficiency in the concentration discrimination (Figure 3). BIONOTE was able to predict the presence of contaminating lower-grade oils up to concentration values lower than 10% (v/v). The Root Mean Square Error in Cross Validation (RMSECV), using the Leave One Out criterion, was slightly different among the four kinds of sophistication. System performance was almost the same for the soybean, sunflower seeds, and peanut oils with RMSECV ranging from 2.1% to 4.4%, while the discrimination of the pomace oil sophistications resulted less precise accounting for an error of 8.3% (v/v) (Figure 3).

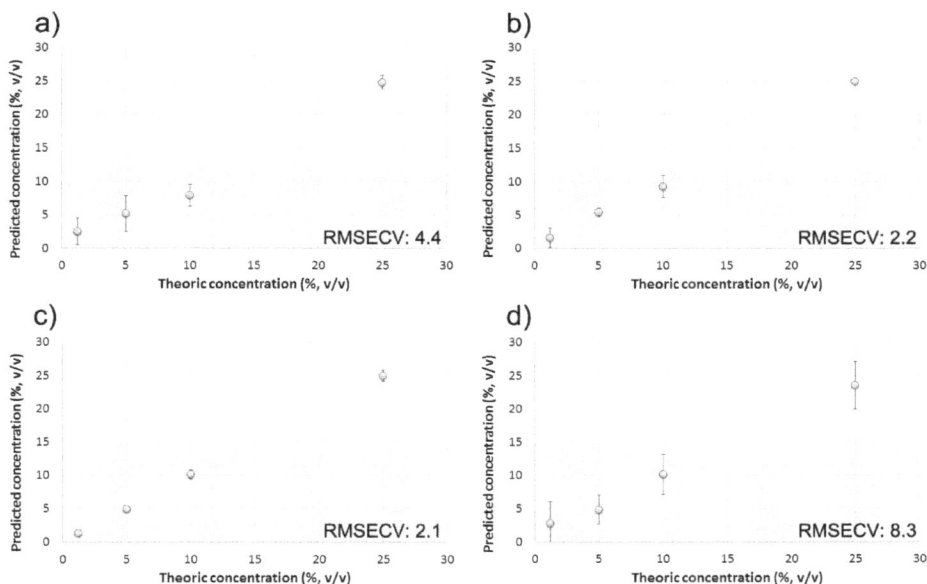

Figure 3. Calculated PLS-DA model for the prediction of contaminating oils concentration. Calibration model has been built using a commercial EVOO sophisticated with 0%–25% (v/v) of (**a**) soybean oil; (**b**) sunflower seeds oil; (**c**) peanut oil; and (**d**) pomace oil. RMSECV associated with the models are reported.

3.4. BIONOTE Relevance to the Chemical Parameters

BIONOTE relevance to the measured chemical parameters have been investigated by calculating four different models to predict polyphenols content, free acidity, peroxide value, and TEAC on the gas and liquid sensor array data. The results obtained are very promising (see Figure 4, panel a: polyphenols; panel b: free acidity; panel c: peroxide value; and panel d: TEAC).

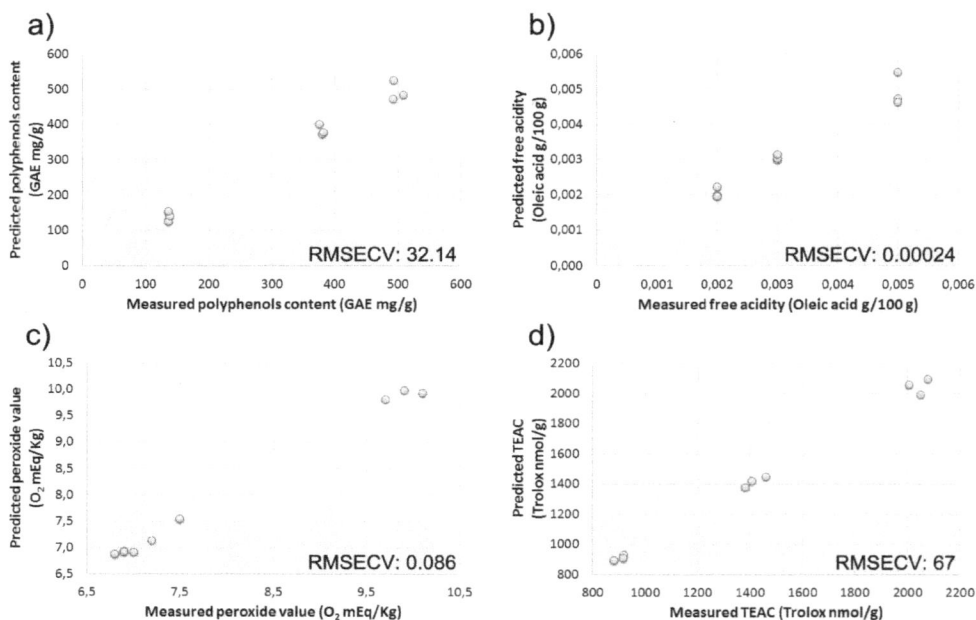

Figure 4. Measured *versus* predicted (PLS-DA model based on BIONOTE data) values of (**a**) polyphenols; (**b**) free acidity; (**c**) peroxide value; and (**d**) TEAC.

4. Discussion

Adulteration is a common problem usually related to high-value products. As a consequence of the fundamental role in the Mediterranean diet and the documented nutraceutical effect [6], EVOO represents a clear target for sophistication aimed to trade. According to recent studies, adulteration is becoming an escalating issue for olive oil in the market with consequences undermining the quality attributes of the product and sometimes even its safety consumption [21]. Although reliable and accurate analyses intended to guarantee olive oil quality in the broadest sense already exist, these are not routinely used. While chemical parameters as free acidity, peroxide value, ΔK, and refractive index are necessary to define if an olive oil fulfills the requirements to be labeled and marketed as EVOO, these constraints are not sufficient for authenticity verification in the most of cases [1,22]. Fraudulent olive oil admixtures are usually chemically corrected to meet international standards, thus requiring more complex analyses to be recognized as adulterations. Nevertheless, even when official analytical methods are applied to screen olive oil samples, olives' biological differences, due to geographical origin and genetic aspects, sometimes generate problems to distinguish between sophistications and authentic EVOOs [23]. So far, numerous modern techniques

164

have been proposed to support or replace official standard methods in the task of olive oil authentication [10,24–28]. However, those do not offer clear advantages yet, because their adulteration detection limits, being usually greater than 10% of contamination, are worse in comparison with chromatographic techniques' ones. In this study, a novel system able to characterize EVOOs in terms of genuineness and authenticity has been presented. The BIONOTE platform takes advantage of either liquid and gas analysis to accomplish a multi-parametric characterization, giving comprehensive information about the sample [17]. The overall sensors' responses are elaborated through multivariate data analysis techniques to highlight similarities and differences, resulting in a correct classification rate of 100%, even when similar EVOOs have been analyzed. Hence, BIONOTE showed the ability to discriminate between twelve Italian EVOOs originating from different Apulian neighboring olive tree orchards. The result highlighted the capability of BIONOTE not only to identify EVOOs against lower grade olive oils, but also to discriminate between EVOOs obtained from different olive cultivars. This is a notable outcome because this issue is usually addressed via more complex genetic approaches. The innovative electronic interface, providing to the system a higher reproducibility and sensitivity comparable to similar devices [29–32], allowed BIONOTE to be also successfully employed in the authenticity verification process, with admixtures percentage thresholds below the best levels reported by literature. BIONOTE was challenged with different kind of EVOO sophistications, covering concentrations lower than 10% (v/v), and in all cases it was able to distinguish authentic oil from an adulterated one. The system detected the presence of fraudulent admixing of extraneous vegetable oils (soybean, sunflower seeds and peanut oils) up to concentrations lower than 5%. However, when the pomace oil was used, system performance decreased. This discrepancy, leading to an increment of the detection limit to about 8%, could be probably explained by the shared origin between EVOO and pomace oil. Considering the demand of EVOO traceability and safety claimed by both producers and consumers, BIONOTE represents a potential solution. In fact, the BIONOTE system is able to address the EVOO authenticity issue focusing not only on the labeling control but also the genuineness of the oil, accounting for geographical origin and olive varieties composition at the same time.

5. Conclusions

Nowadays, global markets and international regulations have increased significantly the number of samples that require validation, raising the necessity of rapid analytical methods. In this context, BIONOTE could represent a real opportunity thanks to its reduced time of analysis. However, due to the profiling approach on which the system is based on, BIONOTE has not been intended to replace the high specificity of the official chromatographic methods. Hence, it

165

is proposed as a rapid tool for preliminary high-throughput screening, aimed to detect samples that require further analytical verifications. This workflow has been designed to reduce the employment of high-value instrumentation and qualified personnel only to specific cases, thus decreasing the costs, while maintaining the elevated number of samples analyzed.

Author Contributions: Marco Santonico and Giorgio Pennazza conceived and designed the experiments and analyzed the data; Simone Grasso and Francesco Genova performed the experiments and wrote the paper; Alessandro Zompanti and Francesca Romana Parente contributed analysis tools.

Conflicts of Interest: The authors declare no conflicts of interest.

References

1. International Olive Council (IOC). *Trade Standards Applying to Olive Oils and Olive-Pomace Oils*; COI/T.15/NC No 3/Rev. 7; International Olive Council (IOC): Madrid, Spain, 2013.
2. European Communities (EC). Commission Implementing Regulation (EU) No 1348/2013. Official Journal of the European Union (Regulation No 1348/2013 L338/31, 16 December 2013). Available online: http://www.beuth.de/en/regulation/euv-1348-2013/200370127 (accessed on 28 August 2015).
3. Aparicio, R.; Harwood, J. *Handbook of Olive Oil Analysis and Properties*, 2nd ed.; Springer: New York, NY, USA, 2013.
4. Angerosa, F.; Campestre, C.; Giansante, G. Analysis and authentication. In *Olive Oil: Chemistry and Technology*, 2nd ed.; AOCS Publishing: New York, NY, USA, 2006.
5. Gómez-Rico, A.; Salvador, M.D.; Moriana, A.; Pérez, D.; Olmedilla, N.; Ribas, F.; Fregapane, G. Influence of different irrigation strategies in a traditional cornicabra cv. Olive orchard on virgin olive oil composition and quality. *Food Chem.* **2007**, *100*, 568–578.
6. Harwood, J.L.; Yaqoob, P. Nutritional and health aspects of olive oil. *Eur. J. Lipid Sci. Technol.* **2002**, *104*, 685–697.
7. Kamal-Eldin, A.; Moreau, R.A. *Gourmet and Health-Promoting Specialty Oils*; AOCS Publishing: New York, NY, USA, 2009.
8. Zhang, J.; Zhang, X.; Dediu, L.; Victor, C. Review of the current application of fingerprinting allowing detection of food adulteration and fraud in china. *Food Control* **2011**, *22*, 1126–1135.
9. Aparicio, R.; Aparicio-Ruiz, R.; García-González, D.; van Amerongen, A.; Barug, D.; Lauwars, M. Rapid methods for testing of oil authenticity: The case of olive oil. *Rapid Methods* **2007**, *10*, 163–188.
10. Aparicio, R.; Morales, M.T.; Aparicio-Ruiz, R.; Tena, N.; García-González, D.L. Authenticity of olive oil: Mapping and comparing official methods and promising alternatives. *Food Res. Int.* **2013**, *54*, 2025–2038.
11. American Oils Chemists' Society (AOCS). *Determination of Fatty Acids in Olive Oils by Capillary GLC*; Official Method Ch 2–91; American Oils Chemists' Society (AOCS): New York, NY, USA, 2009.

12. International Olive Council (IOC). *Determination of Trans-Unsaturated Fatty Acids by Capillary Column Gas Chromatography*; COI/T.20/Doc. No 17/Rev. 1; International Olive Council (IOC): Madrid, Spain, 2001.

13. International Olive Council (IOC). *Determination of Wax Content by Capillary-Column Gas Chromatography*; COI/T.20/Doc. No 18; International Olive Council (IOC): Madrid, Spain, 2003.

14. International Olive Council (IOC). *Determination of Aliphatic Alcohols Content by Capillary Gas Chromatography*; COI/T.20/Doc. No 26; International Olive Council (IOC): Madrid, Spain, 2003.

15. International Olive Council (IOC). *Determination of the Composition and Content of Sterols and Triterpene Dialcohols by Capillary Column Gas Chromatography*; COI/T.20/Doc. No 30; International Olive Council (IOC): Madrid, Spain, 2013.

16. Santonico, M.; Pennazza, G.; Grasso, S.; Amico, A.; Bizzarri, M. Design and test of a biosensor-based multisensorial system: A proof of concept study. *Sensors* **2013**, *13*, 16625–16640.

17. Singleton, V.L.; Rossi, J.A. Colorimetry of total phenolics with phosphomolybdic-phosphotungstic acid reagents. *Am. J. Enol. Vitic.* **1965**, *16*, 144–158.

18. American Oils Chemists' Society (AOCS). *Acid Value*; Official Method Cd 3d-63; American Oils Chemists' Society (AOCS): New York, NY, USA, 2009.

19. American Oils Chemists' Society (AOCS). *Peroxide Value Acetic Acid-Isooctane Method*; Official Method Ch Cd 8b-90; American Oils Chemists' Society (AOCS): New York, NY, USA, 2011.

20. American Oils Chemists' Society (AOCS). *Determination of Specific Extinction of Oils and Fats, Ultraviolet Absorption*; Official Method Ch 5–91; American Oils Chemists' Society (AOCS): New York, NY, USA, 2009.

21. García-González, D.L.; Aparicio, R. Research in olive oil: Challenges for the near future. *J. Agric. Food Chem.* **2010**, *58*, 12569–12577.

22. International Olive Council (IOC). *Global Method for the Detection of Extraneous Oils in Olive Oils*; COI/T.20/Doc. No. 25; International Olive Council (IOC): Madrid, Spain, 2006.

23. Mignani, A.G.; García-Allende, P.B.; Ciaccheri, L.; Conde, O.M.; Cimato, A.; Attilio, C.; Tura, D. Comparative Analysis of Quality Parameters of Italian Extra Virgin Olive Oils According to Their Region of Origin. In Proceedings of the SPIE, San Francisco, CA, USA, 3 March 2008.

24. Agiomyrgianaki, A.; Petrakis, P.V.; Dais, P. Detection of refined olive oil adulteration with refined hazelnut oil by employing nmr spectroscopy and multivariate statistical analysis. *Talanta* **2010**, *80*, 2165–2171.

25. Calvano, C.D.; Ceglie, C.D.; D'Accolti, L.; Zambonin, C.G. Maldi-tof mass spectrometry detection of extra-virgin olive oil adulteration with hazelnut oil by analysis of phospholipids using an ionic liquid as matrix and extraction solvent. *Food Chem.* **2012**, *134*, 1192–1198.

26. Chiavaro, E.; Vittadini, E.; Rodriguez-Estrada, M.T.; Cerretani, L.; Bendini, A. Differential scanning calorimeter application to the detectionof refined hazelnut oil in extra virgin olive oil. *Food Chem.* **2008**, *110*, 248–256.

27. Maggio, R.M.; Cerretani, L.; Chiavaro, E.; Kaufman, T.S.; Bendini, A. A novel chemometric strategy for the estimation of extra virgin olive oil adulteration with edible oils. *Food Control* **2010**, *21*, 890–895.

28. Poulli, K.I.; Mousdis, G.A.; Georgiou, C.A. Rapid synchronous fluorescence method for virgin olive oil adulteration assessment. *Food Chem.* **2007**, *105*, 369–375.

29. Apetrei, C.; Apetrei, I.M.; Villanueva, S.; de Saja, J.A.; Gutierrez-Rosales, F.; Rodriguez-Mendez, M.L. Combination of an e-nose, an e-tongue and an e-eye for the characterization of olive oils with different degree of bitterness. *Anal. Chim. Acta* **2010**, *663*, 91–97.

30. Cosio, M.S.; Ballabio, D.; Benedetti, S.; Gigliotti, C. Geographical origin and authentication of extra virgin olive oils by an electronic nose in combination with artificial neural networks. *Anal. Chim. Acta* **2006**, *567*, 202–210.

31. Escuderos, M.E.; Sánchez, S.; Jiménez, A. Virgin olive oil sensory evaluation by an artificial olfactory system, based on quartz crystal microbalance (QCM) sensors. *Sens. Actuators B Chem.* **2010**, *147*, 159–164.

32. Haddi, Z.; Alami, H.; El Bari, N.; Tounsi, M.; Barhoumi, H.; Maaref, A.; Jaffrezic-Renault, N.; Bouchikhi, B. Electronic nose and tongue combination for improved classification of moroccan virgin olive oil profiles. *Food Res. Int.* **2013**, *54*, 1488–1498.

Analysis of a Lipid/Polymer Membrane for Bitterness Sensing with a Preconditioning Process

Rui Yatabe, Junpei Noda, Yusuke Tahara, Yoshinobu Naito, Hidekazu Ikezaki and Kiyoshi Toko

Abstract: It is possible to evaluate the taste of foods or medicines using a taste sensor. The taste sensor converts information on taste into an electrical signal using several lipid/polymer membranes. A lipid/polymer membrane for bitterness sensing can evaluate aftertaste after immersion in monosodium glutamate (MSG), which is called "preconditioning". However, we have not yet analyzed the change in the surface structure of the membrane as a result of preconditioning. Thus, we analyzed the change in the surface by performing contact angle and surface zeta potential measurements, Fourier transform infrared spectroscopy (FTIR), X-ray photon spectroscopy (XPS) and gas cluster ion beam time-of-flight secondary ion mass spectrometry (GCIB-TOF-SIMS). After preconditioning, the concentrations of MSG and tetradodecylammonium bromide (TDAB), contained in the lipid membrane were found to be higher in the surface region than in the bulk region. The effect of preconditioning was revealed by the above analysis methods.

Reprinted from *Sensors*. Cite as: Yatabe, R.; Noda, J.; Tahara, Y.; Naito, Y.; Ikezaki, H.; Toko, K. Analysis of a Lipid/Polymer Membrane for Bitterness Sensing with a Preconditioning Process. *Sensors* **2015**, *15*, 22439–22450.

1. Introduction

Various types of sensor for mimicking the five human senses have been realized. These sensors are categorized into physical and chemical sensors. For example, imaging devices, microphones, and pressure sensors are categorized as physical sensors corresponding to the visual sense, auditory sense and tactile sense, respectively. In particular, rapid advances have been achieved in the performance of imaging devices, which has had an impact on society because imaging devices with high performance are readily available at a low cost. On the other hand, there are few commercialized chemical sensors. There are a number of methods for measuring chemical substances such as gas chromatography mass spectrometry (GCMS), liquid chromatography mass spectrometry (LCMS), high performance liquid chromatography (HPLC), FTIR and nuclear magnetic resonance (NMR). However, these methods should be carried out by specialists with high technical knowledge because the obtained data are complicated, whereas it is desirable for users to be able to easily discriminate and quantify chemical substances using a sensor device.

A taste sensor, a kind of electronic tongue, has been developed and commercialized as a chemical sensor to evaluate the taste of substances [1–14]. It is possible to measure the taste of foods or medicines using the taste sensor, which mimics the taste sense of humans, which is categorized into six types of taste; saltiness, sourness, umami, bitterness, astringency and sweetness. The taste sensor has global selectivity, which means that each of its membranes respond to substances that are categorized in each tastes. Several types of lipid/polymer membrane are used for the taste sensor, which consist of a polymer, plasticizer and lipid. The polymer is used to support the membrane and the plasticizer and lipid are used to control the hydrophobicity and electrical properties of the membrane, respectively. Then, the lipid/polymer membrane transduces the response to a taste into an electrical signal [6].

The measurements performed by the taste sensor are carried out using several types of lipid/polymer membrane designed to detect saltiness, sourness, umami, bitterness, astringency and sweetness. Two types of value are obtained from the measurements. First, a relative value is obtained by immersing the sensor electrodes in a sample solution. Second, a change in the membrane potential caused by adsorption (CPA) value is obtained by immersing the sensor electrodes in another standard solution after the measurement of the sample solution. The two values are used to evaluate the initial taste and aftertaste. The CPA can be obtained using lipid/polymer membranes for umami, bitterness and astringency. It is known that the CPA is related to the adsorption of taste substances on the membrane [15–17]. In this study, we focused on the lipid/polymer membrane for sensing bitterness, which is called "C00". It was previously reported that the CPA became stable after the C00 membrane was immersed in monosodium glutamate (MSG) solution for several days [18]. In this study, we analyzed the change in the surface structure of the lipid/polymer membrane for bitterness sensing by immersing it in MSG solution.

2. Experimental Section

2.1. Materials

Polyvinyl chloride (n about 1100, PVC) was purchased from Wako Pure Chemical Industry (Osaka, Japan) as a polymer material. 2-Nitrophenyl octyl ether (NPOE) was obtained from Dojindo Laboratories (Kumamoto, Japan) as a plasticizer. Tetradodecylammonium bromide (TDAB) was purchased from Sigma Aldrich (St. Louis, MO, USA) as a lipid reagent. Tetrahydrofuran (THF) was obtained from Sigma Aldrich as a solvent. Sodium hydrogen L-glutamate monohydrate (monosodium glutamate, MSG), potassium chloride (KCl) and L(+)-tartaric acid were purchased from Kanto Chemical, Co., Inc. (Tokyo, Japan). Iso-alpha acid was obtained from Intelligent Sensor Technology Inc. (Kanagawa, Japan). Hydrogen peroxide (30%, Wako Pure Chemical Industry) and sulfuric acid (Kanto Chemical, Co., Inc.) were purchased for use as piranha solution. The standard solution in this study was made from 30 mM KCl and 0.3 mM tartaric acid in water. All aqueous solutions were prepared from Milli-Q water obtained from a Milli-Q system (Millipore, Billerica, MA, USA).

2.2. Fabrication of Lipid/Polymer Membrane for Bitterness Sensing

The lipid/polymer membrane for bitterness sensing was fabricated by the following process: first, TDAB was dissolved in THF. Next, NPOE and the TDAB solution were mixed. Then, PVC was added to the mixed solution. After that, the mixed solution was dried on a Petri dish at room temperature for 3 days to obtain a membrane sample. Next, the membrane sample was immersed in MSG solution (30 mM MSG in standard solution) at room temperature for about 7 days (the MSG solution should be remade each day because bacteria breed in the solution). Finally, the membrane sample was cut and attached to the sensor surface using PVC in THF solution to form a sensor electrode. This fabrication process was used in our previous study [6].

2.3. Measurement Procedure of Taste Sensor

The relative value and CPA value were measured using a taste sensor system (TS-5000Z, Intelligent Sensor Technology, Inc.). Several electrodes were prepared for the taste sensor system. One of them was a reference electrode, which was a Ag/AgCl electrode immersed in saturated KCl solution. The others were sensor electrodes that were made from the lipid/polymer membrane. The potentials of these electrodes were generated by immersing them in a sample solution. The output was the electrical potential difference between the sensor electrode and the reference electrode.

The measurement procedure was carried out by the following steps. First, the electrodes were immersed in the standard solution and the sensor output was measured as V_r. Next, the electrodes were immersed in the sample solution and the output was measured as V_s. Then, the electrodes were immersed in another standard solution and the output was measured as V_r'. The relative value and CPA value were calculated from the following equations:

$$\text{Relative value} = V_s - V_r \tag{1}$$

$$\text{CPA value} = V_r' - V_r \tag{2}$$

Finally, the electrodes were rinsed with a cleaning solution (100 mM KCl, 10 mM KOH, 30 vol% ethanol) and another standard solution. The measurement procedure was performed five times for each sample. The averages and standard deviations were calculated using the results excluding those for the first measurement.

2.4. Evaluation of Surface Conditions from Surface Zeta Potential and Contact Angle

The surface zeta potential and contact angle were measured to evaluate the change in the surface conditions due to the preconditioning process. First, the zeta potential of the lipid polymer membrane surface was measured by the streaming potential method using a SurPASS analyzer (Anton Paar GmbH, Graz, Austria) with an adjustable gap cell accessory. The solution used for the measurement was 1 mM $KCl_{(aq)}$. A small amount of 100 mM KOH was added to the solution to control the pH to about 7 because the pH of the fresh KCl solution was about 5.8. The measurements were carried out while the solution was titrated with a small amount of 100 mM HCl. The measurements were conducted four times to obtain each data point. Second, the contact angle of the lipid/polymer membrane surface was measured by a DM500 contact angle meter (Kyowa Interface Science Co., Ltd., Saitama, Japan). The measurement was performed with a 2 µL water droplet. The measurements were conducted three times to obtain each data point.

2.5. Analysis of Substances on Lipid/Polymer Membrane Surface by FTIR-RAS

FTIR is one of the most convenient methods for the analysis of organic compounds. However, it is difficult to analyze thin film samples by FTIR, even if the attenuated total reflection (ATR) method is used, because the penetration depth of infrared light is about 500 nm using a germanium prism (or about 1000 nm using a diamond prism, which is commonly used). Therefore, we analyzed the organic substances on the surface of the lipid/polymer membrane by FTIR with reflection absorption spectroscopy (RAS) after the organic substances had been transferred to a gold surface.

The RAS method is used for the analysis of organic thin films on metal substrates. RAS is normally not able to analyze organic film surfaces (the substrate must be a flat metal). Therefore, after the substances on the lipid/polymer surface were transferred to a thin gold film by the following steps, we analyzed the organic layer on the gold film by FTIR-RAS. First, a thin gold film on a glass substrate was cleaned by piranha solution ($H_2SO_4:H_2O_{2 (aq)}$ = 4:1, 120 °C, 10 min) during which the surface of the gold film became clean and hydrophilic. Next, the surface of the lipid/polymer membrane was attached to the gold surface. Then, the lipid/polymer membrane was pressed from its rear side using a spatula. Next, the lipid/polymer membrane was lifted off from the gold surface. Finally, we analyzed the substances adsorbed on the gold surface using an FTIR system (Frontier Gold FTIR, PerkinElmer, Waltham,

MA, USA) RAS with an Advanced Grazing Angle accessory including a polarizer (PIKE Technologies, Fitchburg, WI, USA) using an MCT detector.

2.6. Chemical Analyses of Surface of Lipid/Polymer Membrane

We used two methods for the analysis of organic substances on the surface in addition to FTIR-RAS. One of them was x-ray photoelectron spectroscopy (XPS), which is often used for surface analysis. For the analysis of organic substances, it is possible to recognize some chemical substances from the chemical shifts of nitrogen, oxygen and other elements. The analysis is highly sensitive within the surface region (within 2 nm). We evaluated the surface of the lipid/polymer membrane using an ESCA5800 system (ULVAC-PHI, Inc., Kanagawa, Japan) with a neutralizer. However, there were two problems in our XPS measurements. One of them was the possibility that some substances disappear by evaporation because the measurements were performed under ultrahigh vacuum (below 1×10^{-9} torr). Actually, the measurements were carried out after the lipid/polymer membrane samples had been stored under high vacuum (about 1×10^{-6} torr) for 2–3 weeks because it was impossible to measure the fresh membrane samples owing the vacuum. This suggests that some substances with low boiling points evaporated from the samples (for example, NPOE is a liquid at room temperature, whose boiling point is about 200 °C.). The other problem was contamination. It is difficult to distinguish target compounds and contaminants because it is difficult to identify a substance from information on the atomic and chemical shifts if not all the substances on the surface are known. We carried out gas cluster ion beam (GCIB) time of flight (TOF) secondary ion mass spectrometry (SIMS) with a cooled sample stage to solve these problems.

GCIB-TOF-SIMS is a method of analysis used to determine the depth profile of substances in an organic material. The analysis is performed by TOF-SIMS while etching the surface with an argon GCIB. The merit of etching with a GCIB is less damage to the substances than etching with an argon ion beam. If the etching is performed with an argon ion beam, it is difficult to identify the compounds because fragments of the compounds are generated during the etching. We evaluated the depth profile of substances in the lipid/polymer membrane starting from the surface, which was carried out by Toray Research Center (Tokyo, Japan) using a TOF.SIMS 5 (ION-TOF GmbH, Münster, Germany) instrument. It was possible to distinguish the target compounds and contaminants because the substances could be partially identified from the mass spectrum data. In addition, the effect of evaporation was reduced by cooling the samples to below -140 °C. In fact, it was possible to measure the fresh membrane samples without storing samples in a vacuum. However, there were two problems in our GCIB-TOF-SIMS measurement. One of them was that it was impossible to compare the concentrations of substances because the normalization of data could not be carried out owing to the lack of a standard sample. Such normalization is required to determine the absolute concentrations of the substances because the sensitivity of the mass detector is considerably different for each compounds. The other problem is that impurities in the materials affect the result because the mass detector has high sensitivity. In summary, to reveal the chemical substances on the lipid/polymer surface, we carried out XPS measurements with GCIB-TOF-SIMS measurement to support the XPS data.

3. Results and Discussion

3.1. Effect of Preconditioning of Lipid/Polymer Membrane

The CPA value of the fresh lipid/polymer membrane was low. However, it became high when the membrane was immersed in MSG solution (30 mM in standard solution) for 4–5 days. The immersion process is called "MSG preconditioning". Figure 1 shows the result of preconditioning. The five measurements for a bitterness sample (0.01 vol% iso-a-acid in standard solution) were carried out in one day. After the measurements, the lipid/polymer membrane was immersed in the MSG solution until the next day. The measurements were carried out for 13 days (0–12 days: total number of measurements = 65).

First, the CPA value (shown in Figure 1a) is low until about 3 days. After that, it starts to change then becomes stable at about -80 mV. The electrode potential, which is V_r in Equations (1) and (2), is shown in Figure 1b. The transient behavior of the electrode potential is similar to that of the CPA value. Several similar experiments were carried out using other samples. Although the transients of the results were slightly different from each other, the stable CPA value and electrode potential were about -80 and $+100$ mV, respectively. This CPA value agrees with previously reported values [6]. In summary, the electrical property of the lipid/polymer membrane for bitterness sensing was changed by immersion (preconditioning) in MSG solution.

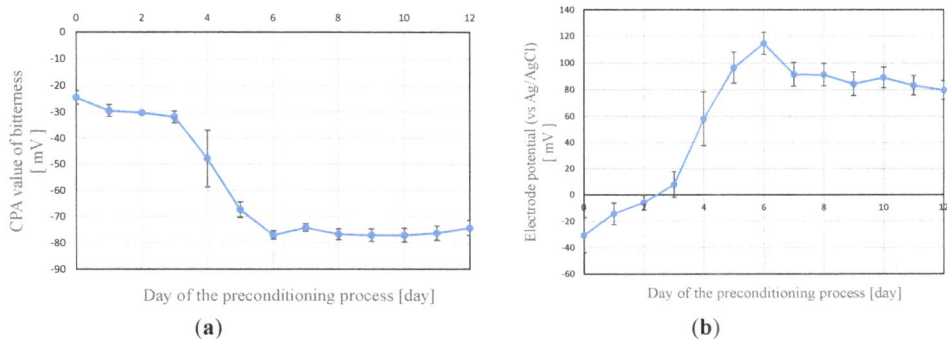

Figure 1. The result of measurements for bitterness sample (0.01 vol% iso-α-acid) depending on the preconditioning time. The five measurements were carried out at each day. The error bar is standard deviation of the data. (**a**) CPA value; (**b**) The electrode potential (*vs.* Ag/AgCl at room temperature) when the electrode was immersed in the standard solution.

3.2. Surface Conditions of Lipid/Polymer Membrane

The electrode potential was affected by the preconditioning process. Figure 2 shows the change in the surface zeta potential with the progress of preconditioning. The potential became increasingly negative with increasing preconditioning time. This result suggests that the MSG molecules are adsorbed on the membrane surface because the molecules are negatively charged at pH higher than 3.22, which is the isoelectric point of MSG.

The contact angle was measured to confirm the result of the surface zeta potential measurements during preconditioning process. Figure 3 shows the result. The contact angle decreased during preconditioning. This means that the surface became hydrophilic as a result of preconditioning. This result is consistent with that for the surface zeta potential because the surface becomes hydrophilic if the MSG molecules are adsorbed on the surface. In conclusion, the surface of the lipid/polymer membrane was negatively charged by the preconditioning and became hydrophilic.

175

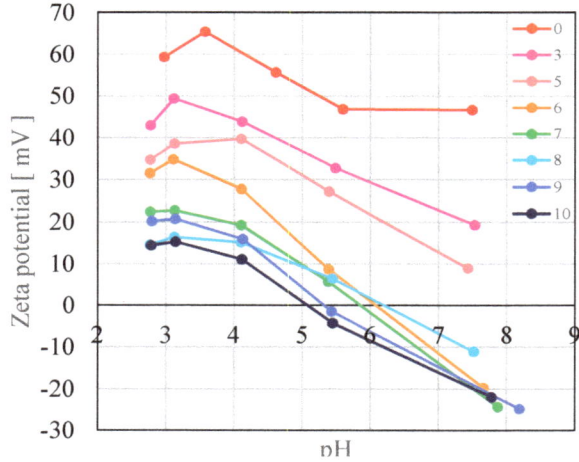

Figure 2. The dependence of the surface zeta potential during preconditioning. For example, the red graph is data of the membrane without the preconditioning. The dark blue graph is data of the membrane with the preconditioning for 10 days.

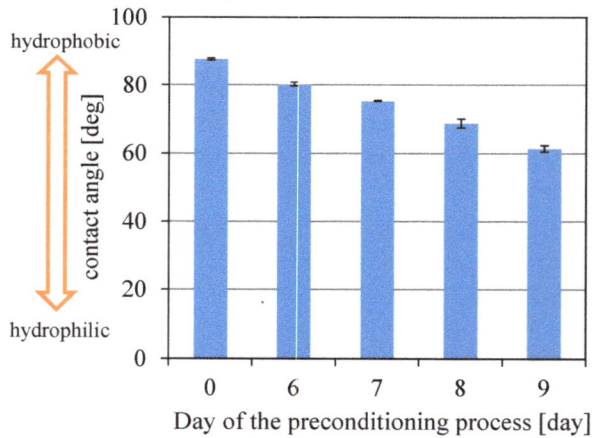

Figure 3. The contact angle of a water droplet with the lipid/polymer membrane.

3.3. FTIR-RAS Analysis of Substances on Lipid/Polymer Membrane Surface

We found that the surface was changed by the preconditioning as reported in the previous section. Chemical analyses of the surface were carried out to reveal the reason for the change. First, FTIR-RAS by the transfer method was carried out in accordance with the procedure in Section 2.4. Figure 4 shows the results for samples with and without preconditioning for 7 days. Several peaks appear for the sample with the preconditioning, and that do not appear for the sample without the preconditioning. The peaks at 2860 and 2920 cm^{-1} result from C-H stretching, the peak at 1610 cm^{-1} results from C=C stretching in the aromatic ring, the peak at 1530 cm^{-1} results from NO$_2$ asymmetric stretching, and the peaks at 1350 and 1280 cm^{-1} result from C–O–C asymmetric stretching. The chemical structure of the substance as below contained in the lipid/polymer membrane is shown in Figure 5. The above peaks suggest the chemical structure of NPOE. However, it would be difficult to recognize TDAB by FTIR even if it exists on the surface because the characteristic peaks of TDAB with high strength do not appear in IR data except for the peak corresponding to C–H stretching (we obtained IR data for the TDAB used in this study by FTIR-ATR in another experiment). In summary, NPOE is exuded onto the surface during the preconditioning.

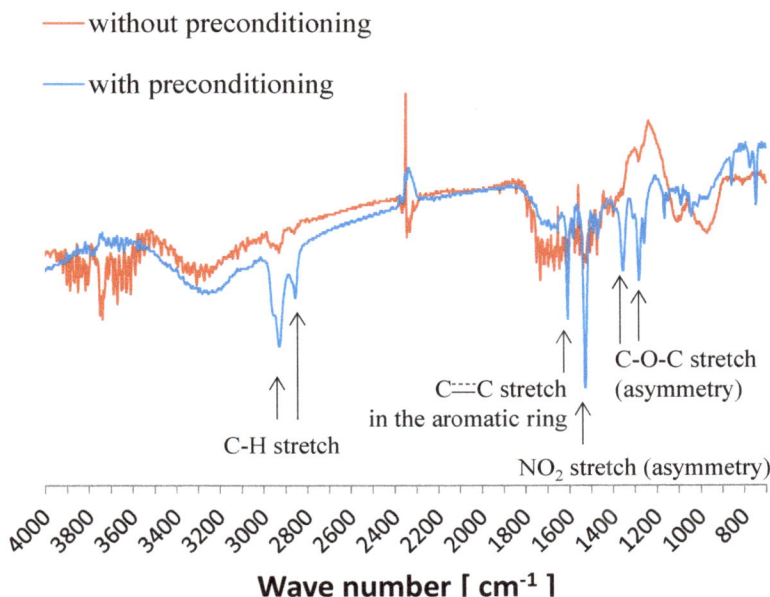

Figure 4. FTIR-RAS with transfer method for samples with and without the preconditioning for 7 days.

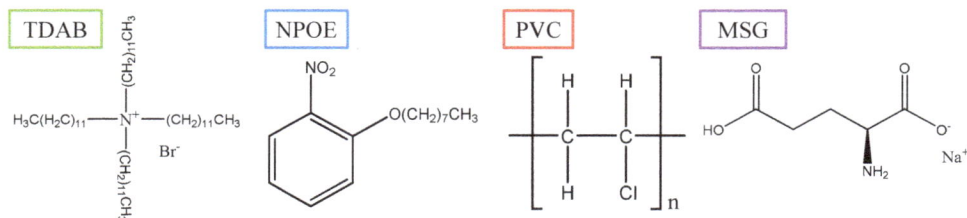

Figure 5. The chemical structures of TDAB, NPOE, PVC and MSG.

3.4. Chemical Analysis of Surface of Lipid/Polymer Membrane by XPS

The analysis of samples with various NPOE and TDAB concentrations with and without preconditioning for 7 days was carried out by XPS. Table 1 shows the conditions of the samples. The samples that were analyzed by FTIR or other methods were samples 1-1-0 (without preconditioning) and 1-1-7 (with preconditioning). Sample 1-1-7 is used as a bitterness sensor. Figure 6 shows the result of the XPS measurements. First, we discuss the samples except for sample 1000-0-0 to simplify the explanation. As shown in the lower graph, the samples not subjected to preconditioning have no peak, whereas the samples subjected to preconditioning have a peak at 402 or 400 eV. This means that TDAB or NPOE was concentrated on the membrane surface by the preconditioning or that MSG was adsorbed on the surface by the preconditioning because TDAB, NPOE and MSG have a nitrogen atom. Next, sample 0-0-7, which has no TDAB or NPOE and was immersed in MSG solution, has a peak at 400 eV. Thus, the peak at 400 eV originates from the MSG. Next, the intensity of the peak at 402 eV for samples 0-0-7, 0-1-7, 1-1-7 and 100-1-7 depends on the amount of TDAB. In addition, sample 1000-0-0, which has only TDAB, has a peak at 402 eV. Therefore, the peak at 402 eV originates from TDAB. Moreover, the intensities of the peak at 402 eV for samples 1-1-7 and 100-1-7 are higher than those of samples 1-1-0 and 100-1-0, which were not subjected to preconditioning. This means that the TDAB on the surface was concentrated by the preconditioning. In conclusion, the preconditioning process concentrated the TDAB on the lipid/polymer surface and caused the adsorption of MSG on the surface.

Table 1. The samples for XPS measurement. The first and second numbers are the amount of TDAB and NPOE, respectively. The last number is the preconditioning time. For example, sample 100-1-7 means that the amount of TDAB is 100 times of sample 1-1-7, which is used as a bitterness sensor.

Sample Name	TDAB (times)	NPOE (times)	Preconditioning Time [day]
0-0-0	0x	0x	0
0-0-7	0x	0x	7
0-1-0	0x	1x	0
0-1-7	0x	1x	7
1-1-0	1x	1x	0
1-1-7	1x	1x	7
100-1-0	100x	1x	0
100-1-7	100x	1x	7
1000-0-0	1000x	0x	0

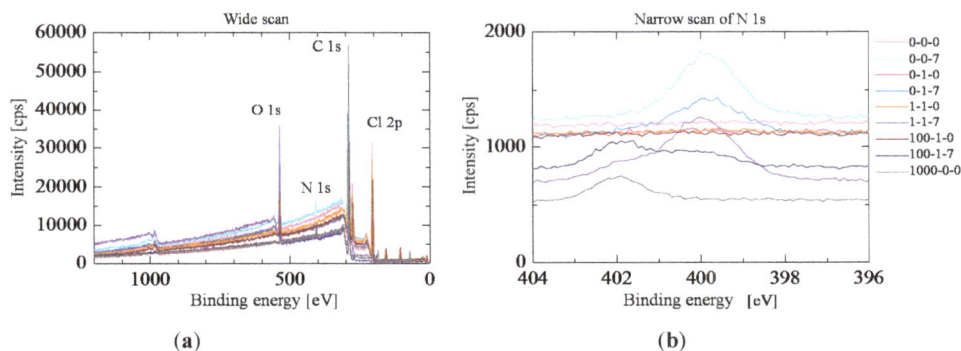

Figure 6. The result of XPS measurement. The upper graph is the result of wide scan. The lower graph is the result of narrow scan for signal of nitrogen 1s. The red lines are the results of samples without preconditioning. The blue lines are the result of samples with preconditioning. (**a**) Wide scan; (**b**) Narrow scan of Nitrogen 1s.

3.5. Analysis of Depth Profile of Substances in Lipid/Polymer Membrane

The depth profile of the concentration of substances in the lipid/polymer membrane was measured by GCIB-TOF-SIMS to confirm the result of XPS. Figure 7 shows the result of GCIB-TOF-SIMS, which was carried out starting from the surface. The concentrations of TDAB and MSG in the surface region are higher than those in the deep region after the preconditioning. TDAB was concentrated in the surface region, enabling it to be trapped at the interface between the MSG aqueous solution and the membrane via the quaternary amino group of TDAB, which is hydrophilic. MSG diffused from the solution into the membrane. This result supports the result of XPS analysis. In addition, we found that the concentrations of TDAB and MSG after preconditioning are higher up to a depth of 1000 nm, although we expected that they would have only been concentrated at the surface. However, we could not explain the data for NPOE because the results for positive and negative ions were different. There might be problem in measuring NPOE by GCIB-TOF-SIMS because the result of FTIR-RAS in Figure 4 suggests that the amount of NPOE increased on the surface. In conclusion, the concentrations of TDAB and MSG were higher in the surface region than in the bulk region.

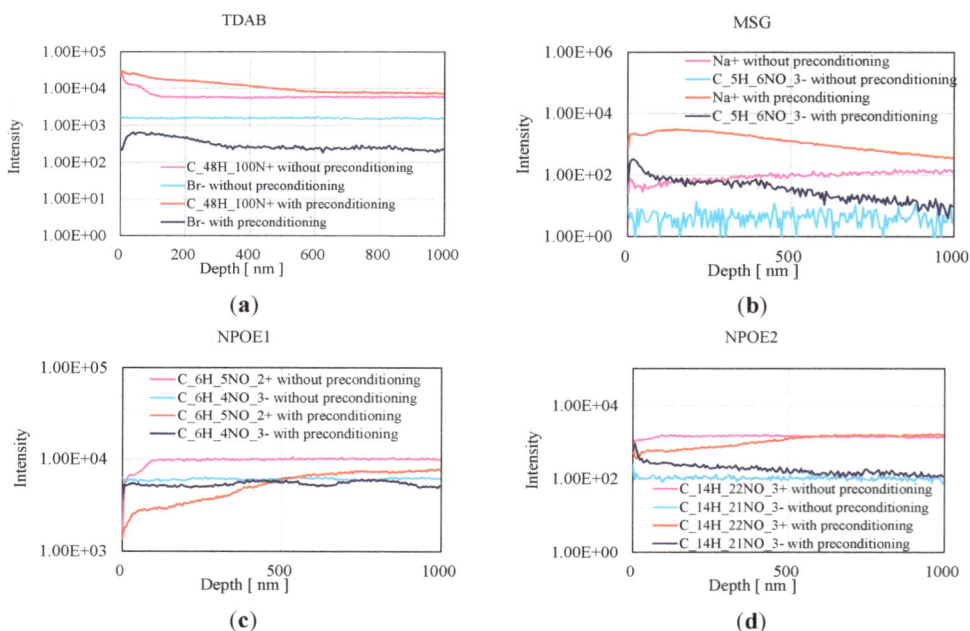

Figure 7. The result of GCIB-TOF-SIMS measurement. Four lines indicate the data of samples with and without the preconditioning using positive and negative ion. ATTENTION: The intensities cannot compare with each other samples because of no normalization. (**a**) $^{69\circ}C_{48}H_{100}N^+$ and $^{81}Br^-$ of TDAB; (**b**) $^{23}Na^+$ and $^{128}C_5H_6NO_3{}^-$ of MSG; (**c**) $^{123}C_6H_5NO_2{}^+$ and $^{138}C_6H_4NO_3{}^-$ of NPOE; (**d**) $^{252}C_{14}H_{22}NO_3{}^+$ and $^{251}C_{14}H_{21}NO_3{}^-$ of NPOE.

4. Conclusions

We analyzed the surface of a lipid/polymer membrane for bitterness sensing to determine the effect of the MSG preconditioning process. The purpose of this study was to explain why it becomes possible to measure bitterness by CPA using the lipid/polymer membrane after the preconditioning process. We evaluated the surface conditions using electrical and chemical methods. First, the contact angle and the surface zeta potential were measured as electrical methods because those have a relationship with the response of the taste sensor. The results indicated that the surface became hydrophilic and was negatively charged by the preconditioning. Next, FTIR-RAS, XPS and GCIB-TOF-SIMS were carried out as chemical methods to identify substances on the surface. Their results indicated that the adsorption of MSG on the surface, and that TDAB was concentrated on the surface. However, the results did not support that of FTIR-RAS measurement, which showed that NPOE was exuded on the surface. In this study, we revealed the change in the surface structure caused by MSG preconditioning. In our future work, we will clarify the relationship between the mechanism of CPA measurement and the change in the surface structure.

Acknowledgments: A part of this work was conducted in Kyushu University, supported by Nanotechnology Platform Program (Molecule and Material Synthesis) of the Ministry of Education, Culture, Sports, Science and Technology (MEXT), Japan and this work was supported by JSPS KAKENHI Grant Number 26820129.

Author Contributions: This work presented here was carried out in collaboration between all authors. Rui Yatabe, Junpei Noda, Yusuke Tahara, Yoshinobu Naito, Hidekazu Ikezaki and Kiyoshi Toko define the research theme. Rui Yatabe and Junpei Noda carried out the experiments and analyzed the data. Rui Yatabe interpreted the results and wrote the paper. Rui Yatabe, Yusuke Tahara, Yoshinobu Naito, Hidekazu Ikezaki and Kiyoshi Toko provided directions for experimental methods, analysis of data, interpretation of the results and writing of the paper. All the authors have contributed to, seen and approved the manuscript.

Conflicts of Interest: The authors declare no conflict of interest.

References and Notes

1. Toko, K., Ed.; *Biochemical Sensors: Mimicking Gustatory and Olfactory Senses*; Pan Stanford Publishing: Singapore, Singapore, 2013.
2. Habara, M.; Toko, K. Taste Sensor. In *Encyclopedia of Sensors*; American Scientific Publishers: Stevenson Ranch, CA, USA, 2006; Volume 10, pp. 107–119.
3. Anand, V.; Kataria, M.; Kukkar, V.; Saharan, V.; Choudhury, P.K. The Latest Trends in the Taste Assessment of Pharmaceuticals. *Drug Discov. Today* **2007**, *12*, 257–265.
4. Citterio, D.; Suzuki, K. Smart Taste Sensors. *Anal. Chem.* **2008**, *80*, 3965–3972.
5. Winquist, F. Voltammetric Electronic Tongues—Basic Principles and Applications. *Microchimica. Acta.* **2008**, *163*, 3–10.
6. Kobayashi, Y.; Hamada, H.; Yamaguchi, Y.; Ikezaki, H.; Toko, K. Development of an Artificial Lipid-Based Membrane Sensor with High Selectivity and Sensitivity to the Bitterness of Drugs and with High Correlation with Sensory Score. *IEEJ Trans. Electr. Electron. Eng.* **2009**, *4*, 710–719.
7. Harada, T.; Uchida, T.; Yoshida, M.; Kobayashi, Y.; Narazaki, R.; Ohwaki, T. A New Method for Evaluating the Bitterness of Medicines in Development using a Taste Sensor and a Disintegration Testing Apparatus. *Chem. Pharm. Bull.* **2010**, *58*, 1009–1014.

8. Kobayashi, Y.; Habara, M.; Ikezazki, H.; Chen, R.; Naito, Y.; Toko, K. Advanced Taste Sensors Based on Artificial Lipids with Global Selectivity to Basic Taste Qualities and High Correlation to Sensory Scores. *Sensors* **2010**, *10*, 3411–3443.

9. Riul, A., Jr.; Dantas, C.A.R.; Miyazaki, C.M.; Oliveira, O.N., Jr. Recent Advances in Electronic Tongues. *Analyst* **2010**, *135*, 2481–2495.

10. Uchiyama, Y.; Yamashita, M.; Kato, M.; Suzuki, T.; Omori, M.; Chen, R. Evaluation of the Taste of Tea with Different Degrees of Fermentation using a Taste Sensing System. *Sens. Mater.* **2011**, *23*, 501–506.

11. Woertz, K.; Tissen, C.; Kleinebudde, P.; Breitkreutz, J. A Comparative Study on Two Electronic Tongues for Pharmaceutical Formulation Development. *J. Pharm. Biomed. Anal.* **2011**, *55*, 272–281.

12. Woertz, K.; Tissen, C.; Kleinebudde, P.; Breitkreutz, J. Taste Sensing Systems (Electronic Tongues) for Pharmaceutical Applications. *Int. J. Pharm.* **2011**, *417*, 256–271.

13. Savage, N. Technology: The Taste of Things to Come. *Nature* **2012**, *486*, S18–S19.

14. Tahara, Y.; Toko, K. Electronic Tongues-a Review. *IEEE Sens. J.* **2013**, *13*, 3001–3011.

15. Hara, D.; Fukagawa, T.; Tahara, Y.; Yasuura, M.; Toko, K. Examination of Amount of Astringent Substances Adsorbed Onto lipid/polymer Membrane used in Taste Sensor. *Sens. Lett.* **2014**, *12*, 1172–1176.

16. Toko, K.; Hara, D.; Tahara, Y.; Yasuura, M.; Ikezaki, H. Relationship between the Amount of Bitter Substances Adsorbed Onto Lipid/polymer Membrane and the Electric Response of Taste Sensors. *Sensors* **2014**, *14*, 16274–16286.

17. Harada, Y.; Tahara, Y.; Toko, K. Study of the Relationship between Taste Sensor Response and the Amount of Epigallocatechin Gallate Adsorbed Onto a Lipid-Polymer Membrane. *Sensors* **2015**, *15*, 6241–6249.

18. Insent, Inc. *Highly Durable and Rapidly Measurable Taste Sensor for Quality Control of Products Using Artificial Lipid/polymer Membrane*; Report of Risk-taking Fund for Technology Development, Japan Science and Technology Agency: Tokyo, Japan, 2007; pp. 25–36.

Novel PCR Assays Complement Laser Biosensor-Based Method and Facilitate *Listeria* Species Detection from Food

Kwang-Pyo Kim, Atul K. Singh, Xingjian Bai, Lena Leprun and Arun K. Bhunia

Abstract: The goal of this study was to develop the *Listeria* species-specific PCR assays based on a house-keeping gene (*lmo1634*) encoding alcohol acetaldehyde dehydrogenase (Aad), previously designated as *Listeria* adhesion protein (LAP), and compare results with a label-free light scattering sensor, BARDOT (bacterial rapid detection using optical scattering technology). PCR primer sets targeting the *lap* genes from the species of *Listeria sensu stricto* were designed and tested with 47 *Listeria* and 8 non-*Listeria* strains. The resulting PCR primer sets detected either all species of *Listeria sensu stricto* or individual *L. innocua*, *L. ivanovii* and *L. seeligeri*, *L. welshimeri*, and *L. marthii* without producing any amplified products from other bacteria tested. The PCR assays with *Listeria sensu stricto*-specific primers also successfully detected all species of *Listeria sensu stricto* and/or *Listeria innocua* from mixed culture-inoculated food samples, and each bacterium in food was verified by using the light scattering sensor that generated unique scatter signature for each species of *Listeria* tested. The PCR assays based on the house-keeping gene *aad* (*lap*) can be used for detection of either all species of *Listeria sensu stricto* or certain individual *Listeria* species in a mixture from food with a detection limit of about 10^4 CFU/mL.

Reprinted from *Sensors*. Cite as: Kim, K.P.; Singh, A.K.; Bai, X.J.; Leprun, L.; Bhunia, A.K. Novel PCR Assays Complement Laser Biosensor-Based Method and Facilitate *Listeria* Species Detection from Food. *Sensors* **2015**, *15*, 22672–22691.

1. Introduction

Listeria monocytogenes, a foodborne pathogen, causes fatal systemic infection in immunocompromised hosts including the elderly, infants, pregnant women and their fetuses, HIV infected patients, and patients with malignancy receiving chemotherapy. Alcohol acetaldehyde dehydrogenase (Aad) in *L. monocytogenes* is a house-keeping enzyme and is involved in bacterial adhesion and paracellular translocation through epithelial barrier during intestinal phase of listeriosis [1–4]. Such a housekeeping enzyme with moonlighting function in prokaryotes plays an important role in pathogenesis [5,6]. The Aad (Lmo1634) is also known as *Listeria* adhesion protein (LAP) and its homolog is present in all species of *Listeria sensu stricto* (*i.e.*, in the narrow or strict sense) also known as archetypal *Listeria* species

(*L. monocytogenes*, *L. ivanovii*, *L. seeligeri*, *L. welshimeri*, *L. innocua*, and *L. marthii*) [1,7,8]. Whereas, *L. floridensis*, *L. aquatic*, *L. cornellensis*, *L. riparia*, *L. grandensis*, *L. booriae*, *L. rocourtiae*, *L. newyorkensis*, *L. weihenstephanensis*, *L. fleischmannii* and *L. grayi* [8,9] are considered atypical (*sensu lato*: in the broad sense) and these group are phylogenetically divergent from the species of *Listeria sensu stricto* [9–12].

L. monocytogenes is pathogenic to humans and is responsible for fatal outbreaks involving ready-to-eat meat, dairy, fish, fruits, and vegetables [13]. It was responsible for 57 cases (22 fatalities) from consumption of tainted meat products in Canada [14], 27 cases (8 fatalities) from Quargel sour milk curd cheese [15], 147 cases (33 fatalities) from cantaloupe [16], and most recently in 2015, 35 cases (7 deaths) from caramel apple [17] and 10 cases (3 deaths) from ice cream [18]. The case-fatality rate for listeriosis is 20%–30% [19]. Under the United States Food and Drug Administration (FDA) definition of Current Good Manufacturing Practice, cGMP [21 CFR 110.5(a)], it is mandatory to monitor food for adulterations [21 U.S.C 342(a)] including all poisonous or deleterious substances, which may render food injurious to health. The FDA recommends initial rapid screening of frozen or refrigerated ready-to-eat (RTE) food products for *Listeria* species rather than the lengthy specific test for *L. monocytogenes* [20].

In this study, species of *Listeria sensu stricto*-specific PCR primer sets targeting *lap* (*aad*), a house-keeping gene were developed that detected all species of *Listeria sensu stricto* tested. House-keeping genes are integral and essential for bacterial metabolic function and survival [21], thus they provide an attractive target for detection. This molecular assay based on *lap* could be used as a screening tool to address the needs of food safety and the regulatory agency. These PCR primer sets were further used to detect *Listeria* species from inoculated food samples. In addition, the light scattering sensor, BARDOT (bacterial rapid detection using optical scattering technology) [22–24] was also employed to verify the presence of *L. monocytogenes* and *L. innocua* from a mixed culture (*Listeria* plus *Lactobacillus casei* and *Escherichia coli* O157:H7) inoculated food samples. In BARDOT, a red-diode laser (635 nm; 1 mW; 1 mm diameter) passes through the center of a bacterial colony on an agar plate and generates a 2-dimensional forward scatter fingerprint of each colony within 3–5 s [23]. Organism-specific features are extracted from scatter patterns and are used to identify unknown bacteria using the scatter image library [25]. Scatter image libraries for the thirteen serotypes of *L. monocytogenes* (1/2a, 1/2b, 1/2c, 3a, 3b, 3a, 4a, 4b, 4ab, 4c, 4d, 4e and 7) were also developed for the BARDOT-based detection in future studies.

2. Experimental Section

2.1. Bacterial Cultures, Growth and Ribotyping

All bacterial cultures (Table S1) used in this study are from our collection. All cultures were stored at −80 °C as 10% frozen glycerol stocks, and fresh cultures were obtained by propagating in Brain Heart Infusion broth (BHI) or Tryptic soy broth with 0.6% yeast extract (TSB-YE) at 37 °C for 16–18 h, with the exception of *L. rocourtiae*, which was grown at 32 °C. The bacterial cultures were plated on Brain Heart Infusion Agar (BHIA), Luria-Bertani Agar (LBA) to capture the colony scatter patterns. The majority of ribopattern information for cultures was obtained from a previous study from our lab [26]. Additional cultures were ribotyped using the automated Riboprinter® (Qualicon, Inc., Wilmington, DE, USA) as described in our previousstudy [27]. For the food sample study, Fraser Broth (FB) containing 10 mL antimicrobial supplement (25 mg acriflavin, 20 mg nalidixic acid and 500 mg ammonium ferric citrate per liter) was used. Dehydrated media or media components were purchased from BD (Sparks, MD, USA) and FB was purchased from Acumedia (Neogen, Lansing, MI, USA).

2.2. Design of Lap Gene-Specific Primer Sets for Listeria Species

The *lap* sequences in *L. monocytogenes* F4244 (Acc. No. AY561824), *L. innocua* F4248 (Acc. No. AY561825), *L. welshimeri* ATCC35897 (Acc. No. AY561828), *L. seeligeri* SE31 (Acc. No. AY561827) and *L. ivanovii* SE98 (Acc. No. AY561826) were reported previously [1]. In addition, the complete sequences of the *lap* gene from *L. monocytogenes* EGD (Acc. No. NC_003210), *L. innocua* CLIP11262 (Acc. No. NC003212) and *L. marthii* (Acc. No. NZ_CM001047) [28] were obtained from NCBI GenBank [29]. To identify a species-specific DNA sequence region, the MultAlin [30] program was used to align and compare the sequences of the *lap* gene. The scheme for *Listeria* genus/species-specific primer binding sites on the *lap* gene are represented in Figure 1.

Two conserved sequence regions (1–54 and 1294–1401) were found in different *Listeria* species (*L. monocytogenes* EGD, F4244; *L. innocua* CLIP11262, F4248; *L. welshimeri*; *L. ivanovii*; *L. seeligeri* and *L. marthii*), and these regions were used to design the species of *Listeria sensu stricto*-specific primer set, ELAP-F2 and LIS-R1 (Table 1). Other *Listeria* species-specific primer sets were developed based on the rule that the 3′-end of primer should be unique to the target species. *L. innocua*-specific primers are designated as Inn-F1 and Inn-R1, and *L. welshimeri*-specific primers are named as Wel-F1 and Wel-R1. A primer set IvaSee-F1 and IvaSee-R1 was specific for both *L. ivanovii* and *L. seeligeri* as they represent close genetic relatedness [31]. In addition, primers Mar-F1 and Mar-R3 were specific for *L. marthii*. Specific primers for either *L. monocytogenes*, *L. ivanovii* or *L. seeligeri* could not be obtained, possibly due to their highly conserved *lap* gene sequence motifs [1].

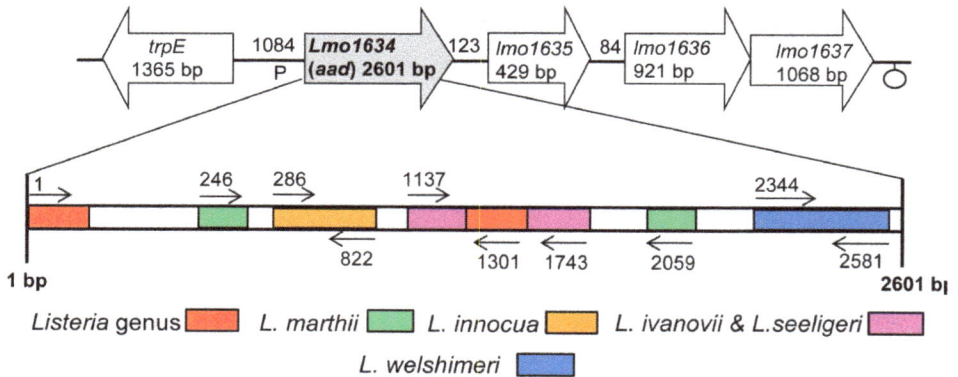

Figure 1. Schematic representation of the *lap* (*aad*) gene-specific primer binding sites for PCR-based detection of *Listeria*. Open block arrow represents genes flanking the *lap*. Abbreviations represents (similar proteins to): *trpE*, anthranilate synthase alpha subunit; *lmo1634*, alcohol-acetaldehyde dehydrogenase (LAP); *lmo1635*, unknown protein; *lmo1636*, ABC transporter (ATP-binding protein); *lmo1637*, membrane protein. The *Listeria lap* primers were specific for all species of *Listeria sensu stricto* tested, but not the atypical listeriae; *L. grayi* and *L. rocourtiae*. Colored boxes indicate primer binding sites for *Listeria* species (see Table 1 for PCR product size).

Table 1. Sequences of species-specific primers based on *lap* sequence used in this study.

Primer	Sequence [a]	Location in *lap* Gene	Product Size (bp)	Specificity
ELAP-F1 LIS-R1	5'CGGTCCCCGGGTACCATGGCAA 5'TTAAAGAAAATGCGGCC3'	1–1301	1301	*Listeria* spp. (except *L. grayi*, *L. rocourtiae*)
Inn-F1 Inn-R1	5'GGAGTTATTAACGAAGATACT3' 5'TTCTGCTTTTACTTCTTTAGCA3'	286–822	536	*L. innocua*
IvaSee-F1 IvaSee-R1	5'AAGCTGCAGTTATTCATTCC3' 5'ATCTAAGAATTTTTGTTTTAGT3'	1137–1743	606	*L. ivanovii*, *L. seeligeri*
Wel-F1 Wel-R1	5'TTCTCGTATTATCGGTTTACCA3' 5'GCTTCAAGATAGATTTCTTTCAA3'	2344–2581	237	*L. welshimeri*
Mar-F1 Mar-R1	5'AGAATATATTTGGAACAGCATC3' 5'GTTCGATTGCACGGATGGAAAG3'	246–268 2038–2059	1813	*L. marthii*

[a] Underlined indicates artificial nucleotide addition sites; translation start codon is indicated in bold character.

2.3. PCR Conditions, Primers and DNA Extraction

For PCR, 100 ng of template DNA, 25 pmol of each primer, 0.2 μL of Go*Taq* polymerase (5 U/μL stock; Promega), 1× Go*Taq* flexi colored buffer (5x stock, Promega), 2 mmol/L MgCl$_2$ (25 mmol/L stock, Promega), and 200 μM of dNTPs (10 mmol/L stock, Promega) were mixed for a 25 μL final volume. PCR amplification was done using a thermocycler (GeneAmp PCR System 9700, Applied Biosystems) as follows: Hot start at 95 °C for 5 min; 30 cycles with denaturation at 95 °C for 1 min, annealing at 54 °C for 1 min, and extension at 72 °C for 1.5 min; final extension at 72 °C for 10 min. The amplified DNA was resolved in 1.2% agarose gel and visualized by ethidium bromide staining with a ChemiDoc XRS gel documentation system (Bio-Rad). The species of *Listeria sensu stricto*-specific primer set and individual *Listeria* species-specific primer sets (Table 1) were used for identification of each *Listeria* species in a pure culture and in the model foods. List of primers and their binding locations on *lap* gene is presented in Table 1. To further verify the *lap*-gene specific PCR results, two sets of cell wall hydrolase; CWH or p60 (*iap*) gene-specific primers, Lis1B/MonoA and Lis1B/Ino2 [32], were applied to verify *L. monocytogenes* and *L. innocua* cultures, respectively. The sequence of Lis1B, MonoA and Ino2 primers are 5′-TTATACGCGACCGAAGCCAAC-3′, 5′-CAAACTGCTAACACAGCTACT-3′ and 5′-ACTAGCACTCCAGTTGTTAAAC-3′, respectively.

The genomic DNA from reference cultures or enriched food samples were extracted with DNeasy Tissue Kit (Qiagen) following manufacturer's protocol. Briefly, the cultures were pretreated with lysozyme solution (10 mg/mL in TE buffer (pH 7.0) containing 10 mmol/L Tris-Cl, pH 7.0 and 1 mmol/L EDTA) at 37 °C for 30 min prior to cell lysis. The total DNA was also extracted following the published protocol [33]. The concentration and purity of genomic DNA was determined using NanoDrop 2000C (Thermo Scientific, Franklin, MA, USA).

2.4. Specificity and Sensitivity of Lap Gene Primers for Listeria Detection

A total of 55 *Listeria* (n = 47) and non-*Listeria* (n = 8) cultures were tested to determine the specificity of *lap* gene primer sets for the species of *Listeria sensu stricto* or individual species: *L. monocytogenes* (n = 13), *L. ivanovii* (n = 12), *L. innocua* (n = 10), *L. seeligeri* (n = 5), *L. welshimeri* (n = 3), *L. grayi* (n = 2), *L. marthii* (n = 1) and *L. rocourtiae* (n = 1). Non-*Listeria* cultures included *Enterobacter aerogenes*, *Serratia marcescens*, *Hafnia alvei*, *Lactobacillus casei*, *Lactobacillus acidophilus*, *Bacillus cereus*, *Escherichia coli* O157:H7, and *Salmonella enterica* serovar Typhimurium (Table S1).

To determine the sensitivity (limit of detection) of the *lap* gene primer sets for detection of *Listeria* species, pure cultures of *L. monocytogenes* F4244 and *L. innocua* F4248 cells were plated on modified oxford (MOX) agar for enumeration. Simultaneously, total DNA was extracted in 200 μL of PBST (20 mM phosphate buffered saline (PBS), pH 7.2, with 0.05% Tween 20) from 1 mL of pure cultures using the boiling method [33]. A relationship between the number of cells and the corresponding DNA/genomic equivalents was established before performing the *lap* gene-based PCR to establish the sensitivity of the reaction. Overnight (16 h) grown cells of *L. monocytogenes* F4244 ($8.02 \pm 0.11 \log_{10}$ CFU/mL) yielded 96.1 ± 2.1 ng/µL DNA, and 1 ng DNA was equivalent to $5.5 \log_{10}$ genomic equivalent (GE), whereas *L. innocua* F4248 ($8.15 \pm 0.24 \log_{10}$ CFU/mL) yielded 98.4 ± 0.6 ng/µL DNA, and 1 ng DNA was equivalent to $5.5 \log_{10}$ GE. PCR was performed with the diluted DNA and "GE was calculated. The genome size of *L. monocytogenes* and *L. innocua* 2.9 and 3.0 Mbp, "respectively, thus yielded $5.5 \log_{10}$ GE for 1 ng of DNA. The amplified PCR products obtained from different cell concentrations were quantified using the NIH ImageJ tool, an image processing and analysis software [34].

2.5. Laser Optical Sensor and Scatter Image Analysis

Laser optical sensor, also designated BARDOT, works on the biophysical principles (refraction, diffraction, interference) of forward light scattering. An external design of BARDOT and its internal scheme has been described previously [23,35]. The detection time (sample-to-result) for the BARDOT-based detection of *Listeria* spp. colonies on BHI agar plate (BHIA) is about 22 h except for *L. rocourtiae*, which took about 48 h to generate the colony scatter pattern.

To find the optimal incubation time that generates maximal scatter features and distinguishing scatter patterns, a time-lapse study was performed to capture the scatter pattern of *Listeria* species at 17, 22 and 25 h. Scatter patterns were acquired for *Listeria* colonies after plating on BHIA, and each colony (~1 mm diameter) contained about 2.5×10^8 *Listeria* cells. The scatter patterns were captured when the colony size reached close to 1 mm in diameter. A total of 1,884 scatter images from pure cultures of eight *Listeria* species were captured on BHIA, where 677 scatter patterns were used to build the scatter image library and the rest of the scatter patterns were generated to find the optimal incubation time [24]. The scatter image library of eight *Listeria* species (*L. monocytogenes*, *L. innocua*, *L. grayi*, *L. seeligeri*, *L. welshimeri*, *L. marthii*, *L. ivanovii*, and *L. rocourtiae*) consisted of an average of 80 scatter patterns per species. This *Listeria* species library was used to differentiate *L. monocytogenes* and *L. innocua* inoculated in the food sample. Another scatter image library of *L. monocytogenes* and *L. innocua* (110 scatter images) was also built to specifically differentiate the two species. The scatter images were further processed and analyzed using a built-in image analysis software [25]. This analysis generated the cross validation matrix for the principal component analysis of the scatter images of the *Listeria* species. Scatter image libraries for 13 serovars of *L. monocytogenes* were also generated after growth on BHIA or LB agar (LBA) plates.

2.6. Detection and Identification of Listeria in Artificially Inoculated Food Samples

Two types of food samples, ready-to-eat hotdogs (franks) and cantaloupes, were procured from a local grocery store (West Lafayette, IN, USA). To test the application of designed primer sets, food samples were artificially inoculated and tested in three independent experimental replicates. Twenty-five grams of each hotdog and cantaloupe rinds (each piece was about 2 × 3 cm) were artificially inoculated with 100 CFU of single culture (*L. monocytogenes* F4244 or *L. innocua* F4248) or 100 CFU of a mixed culture (50 CFU of *L. monocytogenes* F4244 and 50 CFU of *L. innocua* F4248). Since we did not find any background microbial load in hotdogs, samples were inoculated with *Lactobacillus casei* (100 CFU/25g) and *Escherichia coli* (100 CFU/25g) as background contaminants. Four sets of food samples: (i) food alone; food inoculated with (ii) *L. monocytogenes*; (iii) *L. innocua*; and (iv) *L. monocytogenes* and *L. innocua*, were enriched in FB according to the USDA-FSIS protocol [36]. All inoculated food samples (25 ± 2 g) were enriched in 225 ± 2.5 mL of FB at 37 °C in a shaking incubator (140 rpm) for 24 h. One milliliter of enriched broth was used for the DNA extraction (as described before) and plated on MOX agar plates for enumeration and BHIA for identification by BARDOT [23,37]. Briefly, for BARDOT analysis, FB enriched samples were decimally diluted in 20 mmol/L PBS (pH 7.2), plated on BHIA, and incubated at 37 °C for 24 h or until the colony diameter reached 1.1 ± 0.2 mm. The scatter patterns of colonies were compared with the scatter pattern library of *Listeria* species for identification [23,24].

3. Results

3.1. Specificity and Sensitivity of Lap Gene Primers for Listeria Detection

A total of 55 different *Listeria* (*n* = 47) and non-*Listeria* (*n* = 8) cultures were analyzed (Table S1). When the general *Listeria* primer set, ELAP-F2 and LIS-R1, was used, all the tested species of *Listeria sensu stricto* produced a 1301 bp band (Figure 2A, Table S1). When the *L. innocua*-specific primer set, Inn-F1 and Inn-R1, was used, a 536 bp band was amplified only in the *L. innocua* strains (Table S1) Likewise, the IvaSee-F1 and IvaSee-R1 primer set produced *L. ivanovii* and *L. seeligeri*-specific 606 bp band. The primer set, Wel-F1 and Wel-R1, generated 237 bp band only in the *L. welshimeri* strains, and a *L. marthii*-specific primer set, Mar-F1 and Mar-R3, produced a 1813 bp band without showing any amplified products from the other *Listeria* species (Figure 2B, Table S1).

These results demonstrated that the general *Listeria* primer set (ELAP-F2 and LIS-R1) could detect all the tested species of *Listeria sensu stricto*. The IvaSee-F1 and IvaSee-R1 were able to differentiate either *L. ivanovii* or *L. seeligeri* from other *Listeria* species or non-*Listeria* organisms without giving any false-positive results.

189

L. ivanovii and *L. seeligeri* contain virulence gene sequences in their genome similar to *L. monocytogenes* [31], and exhibited high sequence homology in *lap* [1].

The *L. innocua*-specific primers (Inn-F1 and Inn-R1) were highly specific and did not give any PCR products with other *Listeria* species, including *L. monocytogenes* strains. This primer set could be used to detect *L. innocua* as a mixed culture with *L. monocytogenes*. Since these two species are usually found together in food and other ecological habitats, the presence of *L. innocua* could be used as an indicator for *L. monocytogenes* [38–41].

Use of the *L. welshimeri*-specific primer set, Wel-F1 and Wel-R1, successfully produced a PCR product in all four *L. welshimeri* strains tested. Since we could not design *L. monocytogenes*, *L. ivanovii* or *L. seeligeri*-specific primer sets, PCR assays for detection of these individual species were not possible with the primers sets used in this study.

Since 2010, eleven new *Listeria* species (a total of 17 species) were added to the genus *Listeria*; *L. marthii* [42], *L. rocourtiae* [11], *L. weihenstephanensis* [12], *L. fleischmannii* [10,43], *L. floridensis*, *L. aquatic*, *L. cornellensis*, *L. riparia*, *L. grandensis* [8], *L. booriae*, and *L. newyorkensis* [9]. The presence of a *lap* homologue in *L. marthii* [7] has been reported and the resulting primer set (Mar-F1 and Mar-R3) is specific, but we were unable to obtain any *lap* gene based primers for *L. rocourtiae*, *L. weihenstephanensis*, *L. fleischmannii* and *L. grayi*. These are considered atypical and are phylogenetically divergent from the species of *Listeria sensu stricto* within the genus *Listeria* [1,11,43]. *L. grayi* and *L. rocourtiae* did not give any amplification with the species of *Listeria sensu-stricto*-specific primer set indicating the possible sequence heterogeneity in the *lap* sequence in these atypical listeriae (Table S1).

The specificity of all primer sets was examined with eight non-*Listeria* cultures and none of them yielded any PCR product (Table S1). We were even able to identify four mislabeled microorganisms: Two with general *Listeria* primer set, one of each with *L. innocua*-specific and *L. welshimeri*-specific primer set (Table S1). Ribotyping identified them as *L. monocytogenes* DUP-1035 and DUP-1039, *L. welshimeri* DUP-1074 and *L. innocua* DUP-1009 (Table S1).

The detection limit (sensitivity) of PCR with the species of *Listeria sensu stricto*-specific primer (ELAP-R1/LIS-R1) was 4.5 $\log_{10}$ genome equivalents for both *L. monocytogenes* and *L. innocua* (Figure 3). PCR for the DNA sensitivity was performed with the total DNA extracted from 1 mL culture of *L. monocytogenes* F4244 (8.02 ± 0.11 $\log_{10}$ CFU/mL) and *L. innocua* F4248 (8.15 ± 0.24 $\log_{10}$ CFU/mL) that also indirectly depicted the PCR sensitivity for the bacterial cell number. One milliliter cultures of *L. monocytogenes* and *L. innocua* yielded 96.1 ± 2.1 ng/μL and 98.4 ± 0.6 ng/μL of DNA concentrations, respectively. The genome equivalents were calculated from the genome size (*L. monocytogenes* size is 2.9×10^6 bp and *L. innocua* is 3.0×10^6 bp), and the molecular weight of nucleotide (1 bp = 650 Da).

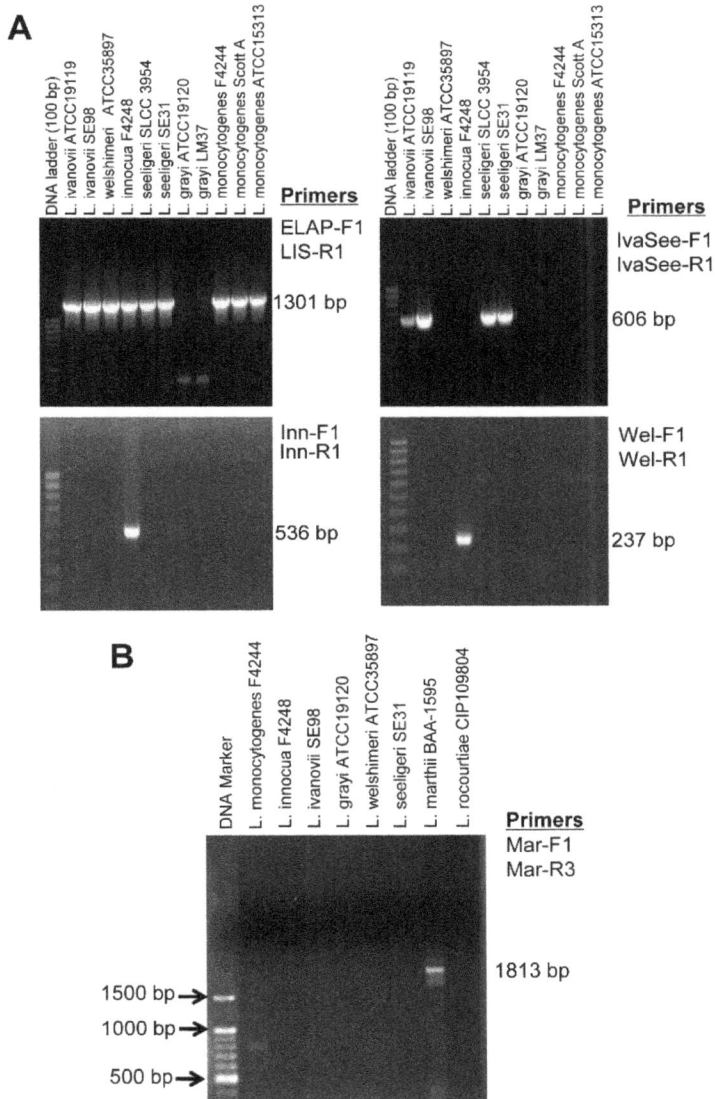

Figure 2. Representative agarose gel showing PCR amplification of selected *Listeria* species by using species-specific primers. (**A**) PCR results based on the species of *Listeria sensu stricto*-specific, and *L. innocua-*, *L. ivanovii-* and *L. seeligeri-* and *L. welshimeri*-specific primers. (**B**) PCR result using *L. marthii*-specific primers.

191

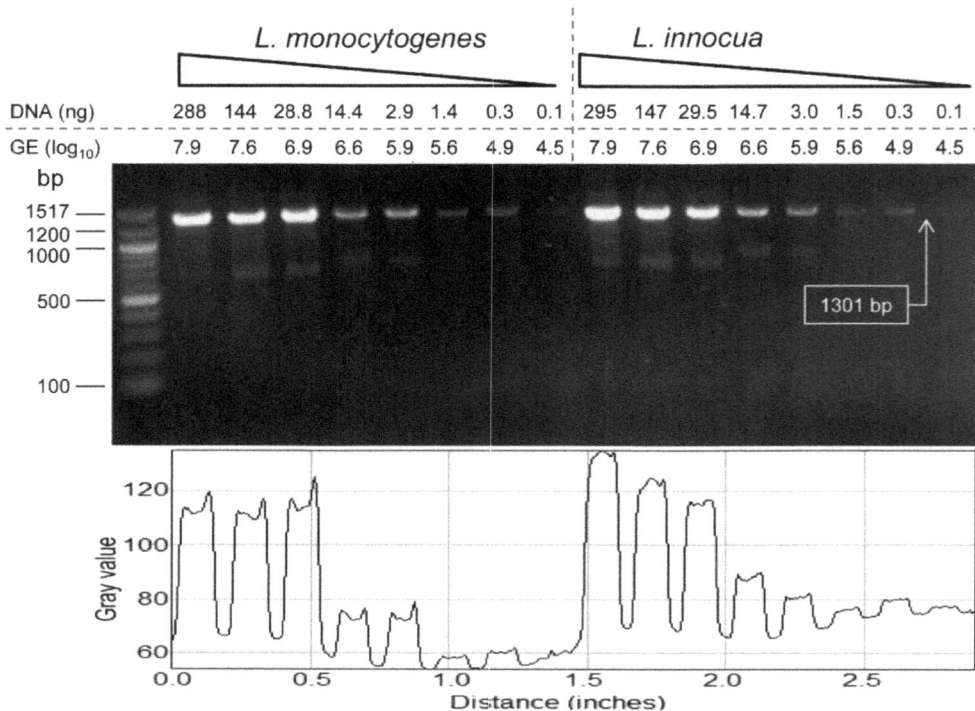

Figure 3. Sensitivity of *lap* gene-based *Listeria* species *sensu stricto*-specific primer (ELAP-F1/LIS-R1) tested against *L. monocytogenes* F4244 and *L. innocua* F4248. Agarose gel showing amplifications with the primers for different concentrations of template DNA of *L. monocytogenes* and *L. innocua* in reaction volume of 25 µL. PCR products in the gel were quantified using the NIH ImageJ image processing and analysis software.

3.2. Scatter Image Library of Listeria Species and Serovars

The light scattering sensor (BARDOT) generated distinguishing forward scattering patterns for colonies of *Listeria* species on BHIA. Time-lapse measurement of the scatter patterns indicated that *Listeria* species generated scatter patterns with maximal differential scatter features at 22 h of incubation (Figure 4A). Principal component analysis performed on the basis of cross validation matrix revealed that *L. innocua*, *L. rocourtiae*, *L. monocytogenes*, *L. marthii*, and *L. seeligeri* can be grouped separately based on the differences in the scatter patterns (Figure 4B) with 100%, 100%, 97.7%, 95.2% and 94.9%, positive predictive value (PPV), also known as classification accuracy, respectively (Table S2). However, *L. grayi*, *L. ivanovii*, and *L. welshimeri* could not be differentiated based on the scatter patterns on BHIA. Application of the *L. monocytogenes* and *L. innocua*-specific image libraries, generated

192

even higher PPVs of 100% for both of the species, and they grouped separately in the principal component analysis (Figure 4C). These image libraries were also used to match scatter images of *L. monocytogenes* and *L. innocua* that were obtained from artificially inoculated food samples mentioned below in the result section.

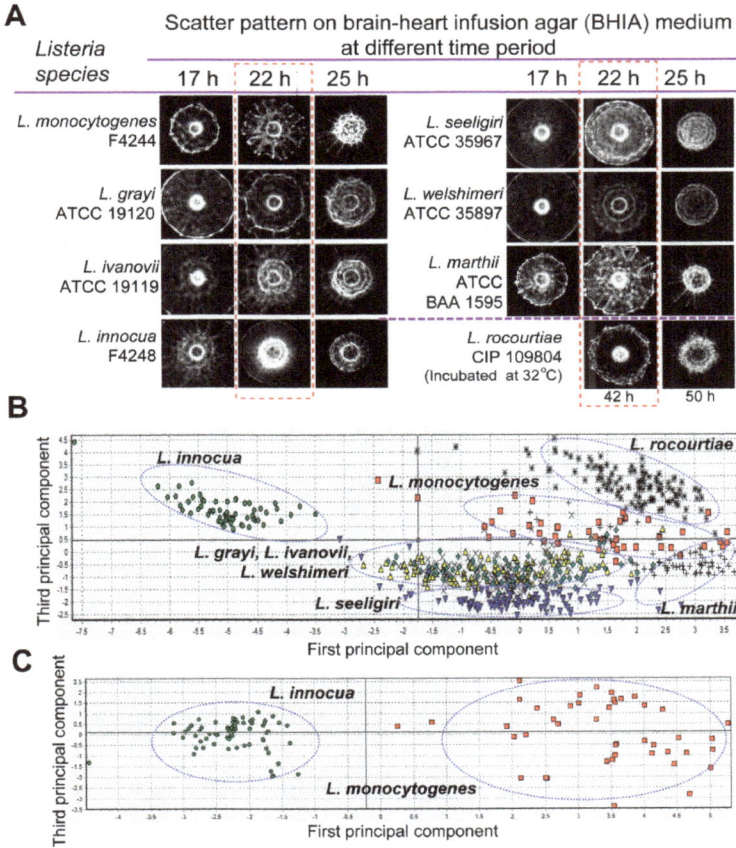

Figure 4. Optical scatter patterns of *Listeria* species and image analysis. (**A**) Colony scatter patterns were captured using BARDOT at different incubation times for eight *Listeria* species on BHI agar plates. Rectangular selection with broken line depicts the optimal incubation time (22 h) that yielded differentiating scatter images when the colony size was 1.1 ± 0.2 mm diameter; (**B**) Principal component analysis of the eight *Listeria* species used to build the scatter image library. Blue oval selections indicate grouping of the *Listeria* species; (**C**) Principal component analysis of *L. monocytogenes* and *L. innocua* colony scatter images that were used to build a two-species scatter image library. The blue oval selections indicate grouping of each *Listeria* species.

In this study we also tested the capabilities of the laser sensor to differentiate *L. monocytogenes* at the serovar level after growth on BHI and LB agar plates (Figure 5). Differences at the serovar level were observed after analysis using the cross validation matrix, where a high PPV average was observed on LB agar compared to the BHI agar media, 90.1% and 82.9%, respectively (Table 2). Scatter pattern analysis for the thirteen serotypes underscores the feasible application of the laser optical sensor to generate a scatter image library with differentiating scatter patterns for *L. monocytogenes* serotypes that can be used for screening and detection of *L. monocytogenes* at the serovar level.

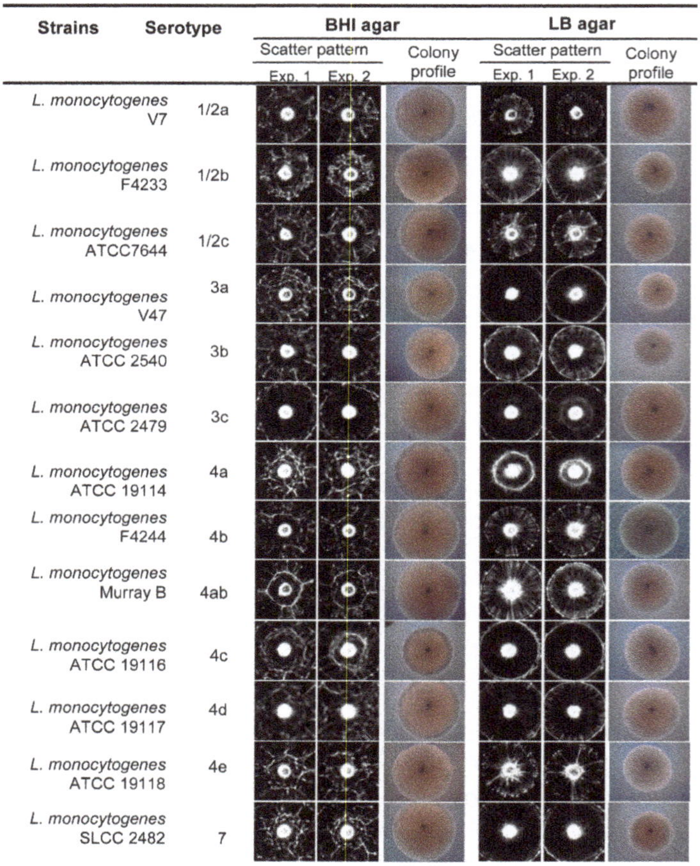

Figure 5. Representative scatter images of colonies of *L. monocytogenes* serotypes grown on BHI and LB agar. Between 50 and 100 colony scatter images for each serovar were collected from each experiment. Colony profiles were measured under phase contrast microscope with 10× objective when the colony size was 1.1 ± 0.2 mm diameter on BHI and LB agar after 21–23 h and 25–27 h of incubation, respectively.

Table 2. Positive predictive value (PPV, precision rate) for the scatter images of colonies obtained from thirteen *L. monocytogenes* serotypes grown on BHIA and LBA media.

Strains	Serotype	% Average Positive Predictive Value (PPV ± SD)	
		BHI	LB
L. monocytogenes V7	1/2a	81.8 ± 2.3	96.2 ± 1.9
L. monocytogenes F4233	1/2b	89.4 ± 1.1	96.8 ± 2.1
L. monocytogenes ATCC7644	1/2c	74.5 ± 3.2	81.3 ± 3.8
L. monocytogenes V47	3a	80.8 ± 1.8	77.6 ± 5.6
L. monocytogenes ATCC 2540	3b	74.2 ± 2.2	97.6 ± 1.7
L. monocytogenes ATCC 2479	3c	99.8 ± 0.9	85.2 ± 2.3
L. monocytogenes ATCC 9114	4a	93.2 ± 1.3	98.0 ± 1.8
L. monocytogenes F4244	4b	46.2 ± 3.5	88.8 ± 2.1
L. monocytogenes Murray B	4ab	92.8 ± 2.9	85.4 ± 3.2
L. monocytogenes ATCC 19116	4c	98.8 ± 1.0	93.6 ± 2.8
L. monocytogenes ATCC 19117	4d	99.0 ± 0.5	88.2 ± 3.1
L. monocytogenes ATCC 19118	4e	65.4 ± 5.6	92.6 ± 1.7
L. monocytogenes SLCC 2482	7	82.4 ± 2.4	90.2 ± 3.1
Average precision rate		82.9 ± 2.2	90.1 ± 2.7

3.3. Detection and Verification of Listeria from Food Samples

The ability of *lap* gene-specific primer sets to detect *L. monocytogenes* and *L. innocua* from inoculated food samples were verified (Figure S1, Table 3). Since a *lap* gene-based *L. monocytogenes* specific primer set could not be designed, we used the combination of species of *Listeria sensu stricto*-specific and *L. innocua* specific primer sets to detect *Listeria* from food. The food samples that revealed positive amplification with ELAP-F1 & LIS-R, but did not show any amplification with the *L. innocua*-specific primer sets (Inn-F1 & Inn-R1), were considered to contain any *Listeria* spp. other than the *L. innocua* (Table 3, Figure S1). Samples with positive amplification for both the primer sets (ELAP-F1 & LIS-R1 and with Inn-F1 & Inn-R1) corroborated the presence of *L. innocua* in the food sample (Table 3). Background microbial colonies obtained from the un-inoculated cantaloupe did not result in any positive amplification (Figure 6). This highlights the specificity and applicability of *lap* gene-specific primers for detection of *Listeria* species even in food samples with background microbiota. We further verified these results by analyzing the enriched food samples by BARDOT.

195

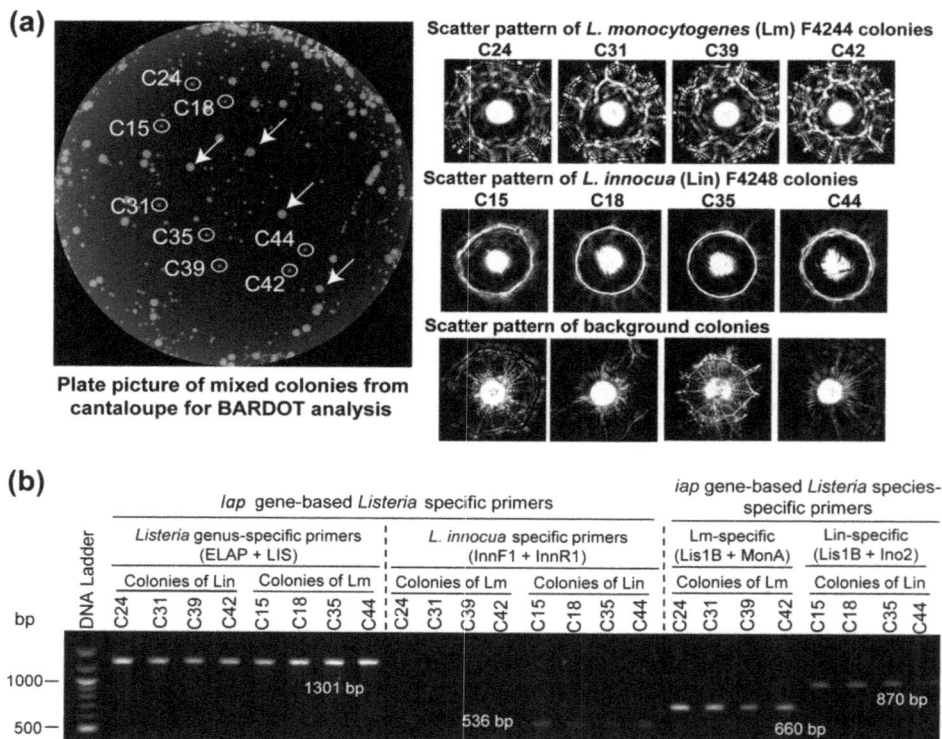

Figure 6. Detection and verification of *L. monocytogenes* (Lm) and *L. innocua* (Lin) in mixture from inoculated cantaloupe samples with BARDOT and PCR. (**A**) Enriched cantaloupe samples containing *L. monocytogenes* and *L. innocua* were plated on BHI agar and colony scatter patterns were obtained. Scatter patterns were matched against the BARDOT scatter image library for *Listeria* identification. The white arrows indicate background bacterial colonies from the cantaloupe; (**B**) BARDOT identified colonies were picked and tested with the primer sets specific for species of *Listeria sensu stricto* (ELAP-F1/LIS-R1), *L. monocytogenes* (Lis1B/MonA), and *L. innocua* (InnF1/InnR1, Lis1B/Ino2) for verification of colonies.

In our previous study, BARDOT generated distinct signature scatter patterns for the colonies of *L. monocytogenes* or *L. innocua* in mixed culture [23]. The distinctive scatter patterns generated with BARDOT facilitated accurate identification of *L. monocytogenes* or *L. innocua* or both in food samples after matching the scatter patterns with the respective image libraries (Figure 6). Colonies # C42, C41, C43, C34, and C29 originated from *L. monocytogenes* and *L. innocua*-inoculated hotdog sample on BHIA were identified as *L. innocua*, while colonies # C28, C25, C23, and C5 were identified as *L. monocytogenes* after comparing scatter images with the library (Figure 6A). These colonies were initially identified as *Listeria* spp. by PCR with the primer set (ELAP-F1/LIS-R1) designed in this study (Figure 6B). Further, these colonies were also confirmed at the species level (*L. monocytogenes* and *L. innocua*) using the primers for the *iap* gene [32,44]. The scatter patterns of *L. innocua*, when matched with the libraries of *Listeria* species as well as *L. monocytogenes* and *L. innocua*, generated 100% match with scatter image library. *L. monocytogenes* colonies from the artificially inoculated hotdog sample revealed low PPV (<80%) when matched with the libraries of *Listeria* species; however, the same *L. monocytogenes* colony scatter pattern generated a high PPV (>90%) when matched with the *L. monocytogenes* and *L. innocua* library. The low PPV of *L. monocytogenes* with the *Listeria* species library could be attributed to the overlapping pattern of *L. monocytogenes* with the scatter pattern of other *Listeria* species. BARDOT-based identification of *L. monocytogenes* and *L. innocua* colonies along with PCR analysis with the *lap* and *iap* gene-specific primers resulted in 100% and 100% identification, respectively, for both the *Listeria* species.

Table 3. *Listeria* detection using *lap* gene-specific primers in food systems.

			PCR [b]			
			Hotdog		Cantaloupe	
Treatment [a]	Inoculation (CFU/25g)	Enrichment Time (h) in Fraser Broth at 37 °C	ELAP-F1/ LIS-R1	Inn-F1/ Inn-R1	ELAP-F1/ LIS-R1	Inn-F1/ Inn-R1
Uninoculated	0	24	−	−	−	−
L. innocua (Lin)	100	24	+	+	+	+
L. monocytogenes (Lm)	100	24	+	−	+	−
Lin and Lm[c]	100	24	+	+	+	+
Lb. casei	100	24	−	−	−	−
E. coli	100	24	−	−	−	−

[a] Three independent experiments were performed for each food sample; [b] DNA extracted from broth enrichment following the published protocol [33] were amplified with the *lap* gene specific primers for species of *Listeria sensu stricto* (ELAP-F1/LIS-R1) and *L. innocua* (Inn-F1/InnR1); [c] Food samples were inoculated with 50 CFU each of *L. monocytogenes* (Lm) and *L. innocua* (Lin) in 25 g of food sample.

4. Discussion

This study reports the application feasibility of primer sets designed from a gene encoding the house-keeping alcohol acetaldehyde dehydrogenase (*aad*), also known as *Listeria* adhesion protein (*lap*), for detection of *Listeria* at the genus and species level. This highly conserved house-keeping enzyme is involved in the pathogenesis of virulent *Listeria* but not avirulent *Listeria* species [1,45] thus providing an attractive target for *Listeria* detection. The *aad* (*lap*) sequence is conserved (97%–98% homology) in the species of *Listeria sensu stricto* and yielded a primer set (ELAP-F1 and LIS-R1) that detected these *Listeria sensu stricto* (archetypal) species, -but not the atypical (*Listeria sensu lato*) listeriae: *L. grayi* and *L. rocourtiae*. Even though both *L. grayi* and *L. rocourtiae* possess a *lap* (*aad*) homolog, they did not give any amplification with the species of *Listeria sensu stricto*-specific primer set indicating a possible sequence heterogeneity in the *lap* gene in these atypical listeriae. Indeed, *lap* gene sequence comparison between *L. monocytogenes* F4244 (AY561824) and *L. grayi* DSM 20601 (NZ_GL538352.1) or *L. rocourtiae* FSL F6-920 c6 (NZ_AODK01000006.1) revealed only 80% homology. The other newly isolated *Listeria* species were not tested in the PCR assay, but we anticipate negative results for these species since they are genetically similar to the two atypical listeriae tested in this study [8,9]. Furthermore, a minor variation in *lap* sequences (97%–98% similarity) among the different species of *Listeria sensu stricto* [1] yielded highly specific primer sets for *L. innocua*, *L. welshimeri*, *L. marthii*, and *L. ivanovii* and *L. seeligeri* together, but none for *L. monocytogenes* (Figure 1, Table S1). These could be useful for specific identification at the species level. *Listeria sensu stricto*-specific and other species-specific sets of primers also did not amplify any non-listerial bacteria tested with pure cultures or in a food matrix (Tables S1 and 3), highlighting the specificity of the *lap* gene-specific primer sets for *Listeria* spp. The detection limit of primer sets with the diluted template DNA revealed an indirect detection limit of about 4.5 $\log_{10}$ genome equivalents for this assay. These primer sets could be used to detect and identify *Listeria* species during screening of frozen or refrigerated ready-to-eat (RTE) food products as recommended by the FDA [20]. Application of PCR-based assay targeting gene encoding house-keeping enzyme (cell wall hydrolase; CWH or p60 encoded by *iap*) in *Listeria* spp. was reported earlier, in which species-specific primer sets successfully detected each *Listeria* spp.; *L. monocytogenes*, *L. ivanovii*, *L. innocua*, *L. seeligeri*, *L. welshimeri* within the genus, except *L. grayi* [32]. Similarly, PCR assay targeting genes encoding aminopeptidase and fibronectin-binding protein were also used for rapid detection of *L. monocytogenes* [46,47].

We developed the light scattering sensor, BARDOT, through collaborative efforts with engineers at the Center for Food Safety Engineering at Purdue University [48]. We have successfully used BARDOT to differentiate and detect *L. monocytogenes*, *L. ivanovii*, *L. innocua*, *L. seeligeri*, *L. welshimeri* and *L. grayi*. The BARDOT system was also successfully applied to differentiate *L. monocytogenes* from other pathogens (*Salmonella enterica* serovar Enteritidis and Typhimurium, *E. coli* O157:H7) based on the scatter patterns from artificially inoculated ready-to-eat hotdog, shredded beef, raw ground beef and chicken, frozen and fresh spinach, and fresh tomato [37]. Recently, we have also optimized the BARDOT-based method for detection and screening of several additional foodborne pathogens including *Bacillus* spp. [49], *Campylobacter* spp. [50], *Salmonella enterica* serovars [24], Shiga-toxigenic *E. coli* [35], and *Vibrio* spp. [51]. In this study, we used BARDOT to differentiate the species of *Listeria* when grown on BHIA. On BHIA, BARDOT successfully differentiated *L. monocytogenes*, *L. innocua*, *L. rocourtiae*, *L. marthii*, and *L. seeligeri*; however, it did not yield satisfactory differential patterns of *L. grayi*, *L. ivanovii*, and *L. welshimeri*

(Figure 4B). In our previous report we have shown that the BHIA and modified Oxford agar prepared without ferric ammonium citrate were able to successfully differentiate the colonies of *L. monocytogenes*, *L. ivanovii*, *L. innocua*, *L. seeligeri*, *L. welshimeri* and *L. grayi*, and colonies of *L. monocytogenes* from *L. innocua*, respectively, based on scatter signature patterns [23,44]. These findings reaffirm the media-dependent generation of scatter signatures for bacterial identification.

L. monocytogenes is the primary human pathogen in the genus *Listeria* and among the 13 serotypes, serotype 1/2a and 4b are responsible for ~75% of all *L. monocytogenes* related outbreaks. Here we have shown that BARDOT can differentiate *L. monocytogenes* serovars 1/2a and 4b with high accuracy on LBA with 96.2% ± 1.9% and 88.8% ± 2.1% PPV, respectively. Observed differences in the scatter pattern of different serovars could be attributed to the O (somatic) antigens expressed on the surface of *L. monocytogenes* [52]. Furthermore, metabolic activity and genomic differences between different species or serotypes of *Listeria* can also contribute to the differential scatter patterns [37]. The genome size for serotype 1/2a (*L. monocytogenes* F6854) is 2.97×10^6 bp, with a total of 3028 genes, of which 2963 are protein coding genes. Serotype 4b (*L. monocytogenes* F2365) has a genome size of 2.91×10^6 bp, with a total of 2933 genes, of which 2848 are protein coding genes. Thus, a difference of 115 protein coding genes in serotype 1/2a to that of 4b could be crucial in generating differential scatter patterns for these two serotypes. In a comparative whole genome sequencing study, it was found that 83 genes were restricted to 1/2a serotype and 51 genes were restricted to 4b serotype [53].

5. Conclusions

In summary, *lap* gene based *Listeria sensu stricto* and individual species-specific primers successfully detected all tested species of *Listeria sensu stricto* (archetypal) while some limitations for individual species level detection. The PCR based assays with the species of *Listeria sensu stricto*-specific primer sets based on *lap* and *iap* genes also successfully detected *L. monocytogenes* and *L. innocua* from mixed culture-inoculated food samples, and each bacterium in food was verified by the light scattering sensor that generated unique scatter signature for each species of *Listeria*. These data emphasize that the BARDOT system could be used to identify *Listeria* spp. on agar plates from a mixed cultures and may serve as a complimentary tool when testing samples with nucleic acid-based molecular methods.

Acknowledgments: This research was supported through a cooperative agreement with the Agricultural Research Service of the U.S. Department of Agriculture project number 1935-42000-072-02G and the Center for Food Safety Engineering at Purdue University. We thank Valerie Ryan and Taylor Bailey for critical reading of the manuscript.

Author Contributions: Arun K. Bhunia and Kwang-Pyo Kim conceived the idea. Kwang-Pyo Kim, Atul K. Singh, Xingian Bai, and Lena Leprun performed the experiments. Kwang-Pyo Kim, Atul K. Singh and Arun K. Bhunia analyzed the results. Kwang-Pyo Kim, Atul K. Singh and Arun K. Bhunia wrote the manuscript.

Conflicts of Interest: The authors declare no conflict of interest.

References

1. Jagadeesan, B.; Koo, O.K.; Kim, K.P.; Burkholder, K.M.; Mishra, K.K.; Aroonnual, A.; Bhunia, A.K. LAP, an alcohol acetaldehyde dehydrogenase enzyme in *Listeria* promotes bacterial adhesion to enterocyte-like Caco-2 cells only in pathogenic species. *Microbiology* **2010**, *156*, 2782–2795.

2. Jagadeesan, B.; Fleishman Littlejohn, A.E.; Amalaradjou, M.A. R.; Singh, A.K.; Mishra, K.K.; La, D.; Kihara, D.; Bhunia, A.K. *N*-Terminal Gly$_{224}$–Gly$_{411}$ domain in *Listeria* adhesion protein interacts with host receptor Hsp60. *PLoS ONE* **2011**, *6*.

3. Burkholder, K.M.; Bhunia, A.K. *Listeria monocytogenes* uses Listeria adhesion protein (LAP) to promote bacterial transepithelial translocation, and induces expression of LAP receptor Hsp60. *Infect. Immun.* **2010**, *78*, 5062–5073.

4. Kim, K.P.; Jagadeesan, B.; Burkholder, K.M.; Jaradat, Z.W.; Wampler, J.L.; Lathrop, A.A.; Morgan, M.T.; Bhunia, A.K. Adhesion characteristics of *Listeria adhesion* protein (LAP)-expressing *Escherichia coli* to Caco-2 cells and of recombinant LAP to eukaryotic receptor Hsp60 as examined in a surface plasmon resonance sensor. *FEMS Microbiol. Lett.* **2006**, *256*, 324–332.

5. Henderson, B.; Martin, A. Bacterial moonlighting proteins and bacterial virulence. *Curr. Top. Microbiol. Immunol.* **2013**, *358*, 155–213.

6. Burkholder, K.M.; Bhunia, A.K. *Listeria monocytogenes* and Host Hsp60—An invasive pairing. In *Moonlighting Cell Stress Proteins in Microbial Infections, Heat Shok Proteins*; Henderson, B., Ed.; Springer Science+Business Media: Dordrecht, Germany, 2013; pp. 267–282.

7. Den Bakker, H.C.; Bundrant, B.N.; Fortes, E.D.; Orsi, R.H.; Wiedmann, M. A population genetics-based and phylogenetic approach to understanding the evolution of virulence in the genus *Listeria*. *Appl. Environ. Microbiol.* **2010**, *76*, 6085–6100.

8. Den Bakker, H.C.; Warchocki, S.; Wright, E.M.; Allred, A.F.; Ahlstrom, C.; Manuel, C.S.; Stasiewicz, M.J.; Burrell, A.; Roof, S.; Strawn, L.K.; *et al.* *Listeria floridensis* sp nov., *Listeria aquatica* sp nov., *Listeria cornellensis* sp nov., *Listeria riparia* sp nov and *Listeria grandensis* sp nov., from agricultural and natural environments. *Int. J. Syst. Evol. Microbiol.* **2014**, *64*, 1882–1889.

9. Weller, D.; Andrus, A.; Wiedmann, M.; den Bakker, H.C. *Listeria booriae* sp nov and *Listeria newyorkensis* sp nov., from food processing environments in the USA. *Int. J. Syst. Evol. Microbiol.* **2015**, *65*, 286–292.

10. Bertsch, D.; Rau, J.; Eugster, M.R.; Haug, M.C.; Lawson, P.A.; Lacroix, C.; Meile, L. *Listeria fleischmannii* sp nov., isolated from cheese. *Int. J. Syst. Evol. Microbiol.* **2013**, *63*, 526–532.

11. Leclercq, A.; Clermont, D.; Bizet, C.; Grimont, P.A.D.; le Fleche-Mateos, A.; Roche, S.M.; Buchrieser, C.; Cadet-Daniel, V.; le Monnier, A.; Lecuit, M.; *et al. Listeria rocourtiae* sp. nov. *Int. J. Syst. Evol. Microbiol.* **2010**, *60*, 2210–2214.

12. Halter, E.L.; Neuhaus, K.; Scherer, S. *Listeria weihenstephanensis* sp. nov., isolated from the water plant Lemna trisulca taken from a freshwater pond. *Int. J. Syst. Evol. Microbiol.* **2013**, *63*, 641–647.

13. Silk, B.J.; Date, K.A.; Jackson, K.A.; Pouillot, R.; Holt, K.G.; Graves, L.M.; Ong, K.L.; Hurd, S.; Meyer, R.; Marcus, R.; *et al.* Invasive listeriosis in the foodborne diseases active surveillance network (FoodNet), 2004–2009: Further targeted prevention needed for higher-risk groups. *Clin. Infect. Dis.* **2012**, *54*, S396–S404.

14. Anonymous. 2008 Listeriosis Outbreak in Ontario Epidemiologic Summary. Available online: http://www.health. gov.on.ca/en/public/publications/disease/docs/listeriosis_outbreak_epi_sum.pdf (accessed on 26 May 2015).

15. Fretz, R.; Pichler, J.; Sagel, U.; Much, P.; Ruppitsch, W.; Pietzka, A.T.; Stöger, A.; Huhulescu, S.; Heuberger, S.; Appl, G.; *et al.* Update: Multinational listeriosis outbreak due to "Quargel", a sour milk curd cheese, caused by two different *L. monocytogenes* serotype 1/2a strains, 2009–2010. *Eurosurveillance* **2010**, *15*, 1–2.

16. CDC. Multistate outbreak of listeriosis associated with Jensen Farms cantaloupe—United States, August–September 2011. *MMWR. Morb. Mortal. Weekly Rep.* **2011**, *60*, 1357–1358.

17. Anonymous. Multistate Outbreak of Listeriosis Linked to Commercially Produced, Prepackaged Caramel Apples Made from Bidart Bros. Apples (Final Update). Available online: http://www.cdc.gov/listeria/outbreaks/ caramel-apples-12-14/ (accessed on 2 September 2015).

18. Anonymous. Multistate Outbreak of Listeriosis Linked to Blue Bell Creameries Products (Final Update). Available online: http://www.cdc.gov/listeria/outbreaks/ice-cream-03-15/ (accessed on 2 September 2015).

19. Wang, S.; Orsi, R.H. Listeria. In *Foodborne Infections and Intoxications*, 4th ed.; Morris, J.G.J., Potter, M.E., Eds.; Elsevier: New York, NY, USA, 2013; pp. 199–216.

20. FDA. Guidance for industry: Control of *Listeria monocytogenes* in refrigerated or frozen ready-to-eat foods; Draft guidance. In *US Department of Health and Human Services: Center for Food Safety and Applied Nutrition*; Rockville, MD, USA, 2008.

21. Gil, R.; Silva, F.J.; Pereto, J.; Moya, A. Determination of the core of a minimal bacterial gene set. *Microbiol. Mol. Biol. Rev.* **2004**, *68*, 518–537.

22. Bae, E.; Banada, P.P.; Huff, K.; Bhunia, A.K.; Robinson, J.P.; Hirleman, E.D. Biophysical modeling of forward scattering from bacterial colonies using scalar diffraction theory. *Appl. Opt.* **2007**, *46*, 3639–3648.

23. Banada, P.P.; Guo, S.; Bayraktar, B.; Bae, E.; Rajwa, B.; Robinson, J.P.; Hirleman, E.D.; Bhunia, A.K. Optical forward-scattering for detection of *Listeria monocytogenes* and other *Listeria* species. *Biosens. Bioelectron.* **2007**, *22*, 1664–1671.

24. Singh, A.K.; Bettasso, A.M.; Bae, E.; Rajwa, B.; Dundar, M.M.; Forster, M.D.; Liu, L.; Barrett, B.; Lovchik, J.; Robinson, J.P.; *et al.* Laser optical sensor, a label-free on-plate *Salmonella enterica* colony detection tool. *mBio* **2014**, *5*.

25. Ahmed, W.M.; Bayraktar, B.; Bhunia, A.K.; Hirleman, E.D.; Robinson, J.P.; Rajwa, B. Classification of bacterial contamination using image processing and distributed computing. *IEEE J. Biomed. Health Inform.* **2013**, *17*, 232–239.

26. Lathrop, A.A.; Jaradat, Z.W.; Haley, T.; Bhunia, A.K. Characterization and application of a *Listeria monocytogenes* reactive monoclonal antibody C11E9 in a resonant mirror biosensor. *J. Immunol. Methods* **2003**, *281*, 119–128.

27. Gray, K.M.; Bhunia, A.K. Specific detection of cytopathogenic *Listeria monocytogenes* using a two-step method of immunoseparation and cytotoxicity analysis. *J. Microbiol. Methods* **2005**, *60*, 259–268.

28. Den Bakker, H.C.; Cummings, C.A.; Ferreira, V.; Vatta, P.; Orsi, R.H.; Degoricija, L.; Barker, M.; Petrauskene, O.; Furtado, M.R.; Wiedmann, M. Comparative genomics of the bacterial genus *Listeria*: Genome evolution is characterized by limited gene acquisition and limited gene loss. *BMC Genomics* **2010**, *11*.

29. Glaser, P.; Frangeul, L.; Buchrieser, C.; Rusniok, C.; Amend, A.; Baquero, F.; Berche, P.; Bloecker, H.; Brandt, P.; Chakraborty, T.; *et al.* Comparative genomics of *Listeria* species. *Science* **2001**, *294*, 849–852.

30. Corpet, F. Multiple sequence alignment with hierarchical clustering. *Nucleic Acids Res.* **1988**, *16*, 10881–10890.

31. Gouin, E.; Mengaud, J.; Cossart, P. The virulence gene cluster of *Listeria monocytogenes* is also present in *Listeria ivanovii*, an animal pathogen, and *Listeria seeligeri*, a nonpathogenic species. *Infect. Immun.* **1994**, *62*, 3550–3553.

32. Bubert, A.; Hein, I.; Rauch, M.; Lehner, A.; Yoon, B.; Goebel, W.; Wagner, M. Detection and differentiation of *Listeria* spp. by a single reaction based on multiplex PCR. *Appl. Environ. Microbiol.* **1999**, *65*, 4688–4692.

33. Longhi, C.; Maffeo, A.; Penta, M.; Petrone, G.; Seganti, L.; Conte, M.P. Detection of *Listeria monocytogenes* in Italian-style soft cheeses. *J. Appl. Microbiol.* **2003**, *94*, 879–885.

34. Schneider, C.A.; Rasband, W.S.; Eliceiri, K.W. NIH Image to ImageJ: 25 years of image analysis. *Nat. Method* **2012**, *9*, 671–675.

35. Tang, Y.; Kim, H.; Singh, A.K.; Aroonnual, A.; Bae, E.; Rajwa, B.; Fratamico, P.M.; Bhunia, A.K. Light scattering sensor for direct identification of colonies of *Escherichia coli* serogroups O26, O45, O103, O111, O121, O145 and O157. *PLoS ONE* **2014**, *9*.

36. *Isolation and Identification of Listeria monocytogenes from Red Meat, Poultry and Egg Products, and Environmental Samples*; USDA-FSIS: Washington, DC, USA, 2013; Volume MLG 8.09, p. 21.

37. Banada, P.P.; Huff, K.; Bae, E.; Rajwa, B.; Aroonnual, A.; Bayraktar, B.; Adil, A.; Robinson, J.P.; Hirleman, E.D.; Bhunia, A.K. Label-free detection of multiple bacterial pathogens using light-scattering sensor. *Biosens. Bioelectron.* **2009**, *24*, 1685–1692.

38. Curiale, M.S.; Sons, T.; Fanning, L.; Lepper, W.; McIver, D.; Garramone, S.; Mozola, M. Deoxyribonucleic acid hybridization method for the detection of Listeria in dairy products, seafoods, and meats: Collaborative study. *J. AOAC Int.* **1994**, *77*, 602–617.

39. Rocourt, J.; Cossart, P. *Listeria Monocytogenes*; ASM Press: Washington, DC, USA, 1997.

40. Besse, N.G.; Barre, L.; Buhariwalla, C.; Vignaud, M.L.; Khamissi, E.; Decourseulles, E.; Nirsimloo, M.; Chelly, M.; Kalmokoff, M. The overgrowth of *Listeria monocytogenes* by other *Listeria* spp. in food samples undergoing enrichment cultivation has a nutritional basis. *Int. J. Food Microbiol.* **2010**, *136*, 345–351.

41. Ryser, E.T. Foodborne listeriosis. In *Listeria, Listeriosis, and Food Safety*; Ryser, E.T., Marth, E.H., Eds.; Marcel Decker: New York, NY, USA, 1999; pp. 299–358.

42. Graves, L.M.; Helsel, L.O.; Steigerwalt, A.G.; Morey, R.E.; Daneshvar, M.I.; Roof, S.E.; Orsi, R.H.; Fortes, E.D.; Milillo, S.R.; den Bakker, H.C.; *et al.* *Listeria marthii* sp. nov., isolated from the natural environment, Finger Lakes National Forest. *Int. J. Syst. Evol. Microbiol.* **2010**, *60*, 1280–1288.

43. Den Bakker, H.C.; Manuel, C.S.; Fortes, E.D.; Wiedmann, M.; Nightingale, K.K. Genome sequencing identifies *Listeria fleischmannii* subsp *coloradonensis* subsp nov., isolated from a ranch. *Int. J. Syst. Evol. Microbiol.* **2013**, *63*, 3257–3268.

44. Koo, O.K.; Aroonnual, A.; Bhunia, A.K. Human heat-shock protein 60 receptor-coated paramagnetic beads show improved capture of *Listeria monocytogenes* in the presence of other *Listeria* in food. *J. Appl. Microbiol.* **2011**, *111*, 93–104.

45. Kim, H.; Bhunia, A.K. Secreted *Listeria* adhesion protein (Lap) influences Lap-mediated *Listeria monocytogenes* paracellular translocation through epithelial barrier. *Gut Pathog.* **2013**, *5*.

46. Winters, D.K.; Maloney, T.P.; Johnson, M.G. Rapid detection of *Listeria monocytogenes* by a PCR assay specific for an aminopeptidase. *Mol. Cell. Probes* **1999**, *13*, 127–131.

47. Gilot, P.; Content, J. Specific identification of *Listeria welshimeri* and *Listeria monocytogenes* by PCR assays targeting a gene encoding a fibronectin-binding protein. *J. Clin. Microbiol.* **2002**, *40*, 698–703.

48. Bhunia, A.K.; Bae, E.; Rajwa, B.; Robinson, J.P.; Hirleman, E.D. Utilization of optical forward scatter image biological database: Foodborne pathogen colony differentiation and detection. In *Omics, Microbial Modeling and Technologies for Foodborne Pathogens*; Yan, X., Juneja, V.K., Fratamico, P.M., Smith, J.L, Eds.; Lancaster, PA, USA, 2012; pp. 553–578.

49. Singh, A.K.; Sun, X.; Bai, X.; Kim, H.; Abdalhaseib, M.U.; Bae, E.; Bhunia, A.K. Label-free, non-invasive light scattering sensor for rapid screening of *Bacillus* colonies. *J. Microbiol. Methods* **2015**, *109*, 56–66.

50. He, Y.; Reed, S.; Bhunia, A.K.; Gehring, A.; Nguyen, L.H.; Irwin, P.L. Rapid identification and classification of *Campylobacter* spp. using laser optical scattering technology. *Food Microbiol.* **2015**, *47*, 28–35.

51. Huff, K.; Aroonnual, A.; Littlejohn, A.E.F.; Rajwa, B.; Bae, E.; Banada, P.P.; Patsekin, V.; Hirleman, E.D.; Robinson, J.P.; Richards, G.P.; *et al.* Light-scattering sensor for real-time identification of *Vibrio parahaemolyticus*, *Vibrio vulnificus* and *Vibrio cholerae* colonies on solid agar plate. *Microb. Biotechnol.* **2012**, *5*, 607–620.

52. Liu, D. Identification, subtyping and virulence determination of *Listeria monocytogenes*, an important foodborne pathogen. *J. Med. Microbiol.* **2006**, *55*, 645–659.

53. Nelson, K.E.; Fouts, D.E.; Mongodin, E.F.; Ravel, J.; de Boy, R.T.; Kolonay, J.F.; Rasko, D.A.; Angiuoli, S.V.; Gill, S.R.; Paulsen, I.T.; *et al.* Whole genome comparisons of serotype 4b and 1/2a strains of the food-borne pathogen Listeria monocytogenes reveal new insights into the core genome components of this species. *Nucleic Acids Res.* **2004**, *32*, 2386–2395.

Fluorescence-Based Bioassays for the Detection and Evaluation of Food Materials

Kentaro Nishi, Shin-Ichiro Isobe, Yun Zhu and Ryoiti Kiyama

Abstract: We summarize here the recent progress in fluorescence-based bioassays for the detection and evaluation of food materials by focusing on fluorescent dyes used in bioassays and applications of these assays for food safety, quality and efficacy. Fluorescent dyes have been used in various bioassays, such as biosensing, cell assay, energy transfer-based assay, probing, protein/immunological assay and microarray/biochip assay. Among the arrays used in microarray/biochip assay, fluorescence-based microarrays/biochips, such as antibody/protein microarrays, bead/suspension arrays, capillary/sensor arrays, DNA microarrays/polymerase chain reaction (PCR)-based arrays, glycan/lectin arrays, immunoassay/enzyme-linked immunosorbent assay (ELISA)-based arrays, microfluidic chips and tissue arrays, have been developed and used for the assessment of allergy/poisoning/toxicity, contamination and efficacy/mechanism, and quality control/safety. DNA microarray assays have been used widely for food safety and quality as well as searches for active components. DNA microarray-based gene expression profiling may be useful for such purposes due to its advantages in the evaluation of pathway-based intracellular signaling in response to food materials.

Reprinted from *Sensors*. Cite as: Nishi, K.; Isobe, S.-I.; Zhu, Y.; Kiyama, R. Fluorescence-Based Bioassays for the Detection and Evaluation of Food Materials. *Sensors* **2015**, *15*, 25831–25867.

1. Introduction

Fluorescent dyes or fluorophores have been widely used as probes (for physical and structural parameters), indicators (e.g., for molecular concentrations) or labels/tracers (e.g., for visualization and localization of biomolecules) in various bioassays [1]. While the development of fluorescent dyes has a history many centuries long, their importance has increased due to the recent advancement of new fluorescent dyes [2], which have been developed along with the development of new biotechnological tools and devices. For example, Laurdan, a naphthalene-based amphiphilic fluorescent dye having as characteristics the ability to penetrate membranes and a large Stokes shift, was developed to study membrane fluidity and dynamics, and its usage was made quite effective by the development of two-photon fluorescent microscopy, a microscope system with two-photon excitation, which enables the detection of signals with less background, less photodamage and more depth discrimination [3–5].

Therefore, the development of fluorescent dyes has had quite an impact when accompanied by the development of suitable devices and their applications. One of the most important currently emerging research fields is the development and application of technologies for new functional foods and quality control and safety of its production. Along with technological innovations, the effective usage of gene/genome information in the pathway-based evaluation of materials is crucial. We summarize here recent progress in fluorescence-based bioassays, including genomic and transcriptomic assays, by focusing on their applications in the study of food safety, quality and efficacy.

1.1. Overview of Fluorescent Dyes

Fluorescent dyes are generally polyaromatic or heterocyclic hydrocarbons, which undergo a three-stage process of fluorescence: excitation, excited-state lifetime and fluorescence emission [6]. Fluorescent dyes are characterized by key properties, such as those revealed by the absorption maximum (λ_{max}), the emission maximum (λ_{em}), the extinction coefficient (ε) and the fluorescence quantum yield (Φ) [2]. For example, the "Stokes shift", defined by the difference between λ_{max} and λ_{em}, is an important property of a fluorescent dye, and a large Stokes shift helps to avoid the reabsorption of emitted photons, giving higher contrast in fluorescent imaging [7].

New technologies, materials and devices have been developed for the efficient detection and utilization of the fluorescence signals in a biological specimen. For example, fluorescence-activated cell sorting (FACS) is an example of the successful application of fluorescence technologies for flow cytometry, and is now used in basic as well as industrial fields of life science [8,9]. Flow cytometry is a technique used for cell counting, cell sorting and biomarker detection, by passing a cell suspension in a stream of fluid through an electronic detection apparatus, allowing simultaneous multiparametric analyses of many thousands of micrometer-sized particles per second. Its applications include food study, such as water testing, milk analysis, brewing/wine production and food microbiology [10]. Meanwhile, fluorescence *in situ* hybridization (FISH) is a cytogenetic technique in which fluorescently labeled probes are hybridized with parts of DNA on chromosomes or specific RNA targets (e.g., mRNA and miRNA), and signals are detected by fluorescence microscopy. After a 30-year history, the original FISH protocol has been diversified into a number of new protocols with improved sensitivity, specificity and resolution [11]. For example, chromosome orientation-FISH, or CO-FISH, can detect strand-specific target DNA, and thus is useful to detect chromosomal abnormalities, such as Robertsonian translocations, chromosomal inversion and telomeric alterations [12].

A number of fluorescent techniques utilize Förster resonance energy transfer (FRET), a mechanism of energy transfer from a donor dye to a different acceptor dye, which is used to analyze conformations, interactions and concentrations of proteins and nucleic acids [6]. Protein-protein interactions can be detected by other fluorescent techniques, such as bioluminescence resonance energy transfer (BRET) assay, a modification of FRET, and biomolecular fluorescence complementation (BiFC) assay. BiFC assay is based on structural complementation between two non-fluorescent N- and C-terminal fragments of a fluorescent protein, and has contrasting advantages and disadvantages compared with FRET [13,14].

Other than aromatic hydrocarbons, several unique materials have also been utilized for fluorescence applications. Quantum dots are fluorescent semiconductor nanoparticles that have potential in biology, such as specific labeling of cells and tissues, long-term imaging, lack of cytotoxicity, *in vivo* multicolor imaging and FRET-based sensing [15]. A variety of fluorescent colors are available, depending on the size and shape of the particles. Additionally, some lanthanide ions are useful for bioassays due to their superior characteristics, such as long fluorescent lifetimes, large Stokes shifts and sharp emission profiles [16]. These materials have been used to study food safety, quality and efficacy (see Section 2).

1.2. Fluorescent Dyes for Bioassays

Fluorescent probes are required to match certain conditions for experiments, such as wavelength range, Stokes shift and spectral bandwidth, which are partly imposed by the instrumentation and the requirements of multicolor labeling experiments [6]. To design fluorescent experiments, the fluorescent output of a dye judged by the extinction coefficient and the fluorescence quantum yield needs to be considered. Additionally, under high-intensity illumination conditions, the irreversible destruction or photobleaching of fluorescent dyes is an important factor. Polyaromatic fluorescent dyes with extended π-conjugated systems could thus be ideal for designing dyes with longer Stokes shifts [7], which may improve the performance of fluorescent dyes. Here, we summarize the fluorescent dyes frequently used for bioassays.

Since its first synthesis in 1871, fluorescein, along with its derivatives, has been used as a powerful tool in various fields of life science [17]. Fluorescein is composed of two parts of xanthene, the chromophore part, and benzene, and exhibits excitation at 490 nm and emission at 514 nm ($\lambda_{max}/\lambda_{em} = 490/514$ nm), with fluorescent properties of $\varepsilon = 9.3 \times 10^4$ $M^{-1} \cdot cm^{-1}$ and $\Phi = 0.95$ [2]. A variety of fluorescein derivatives have been synthesized to improve its chemical, fluorescent and biological properties, and its stability, such as Oregon Green, fluorescein isothiocyanate (FITC), fluorescein diacetate and carboxyfluorescein (FAM). These dyes and fluorescein have been used in various bioassays/biomaterials, such as cell assays (flow cytometry, suspension arrays, fluorescent microscopy, fluorescent cell assay and fluorescent cytomics), FRET-based assays, probing (CO-FISH, fluorescent caspase assay, fluorescent hybridization, fluorescent nanoparticle assay, fluorescent nucleic acid assay and small-molecule fluorochrome assay) and microarray/biochip assays (see Section 2.1).

Rhodamines are isologs of fluorescein, having two amino groups, one of which is positively charged, and have properties similar to fluorescein, such as $\lambda_{max}/\lambda_{em}$ = 496/517 nm, $\varepsilon = 7.4 \times 10^4\,M^{-1}\cdot cm^{-1}$ and Φ = 0.92 for rhodamine 110 [2]. Rhodamine derivatives were developed for imaging, such as carboxytetramethylrhodamine (TAMRA), tetramethylrhodamine (TMR) and its derivative (tetramethylrhodamine isothiocyanate or TRITC), or to improve photostability and increase brightness, such as Alexa Fluor and DyLight Fluor dyes. Rhodamines (rhodamine, rhodamine B, lissamine rhodamine B, sulforhodamine B, Texas Red, TMR and TRITC) were extensively used in various bioassays/biomaterials, such as cell assays (fluorescent cytomics), probing (fluorescent hybridization, fluorescent nanoparticle assay, fluorescent nucleic acid assay and small-molecule fluorochrome assay) and microarray/biochip assays (see Section 2.1).

Cyanines are composed of two quaternized heteroaromatic bases joined by a polymethine chain, and their colors depend on the number of carbons (3 for Cy3 and 5 for Cy5) in the polymethine chain. Among cyanines, Cy3 and Cy5 have been most utilized, and while Cy3 shows fluorescent properties of $\lambda_{max}/\lambda_{em}$ = 554/568 nm, $\varepsilon = 1.3 \times 10^5\,M^{-1}\cdot cm^{-1}$ and Φ = 0.14, Cy5 shows those of $\lambda_{max}/\lambda_{em}$ = 652/672 nm, $\varepsilon = 2.0 \times 10^5\,M^{-1}\cdot cm^{-1}$ and Φ = 0.18 [2]. Cy3 and Cy5 have been used cooperatively and/or complementarily in multi-parameter fluorescence imaging [18], or as test/reference microarray probes [19] or photoconvertible fluorescent probes [20]. Cyanines have been used in various bioassays/biomaterials, such as probing (CO-FISH, fluorescent nanoparticle assay, fluorescent nucleic acid assay, fluorescent spectroscopy and FRET-based assays), protein/immunological assays (sandwich fluoroimmunoassay) and microarray/biochip assays (see Section 2.1).

Alexa Fluor dyes are synthesized through the sulfonation of coumarin, rhodamine, xanthene and cyanine dyes, and have characteristics of greater photostability and brightness as well as lower pH sensitivity than common dyes with comparable excitation/emission [21]. Among Alexa Fluor dyes, Alexa Fluor 488 (green; $\lambda_{max}/\lambda_{em}$ = 495/519 nm, $\varepsilon = 7.3 \times 10^4\,M^{-1}\cdot cm^{-1}$ and Φ = 0.92), Alexa Fluor 546 (orange; $\lambda_{max}/\lambda_{em}$ = 556/573 nm, $\varepsilon = 1.1 \times 10^5\,M^{-1}\cdot cm^{-1}$ and Φ = 0.79), Alexa Fluor 555 (red-orange; $\lambda_{max}/\lambda_{em}$ = 555/565 nm, $\varepsilon = 1.6 \times 10^5\,M^{-1}\cdot cm^{-1}$ and Φ = 0.10) and Alexa Fluor 647 (far-red; $\lambda_{max}/\lambda_{em}$ = 650/668 nm, $\varepsilon = 2.7 \times 10^5\,M^{-1}\cdot cm^{-1}$ and Φ = 0.33) were frequently used in bioassays [6]. Alexa Fluor dyes have been used in various bioassays/biomaterials, such as biosensing (magnetic modulation biosensing), probing (small-molecule fluorochrome assay) and microarray/biochip assays (see Section 2.1).

Green fluorescent protein (GFP) of the jellyfish Aequorea victoria is a protein composed of 238 amino acid residues, which has an eleven-stranded β barrel with an α helix covalently bonded with a chromophore running through the center [22]. GFP has two excitation peaks, at 395 (major) and 475 (minor) nm, an emission peak at 508 nm and fluorescent quantum yield of 0.77 [23]. To improve brightness, longer wavelengths and FRET, several mutant GFPs were developed, which include blue fluorescent protein (BFP), cyan fluorescent protein (CFP) and yellow fluorescent protein (YFP) [23,24]. Fluorescent proteins have been used in various bioassays/biomaterials, such as biosensing (fluorescent molecular biosensing), cell assays (flow cytometry, suspension arrays, fluorescent microscopy, fluorescent cell assay, fluorescent reporter-gene assay and single live-cell imaging), FRET-based assays, probing (fluorescent caspase assay and fluorescent reporter assay), protein/immunological assays (BiFC) and microarray/biochip assays (see Section 2.1).

Fluolid dyes are organic electroluminescence dyes, which were developed to overcome the inconvenience of currently available fluorescent reagents, and thus have larger Stokes shifts (more than 120 nm), greater photostability (stable for more than 10 years at room temperature) and more fluorescence in a solid state [25]. Their fluorescent properties are as follows: Fluolid-Green ($\lambda_{max}/\lambda_{em}$ = 395/522 nm), Fluolid-Yellow ($\lambda_{max}/\lambda_{em}$ = 410/541 nm), Fluolid-Orange ($\lambda_{max}/\lambda_{em}$ = 440/602 nm) and Fluolid-Red ($\lambda_{max}/\lambda_{em}$ = 525/660 nm). Owing to their extraordinary stability, Fluolid dyes have been used with a fluorescence scanning electron microscope (FL-SEM) [25] as well as in DNA microarray assay [26] and immunohistochemistry [27].

Fluorescent dyes and proteins other than those described above have also been used in bioassays, which include DAPI ($\lambda_{max}/\lambda_{em}$ = 350/450 nm, ε = 1.2 × 10^5 M^{-1}·cm^{-1} and Φ = 0.83) [28], SYBR Green I ($\lambda_{max}/\lambda_{em}$ = 497/520 nm and Φ = 0.8) [6] and RiboGreen ($\lambda_{max}/\lambda_{em}$ = 500/525 nm, ε = 6.7 × 10^4 M^{-1}·cm^{-1} and Φ = 0.65) [29] for staining DNA or RNA; R-phycoerythrin (PE: $\lambda_{max}/\lambda_{em}$ = 546/578 nm, ε = 2.0 × 10^6 M^{-1}·cm^{-1} and Φ = 0.98) [28] for immunofluorescence assays; Texas Red (TxR: $\lambda_{max}/\lambda_{em}$ = 596/620 nm, ε = 8.5 × 10^4 M^{-1}·cm^{-1} and Φ = 0.51) [28,30] for immunohistochemistry; and NanoOrange ($\lambda_{max}/\lambda_{em}$ = 582/605 nm and Φ = 0.36 in the protein complex) [31] for protein quantification. These dyes have been used in various bioassays/biomaterials for food study, such as cell assays (flow cytometry and suspension arrays), FRET-based assays, probing (CO-FISH), protein/immunological assays (fluorescent protein assay, fluorescent amplification catalyzed by T7 polymerase technique or FACTT, and real-time immune-PCR) and microarray/biochip assays (see Section 2.1).

1.3. Fluorescent Dyes Used in DNA Microarray Assay

Fluorescent dyes play important roles in DNA microarray assays due to their detectivity, speed and increased safety [28]. Fluorescent dyes frequently used in DNA microarray assays are phycoerythrin, Alexa Fluor dyes and cyanines. They have been used either as a single dye, such as phycoerythrin, or as two-color fluorescent probes, such as Cy3/Cy5 and Alexa Fluor 555/647. Owing to their superior reliability, Cy3 and Cy5 have been frequently used in gene expression profiling by means of DNA microarray assays since the early days [19]. Cyanines are, however, suggested to have low photostability and to be destabilized by their negative charges [32], as well as being affected by atmospheric ozone in the laboratory [33] and fluorescence quenching [34]. Alexa Fluor dyes, on the other hand, show greater brightness and photostability than cyanines [6]. Phycoerythrin is used in Affymetrix GeneChip assay as a streptavidin-conjugated form to detect biotinylated target cRNA hybridized with the probes on the platform. However, a significant decrease in fluorescent intensity was observed for phycoerythrin [35]. Mitsubishi Rayon developed a hollow fiber array, Genopal, in which fibers are filled with hydrogels attached to oligonucleotide probes and Cy5-labeled target cDNA is hybridized with the probes [36]. Although cyanines were generally used to label probes in the DNA microarray assay developed by GE Healthcare, a fluorescent dye, Amersham HyPer5, was developed and used to label target DNA [37]. Fluorescence-based microarray/biochip assays are summarized below (see Sections 2.3 and 3.1).

2. Application of Fluorescence-Based Bioassays

Fluorescence-based bioassays have been applied in biotechnology and various fields in life science. For example, various fluorescently labeled antibodies have been used to detect specific organelles, cellular activities (e.g., cell morphology, viability and functions) and cellular processes (e.g., transportation, endocytosis and receptor function) [38]. Clinical and pharmacological applications of fluorescent probes have been explored to diagnose leukemia and other cancers [39]. These applications are supported by basic characteristics of fluorophores, such as structural and environmental effects on fluorescence emission, fluorescence polarization and FRET, which are applied for spectrofluorometry, fluorescent microscopy and fluorescence-based chemical sensing to trace and image biological objects [1]. In this section, we summarize first fluorescence-based bioassays and then their applications by discussing representative literature.

2.1. Fluorescence-Based Bioassays

Fluorescence-based bioassays, classified into biosensing, cell assays, energy transfer-based assays, probing, protein/immunological assays and microarray/biochip assays, are summarized in Table 1. Biosensing, such as fluorescent molecular biosensor, fluorometric high-performance liquid chromatography (HPLC) and magnetic modulation biosensing, has been used to detect intermolecular interactions and targets at low concentrations, or to analyze nitrite/nitrate, where fluorescent dyes, such as Alexa Fluor 488, GFP and 2,3-naphthotriazole, have been used. Cell assays, such as flow cytometry (fluorescence-activated cell sorting: FACS), fluorescent cytomics, fluorescence microscopy, fluorescent reporter-gene assay and live-cell imaging, have been used in particle-based flow cytometric assay, drug delivery research and applications of RNA/DNA aptamers to measure cell fluorescence, to screen hormonally active compounds and to examine gene expression/protein interaction, where fluorescent dyes, such as fluoresceins (including FAM and FITC), GFP/GFP-family proteins, lanthanides, phycoerythrin and rhodamines (TMR-C5), have been used. Meanwhile, energy transfer-based assays have been used in live-cell imaging and to analyze protein structure, where fluorescent dyes, such as BFP/GFP, FITC and phycoerythrin, have been used.

A number of technologies have been developed for various types of probe, such as fluorescent calcium indicators, fluorescent caspase substrates, fluorescent nanoparticles, fluorescent nucleic acids, quantum dots and small-molecule fluorochromes, which have been used in fluorescent hybridization (e.g., FISH), reporter-gene assay and fluorescent spectroscopy, often used in combination with FRET, to examine chromosome aberrations/segregations, gene-gene/DNA-protein interactions, calcium signaling and ion channeling; to evaluate fluorescence bioassays, imaging/labeling/sensing, immunoassays/microarray assays, quantitative structure-activity relationship (QSAR), antimycobacterial susceptibility, biological enzymatic reactions, G-protein-coupled receptor (GPCR) ligands and reactive oxygen species; and to screen bladder and other tumor markers, antagonists of GPCRs and anticancer drugs, by using fluorescent dyes, such as Alexa Fluor 546, Cy3/Cy5, FDA, FITC, Fluo-4, fluorescein, FuraRed, GFP/RFP, hydroethidine/hydrocyanines, quantum dots, rhodamine, SpectrumGold/SpectrumOrange, TRITC and Texas Red.

Table 1. Fluorescence-based bioassays. BEBO: 4-[(3-methyl-6-(benzothiazol-2-yl)-2,3-dihydro-(benzo-1,3-thiazole)-2-methylidene)]-1-methyl-pyridinium iodide; BFP: blue fluorescent protein; BiFC: bimolecular fluorescent complementation; ELISA: enzyme-linked immunosorbent assay; FACS: fluorescence-activated cell sorting; FACTT: fluorescent amplification catalyzed by T7 polymerase technique; FAM: carboxyfluorescein; FDA: fluorescein diacetate; FISH: fluorescent *in situ* hybridization; FITC: fluorescein isothiocyanate; FRET: fluorescence resonance energy transfer; GFP: green fluorescent protein; GPCR: G-protein-coupled receptor; PCR: polymerase chain reaction; QSAR: quantitative structure activity relationship; RFP: red fluorescent protein; TMR: tetramethylrhodamine; TRITC: tetramethylrhodamine isothiocyanate; TxR: Texas Red; YFP: yellow fluorescent protein.

Bioassay/Biomaterial	Purpose/Subject	Fluorescent Dye/Molecule (Representative)	Reference
Biosensing			
Fluorescent molecular biosensing	Detection of intermolecular interactions	GFP	Altschuh *et al.*, 2006 [40]
Fluorometric HPLC	Analysis of nitrite/nitrate	2,3-Naphthotriazole	Jobgen *et al.*, 2007 [41]
Magnetic modulation biosensing	Detection of targets at low concentrations	Alexa Fluor 488	Danielli *et al.*, 2010 [42]
Cell Assay			
Flow cytometry/FACS	Particle-based flow cytometric assay	GFP	Vignali, 2000 [43]
Flow cytometry/Suspension array	Measurement of cell fluorescence	GFP/FITC/Phycoerythrin	Edwards *et al.*, 2004 [44]
Fluorescence microscopy	Drug delivery research	GFP/Fluorescein	White & Errington, 2005 [45]
Fluorescent cell assay	High-throughput drug discovery	GFP-family proteins	Wolff *et al.*, 2008 [46]
Fluorescent cell assay	Application in cellular assays	Lanthanides/GFP/FAM	Hanson & Hanson, 2008 [47]
Fluorescent cytomics	Application of RNA/DNA aptamers	Fluorescein/TMR-C5	Ulrich *et al.*, 2004 [48]
Fluorescent reporter-gene assay	Screening of hormonally active compounds	GFP	Svobodová & Cajthaml, 2010 [49]
Single live-cell imaging	Gene expression/Protein interaction	GFP	Mullassery *et al.*, 2008 [50]

212

Table 1. *Cont.*

Bioassay/Biomaterial	Purpose/Subject	Fluorescent Dye/Molecule (Representative)	Reference
Energy Transfer-Based Assay			
FRET	Live-cell imaging	GFP/BFP	Salipalli et al., 2014 [51]
FRET/Flow cytometry	Analysis of protein structure	FITC/Phycoerythrin/GFP	Szöllosi et al., 1998 [52]
Probing			
FISH	Monitoring of chromosome aberrations	(Not shown)	Léonard et al., 2005 [53]
FISH	Screening of bladder tumor markers	SpectrumGold, etc.	Lokeshwar & Selzer, 2006 [54]
FISH	Study of gene-gene/protein interactions	SpectrumOrange, etc.	Chun et al., 2009 [55]
FISH (CO-FISH)	Chromosome segregation study	Cy3/Cy5/FITC/TxR	Falconer & Lansdorp, 2013 [56]
Fluorescent calcium indicator	Calcium signaling for cell functions	Fluo-4	Apáti et al., 2012 [57]
Fluorescent caspase substrate/FRET	Screening of anticancer drugs	FITC/GFP/RFP	Brunelle & Zhang, 2011 [58]
Fluorescent hybridization	Identification of nucleic acids	Fluorescein/Rhodamine, etc.	Marras et al., 2005 [59]
Fluorescent nanoparticle	Synthesis of fluorescent probes	Cyanines/FITC/TRITC	Sokolova & Epple, 2011 [60]
Fluorescent nucleic acid probe	Labeling of nucleic acid probes	Fluorescein/Rhodamine, etc.	Kricka & Fortina, 2009 [61]
Fluorescent reporter assay	Functional study of ion channels	GFP/RFP	Musa-Aziz et al., 2010 [62]
Fluorescent reporter assay/FRET	Antimycobacterial susceptibility testing	FDA/GFP/RFP	Sánchez & Kouznetsov, 2010 [63]
Fluorescent reporter assay/FRET	Detection of gene expression	GFP/RFP	Jiang et al., 2008 [64]
Fluorescent spectroscopy/FRET	Probing biological enzymatic reactions	Cy3/Cy5	Jahnz & Schwille, 2004 [65]
Quantum dot	Fluorescence bioassay	Quantum dot	Liu et al., 2005 [66]
Quantum dot/FRET	Imaging/labeling/sensing	Quantum dot	Medintz et al., 2005 [67]
Quantum dot/FRET	Immunoassay/microarray assay/imaging	Quantum dot	Zhang & Wang, 2012 [68]
Quantum dot/Suspension array	Detection of cancer markers/tumor cells	Quantum dot	Akinfieva et al., 2013 [69]
Small-molecule fluorochrome	Screening of antagonists for GPCRs	FITC/FuraRed/Alexa Fluor 546	Arterburn et al., 2009 [70]
Small-molecule fluorochrome	Detection of reactive oxygen species	Hydroethidine/Hydrocyanines	Maghzal et al., 2012 [71]
Small-molecule fluorochrome	QSAR	FDA	Horobin et al., 2013 [72]
Small-molecule fluorochrome	Fluorescently labeled GPCR ligands	Rhodamine B, etc.	Vernall et al., 2014 [73]

Table 1. *Cont.*

Bioassay/Biomaterial	Purpose/Subject	Fluorescent Dye/Molecule (Representative)	Reference
Protein/Immunological Assay			
BiFC	Protein interaction/modification	GFP/YFP	Kerppola, 2009 [74]
BiFC	Protein-protein interaction	GFP/YFP, *etc.*	Miller *et al.*, 2015 [75]
Chemifluorescent ELISA	Monitoring of kinase activity	(Not shown)	Wu *et al.*, 2010 [76]
Fluorescent dye-based protein assay	Quantitation of protein	NanoOrange	Noble & Bailey, 2009 [77]
Immuno-detection (FACTT)	Quantification of rare blood biomarkers	RiboGreen	Freudenberg *et al.*, 2008 [78]
Lanthanide-doped fluorescent assay	Application for bioassay/therapy	Lanthanides	Guo & Sun, 2012 [79]
Lanthanide fluorescent immunoassay	Time-resolved fluorescence bioassay	$Eu^{3+}/Sm^{3+}/Tb^{3+}/Dy^{3+}$	Yuan & Wang, 2005 [16]
Lanthanide fluorescent immunoassay	Prion disease research	Lanthanides	Sakudo *et al.*, 2007 [80]
Real-time immuno-PCR	Diagnoses of viral antigens/pathogens	SYBR Green I/BEBO	Barletta, 2006 [81]
Sandwich fluoroimmunoassay	Detection/identification of toxins	Cy5	Ligler *et al.*, 2003 [82]
Microarray/Biochip Assay (see Table 2)			

214

Protein/immunological assays, such as BiFC, chemifluorescent enzyme-linked immunosorbent assay (ELISA), fluorescent dye-based protein assay, immuno-detection (FACTT; see Table 1), lanthanide-doped fluorescent assay, lanthanide fluorescent immunoassay, real-time immuno-PCR and sandwich fluoroimmunoassay, have been used to examine bioassay/therapy, kinase activity, protein interaction/modification and time-resolved fluorescence bioassay, to screen/identify rare blood biomarkers and some toxins, to study prion diseases and to identify viral antigens/pathogens, for which fluorescent dyes, such as BEBO (a cyanine dye), Cy5, GFP/YFP, lanthanides (e.g., Eu^{3+}, Sm^{3+}, Tb^{3+} and Dy^{3+}), NanoOrange, RiboGreen and SYBR Green I, have been used. Microarray/biochip assays are discussed below (see Section 2.3). GFP has been used quite often as reporter conjugates in cell assay. For example, estrogen activity was detected by a reporter construct of the human ERα gene fused to yeast enhanced GFP (yEGFP), which was used as a rapid yeast bioassay to screen estrogen activity in calf urine [83]. This construct was validated as a bioassay for hormonal substances in feed [84], and concomitantly improved by the combination with mass-spectrometry techniques [85,86] and by the use of various test samples [87,88], and combined with androgen assay [89], to attain to a level of a standardized multi-hormonal bioassay system.

2.2. Fluorescence-Based Microarrays/Biochips

2.2.1. Antibody/Protein Microarray

Protein microarray assay is a high-throughput method used to study biochemical activities of proteins, by measuring their binding affinities, specificities and quantities [90]. The array has a support surface, such as a slide glass, a nitrocellulose membrane, a bead and a microtiter plate, to which the captured protein is bound as an array, and probe molecules, typically labeled with a fluorescent dye or conjugated with enzymes for chemiluminescent or colorimetric assays, are added to the array. In a fluorescent assay, the reaction between the fluorescence-labeled probe and the immobilized protein causes the emission of a fluorescent signal at a specific position, which is detected by a laser scanner. There are three types of protein microarray used to study the biochemical activities of proteins: analytical microarrays, functional protein microarrays and reverse-phase protein microarrays [90]. Antibody microarrays belong to the category of analytical microarrays, and sometimes use a sandwich format consisting of capture antibodies (e.g., biotinylated antibodies), analytes (e.g., toxins) and reporter molecules (e.g., avidin-conjugated nanoparticles and fluorophore-conjugated secondary antibodies). They have been used to screen foodborne pathogens such as

Escherichia coli O157:H7 and *Salmonella* spp. [91], and to detect multiplex toxins, such as toxins contaminating milk, apple cider and blood samples [92].

2.2.2. Bead/Suspension Array

The detection of bacterial/plant toxins [93], mycotoxins [94] and pesticides [95] in food has been carried out by using bead/suspension array technology, in which fluorescent dye-labeled microspheres/beads are often used. Appropriate molecules or receptors, such as DNA (oligonucleotides), and antibodies and other proteins, are attached to the microspheres differently labeled with fluorescent dyes, for example. Beads are readily suspendable in solution and are used for hybridization between receptors and corresponding reactive biomolecules. Bead arrays have advantages over flat arrays in the array preparation (containing millions of particles per milliliter) and density (containing hundreds of thousands of array elements per microliter), enabling multiparameter detection and high-throughput processing [96]. Since the optical property of each bead is known, target biomolecules hybridized/bound to the beads can be easily differentiated, and quantification can be achieved by comparing the relative intensity of targets in a set of beads with that of markers in another set of beads using fluorescence detection apparatuses, such as a flow cytometer.

2.2.3. Capillary/Sensor Array

A sensor array typically consists of a recognition component, a transducer component and an electronic detection system. The recognition component uses biomolecules to interact with the analyte of interest. This interaction is measured by biotransducers, such as an optical transducer, which outputs a measurable signal proportional to the presence of the target analyte in the sample. Meanwhile, biomolecules are separated first by capillary electrophoresis in an array and then detected by appropriate sensors in capillary arrays. There have been cases of the application of capillary/sensor arrays for food analysis, such as detecting pathogens and toxins, and fluorescent substances are commonly used in their detection systems. Recently, researchers have performed successful analyses of food using improved sensor arrays, such as those with dendritic fluorophores [97] and a fluorescent indicator-displacement sensor array using titania as a host material [98].

2.2.4. DNA Microarray/PCR-Based Array

Using DNA microarray technology, multiple genes can be characterized simultaneously in a single assay. It has been used widely for the analysis of gene expression, but it can also be used for the analysis of microbial pathogens for food safety and environmental applications. A DNA microarray involves the immobilization of numerous probes, such as cDNA and oligonucleotide probes, at a high density on a solid matrix, such as glass, to which fluorescence-labeled

PCR-amplified target DNA fragments can be hybridized. The signal generated by the bound labeled targets on the microarray allows identification based on the known locations of the probes on the array. Applications of DNA microarray technology for the detection of pathogens contaminating food have been reported (detailed in Section 3).

2.2.5. Glycan/Lectin Array

The use of glycan microarrays, comprising multiple different glycans on a single platform, is a technique for the analysis of glycosylation patterns and the screening of a number of glycan-binding proteins for investigation of their roles in biological systems. Recently, a shotgun glycan microarray prepared from isolated human milk glycans was reported, where viruses, antibodies and glycan-binding proteins including lectins were detected in order to examine the diverse recognition functions of human milk glycans [99]. In addition, a lectin microarray, based on the specific affinity of a lectin to a specific glycan, is another useful platform for glycan analysis. Recently, a bead-based multiplex lectin array was developed, where respective lectins were coupled to differentially fluorescent dye-coated microbeads [100]. These beads were incubated with biotin-labeled glycoproteins in suspension, with visualization using the interaction between biotin and streptavidin-R-phycoerythrin. This microarray was applied for glycosylation profiling of hepatocellular carcinoma-associated immunoglobulin G in a rapid, sensitive and reproducible manner.

2.2.6. Immunoassay/ELISA-Based Array

An immunoassay is a test that relies on the inherent ability of an antibody to recognize and bind to a specific antigen, which might exist in a complex mixture, to measure the presence and/or concentration of the antigen. In life science research, immunoassays are often used in studies of the biological functions of proteins, while, in industry, immunoassays are used in various applications, such as to detect contaminants in food and water and to monitor and assess specific molecules during food processing. In immunoassays, antibodies or antigens are conjugated or coupled with fluorescent dyes, or labeled with other materials, such as biotin and horseradish peroxidase, to produce measurable fluorescent, chemiluminescent or chromogenic signals for detection. One of the most popular immunoassays is ELISA, in which antigens in a sample are first attached to the surface of the platform (e.g., a 96-well microtiter plate), which are then detected with a specific antibody linked to an enzyme (for enzymatic reactions) or a fluorescently labeled secondary antibody. In recent years, fluorogenic labels, such as cyanines and phycoerythrin, have been used in immunoassays to detect mycotoxins for food safety [101,102].

217

2.2.7. Microfluidic Chip

Microfluidic chips have been used in many biological fields, such as drug screening and the monitoring of food processing. A microfluidic chip is a set of microchannels molded into a material like glass, silicon or polymer. The microchannels are connected together forming a network, which is connected to the outside by transporting inputs and outputs through the chip platform. The surface patterning of bonded or sealed microchannels in a microfluidic chip can be achieved by technologies such as laminar flow and capillarity, photolithography, microplasmas and electrochemical biolithography [103]. Microfluidic chips have advantages over conventional devices, such as that the assay can be performed on a small scale and thus requires less time and smaller amounts of samples and reagents, and can be performed automatically with high reproducibility [104]. Thus, microfluidic chips have been combined with other systems, such as capillary electrophoresis, PCR and flow cytometry. For example, a simple microfluidic chip system combined with a probe-immobilized fluorescent bead assay was developed for the rapid detection of bacteria associated with food poisoning [105]. Meanwhile, a microfluidic chip system combined with a BRET-based biosensor was developed for real-time, continuous detection with superior sensitivity of maltose in water or beer [106].

2.2.8. Tissue Array

The assay using a tissue array is a high-throughput analysis that utilizes hundreds or up to a thousand separate tissue samples on a single platform. Using this method, tissue samples can be rapidly analyzed by histological analyses, such as immunohistochemistry and FISH, in order to screen genetic or protein markers, or to detect tissues infected with pathogenic/toxigenic factors. Since most dyes currently used for microbial fluorescent staining are toxic or carcinogenic, a tissue array system using brilliant blue FCF, which is a food dye and thus has no toxic effects, was developed and applied for microbial cell fluorescence staining of pathogenic/toxigenic and beneficial fungi and bacteria [107].

2.3. Application of Fluorescence-Based Microarrays/Biochips for Food Study

Fluorescence-based microarrays/biochips for food study are summarized in Table 2. Antibody/protein microarrays have been applied to detect/screen foodborne pathogens and toxins, where fluorescent dyes, such as Cy3, fluorescein and RuBpy, have been used. Bead/suspension arrays, such as cytometric bead arrays, liquid/magnetic bead arrays and suspension arrays, have been used to detect/quantify mycotoxins, pathogens, genetically modified maize, pesticides and bacterial/plant toxins, where fluorescent dyes, such as Alexa Fluor 532, Cy3, FITC and phycoerythrin, have been used. Capillary/sensor arrays, such as capillary arrays, chemical sensor arrays and fluorescent sensor arrays, have been used to analyze carbohydrates, fresh fruit juices and various food materials, where fluorescent dyes, such as sulforhodamine B, lissamine rhodamine B and synthetic dendritic fluorophores, have been used. DNA microarrays/PCR-based arrays, such as direct RNA hybridization/microarrays, DNA/PNA microarrays, laser microdissection/microarrays, oligonucleotide microarrays, PCR/bead arrays, PCR/microarrays (mutant analysis by PCR and restriction enzyme cleavage or MAPREC assay, and nucleic acid sequence-based amplification implemented microarray analysis or NAIMA; see Table 2) and PCR/single-base extension-tag arrays, have been used to detect mycoplasmas, pathogenic bacteria, grapevine viruses, genetically modified cotton, pathogenic *Vibrio* spp., genetically modified soybean and seafood-borne pathogens, to screen hypoxia-inducible genes and recombinant flavivirus vaccine strains, to examine genotypes of beef/chicken and gene expression profiles of fungi, and to evaluate the authenticity of ginseng drugs, along with fluorescent dyes, such as Alexa Fluor 546/647, Cy3/Cy5, phycoerythrin, PolyAn-Green/PolyAn-Red, AmCyan1, NIR Dye 700/800, Oyster-550 and quantum dots.

Glycan/lectin arrays have been used for functional glycomic analysis or glycosylation profiling, where Alexa Fluor 488, Cy5 and phycoerythrin have been used as fluorescent dyes. Immunoassay/ELISA-based arrays, such as ELISA chips and immunoassay microarrays, or those used in competitive immunoassay, fluoroimmunoassay and sandwich fluoroimmunoassay, have been used to detect/quantify food allergens, mycotoxins, ochratoxin A, pathogens/toxins, staphylococcal enterotoxin B or to assess food safety, where fluorescent dyes, such as Alexa Fluor 647, Cy3/Cy5, fluorescein, FluoSpheres, phycoerythrin and RuBpy, have been used. Microfluidic chips have been used to detect food poisoning bacteria or single-base mismatches, or to monitor food processing, along with fluorescent dyes, such as Alexa Fluor 647, FAM and GFP. Tissue arrays have also been used to stain microbial cells using brilliant blue FCF.

Fluorescence-based microarrays/biochips can be categorized by the number of target chemicals; either the characterization of a single chemical, or the screening of multiple chemicals from a number of samples or mixtures of chemicals. Among the assays shown in Table 2, antibody/protein microarrays, DNA microarrays, glycan/lectin arrays and tissue arrays are advantageous for profiling and analyzing a single chemical due to the ability of multiple probing, while bead/suspension arrays, capillary/sensor arrays and immunoassay/ELISA-based arrays, microfluidic chips and PCR-based arrays are useful for screening because of their high-throughput processing ability.

Table 2. Fluorescence-based microarrays/biochips for food study. ELISA: enzyme-linked immunosorbent assay; FAM: carboxyfluorescein; FITC: fluorescein isothiocyanate; GFP: green fluorescent protein; GMO: genetically modified organism; HPLC: high-performance liquid chromatography; MAPK: mitogen-activated protein kinase; MAPREC: mutant analysis by PCR and restriction enzyme cleavage; NAIMA: nucleic acid sequence-based amplification implemented microarray analysis; PNA: peptide nucleic acid; RuBpy: [Ru(bpy)$_3$]Cl$_2$/Tris(bipyridine)ruthenium(II) chloride.

Method/Tool	Purpose/Subject	Fluorescent Dye/Molecule	Reference
Antibody/Protein microarray			
Antibody microarray	Screening of foodborne pathogens	Cy3/Fluorescein	Gehring et al., 2008 [91]
Antibody microarray	Detection of multiplex toxins	Cy3/RuBpy	Lian et al., 2010 [92]
Bead/Suspension array			
Aptamer/Suspension array	Detection of mycotoxins	FITC	Sun et al., 2014 [94]
Cytometric bead array	Detection of pathogens	Alexa Fluor 532/Cy3	Stroot et al., 2012 [108]
Liquid bead array	Genetically modified maize	Phycoerythrin	Han et al., 2013 [109]
Magnetic suspension assay	Quantification of bacterial/plant toxins	Phycoerythrin	Pauly et al., 2009 [93]
Microsphere suspension array	Multiplex mycotoxin detection	FITC	Deng et al., 2013 [110]
Suspension array	Detection of pesticides	Phycoerythrin	Wang et al., 2014 [95]
Capillary/Sensor array			
Capillary array electrophoresis	Carbohydrate analysis	Sulforhodamine B	Khandurina et al., 2004 [111]
Chemical sensor array	Discrimination of fresh fruit juices	Lissamine rhodamine B	Tan et al., 2014 [98]
Fluorescent sensor array	Electronic tongue for food analysis	Dendritic fluorophores	Niammont et al., 2010 [97]

Table 2. *Cont.*

Method/Tool	Purpose/Subject	Fluorescent Dye/Molecule	Reference
DNA Microarray/PCR-Based Array			
Direct RNA hybridization/Microarray	Detection of mycoplasmas	Alexa Fluor 647	Kong et al., 2007 [112]
DNA microarray	Authentication of ginseng drugs	Cy5	Zhu et al., 2008 [113]
DNA microarray	Hypoxia-inducible genes	Phycoerythrin	Otsuka et al., 2010 [114]
DNA microarray	Genotyping of beef/chicken	Cy3/Cy5	Reverter et al., 2014 [115]
DNA microarray	Food safety assessment	PolyAn-Green/PolyAn-Red	Brunner et al., 2015 [116]
Laser microdissection/Microarray	Gene expression profiling of fungi	AmCyan1	Tang et al., 2006 [117]
MAPREC assay	Recombinant flavivirus vaccine strain	NIR Dye 700/800	Bidzhieva et al., 2011 [118]
NAIMA	GMO detection	Oyster-550	Morisset et al., 2008 [119]
Oligonucleotide microarray	Detection of pathogenic bacteria	Quantum dot	Huang et al., 2014 [120]
Oligonucleotide microarray	Detection of grapevine viruses	Cy3	Abdullahi et al., 2011 [121]
PCR/Bead array	Detection of genetically modified cotton	Phycoerythrin	Choi, 2011 [122]
PCR/Microarray	Detection of pathogenic *Vibrio* spp.	Alexa Fluor 546	Panicker et al., 2004 [123]
PCR/Single-base extension-tag array	Seafood-borne pathogens	Cy3	Chen et al., 2011 [124]
PNA microarray	Genetically modified soybean	Cy3/Cy5	Germini et al., 2004 [125]
Glycan/Lectin Array			
Glycan microarray	Functional glycomic analysis	Alexa Fluor 488/Cy5	Yu et al., 2012 [99]
Lectin array	Glycosylation profiling	Phycoerythrin	Wang et al., 2014 [100]
Immunoassay/ELISA-Based Array			
Competitive immunoassay	Detection of ochratoxin A	Cy5	Ngundi et al., 2005 [126]
ELISA chip	Food safety assessment	Fluorescein	Herrmann et al., 2006 [127]
ELISA chip	Staphylococcal enterotoxin B detection	FluoSpheres	Han et al., 2013 [128]
Fluoroimmunoassay	Detection of food allergens	Alexa Fluor 647	Shriver-Lake et al., 2004 [129]
Fluoroimmunoassay	Detection of mycotoxins	Cy5	Ngundi et al., 2006 [130]
Immunoassay microarray	Detection and quantification of toxins	Cy5	Weingart et al., 2012 [131]
Immunoassay microarray	Multiplex mycotoxin detection	Cy3	Hu et al., 2013 [101]
Immunoassay microarray	Detection of mycotoxins	Phycoerythrin	Peters et al., 2014 [102]
Sandwich fluoroimmunoassay	Detection of pathogens/toxins	Cy5	Ngundi & Taitt, 2006 [132]
Sandwich fluoroimmunoassay	Staphylococcal enterotoxin B detection	RuBpy	Zhang et al., 2011 [133]

Table 2. *Cont.*

Method/Tool	Purpose/Subject	Fluorescent Dye/Molecule	Reference
Microfluidic Chip			
Microfluidic chip	Detection of food poisoning bacteria	Alexa Fluor 647	Ikeda *et al.*, 2006 [105]
Microfluidic chip	Detection of single-base mismatches	FAM	Wang *et al.*, 2013 [134]
Microfluidic chip	In-line monitoring of food processing	GFP	Le *et al.*, 2014 [106]
Tissue Array			
Microbial cell fluorescence staining	Microbial staining	Brilliant blue FCF	Chau *et al.*, 2011 [107]

3. DNA Microarray-Based Assay for Food Study

DNA microarray-based assay for food study has been compared with other technologies. For example, foodborne diseases are a major issue among global public health problems and the development of rapid detection methods is crucial for their prevention and treatment. Law *et al.* summarized rapid methods for the detection of foodborne bacterial pathogens, such as PCR-based methods, PCR-independent methods, DNA microarray assay, biosensor-based assays and immunological methods, and discussed their principles, applications, advantages and limitations [135]. Nucleic acid-based methods generally give high sensitivity, although they require trained personnel and specialized instruments. Biosensor-based assays, on the other hand, can be used without sample pre-enrichment, although they need improvements for on-site detection. Immunological methods, such as ELISA and flow immunoassay, are currently widely used, but have difficulties when interfering molecules are included in the samples. Josefsen *et al.* compared assays for the rapid monitoring of Campylobacter bacteria in poultry production, and real-time PCR is currently closest to a realistic monitoring system, although other methods, such as microarray PCR, miniaturized biosensors, chromatographic techniques and DNA sequencing, could be considered in the future when cost-effective on-site/at-line monitoring capability is achieved [136]. Gui and Patel discussed the merits of DNA testing, such as DNA microarray assay and next-generation sequencing, to detect *Yersinia* and other foodborne pathogens [137]. DNA testing is generally a high-sensitivity and high-throughput assay, allowing the detection of a single molecule in multiple reactions to be performed at once, thus allowing a range of characteristics to be rapidly and simultaneously determined. However, improvements in sample preparation, data analysis and molecular detection techniques are still needed. Lauri and Mariani compared potentials and limitations among four molecular diagnostic methods: PCR, nucleic acid sequence-based amplification (NASBA), oligonucleotide DNA microarray and ligation detection reaction (LDR), in food safety assessment [138]. While DNA microarrays can be used to detect quite a number of DNA species simultaneously, they are expensive and need more time for processing. DNA-based technologies have been used to assess the safety and quality of food, animal feed and environmental samples, by providing traceability to prevent foodborne diseases and markers to monitor genetically modified organisms [139].

3.1. DNA Microarray Assay Protocols

Among microarrays and biochips, DNA microarrays have been developed most extensively and some have already been used to diagnose cancer and other diseases or symptoms [140]. While the traditional solid-phase microarrays contain specific DNA probes attached to the surface of glass, plastic or silicon chips, other types have been developed, which include bead, fiber and electric arrays, where DNA is attached on the surface of latex or polystyrene beads (bead arrays) or attached to gels within plastic hollow fibers (fiber arrays), or an electrical current is generated by redox recycling upon target/probe hybridization (electric arrays). While a variety of DNA microarray assays have been developed, they can be classified into two major types: those for genotyping (e.g., for comparative genomic hybridization, identifying mutations and single-nucleotide polymorphisms and chromatin-immunoprecipitation on a chip) or gene expression analyses (e.g., for gene expression profiling, screening expression marker genes and identifying splice variants). Genotyping is used to detect the contamination of microbes in food, to identify pathogenic/toxic microbial strains/subtypes and to examine the authenticity of plants or the presence of genetically modified organisms by using 16S rRNA genes and/or genomic DNA markers specific to the microbes or the plants. Gene expression profiling, on the other hand, has been used to identify contaminated pathogenic/toxic bacterial strains, to detect specific stress responses and to examine the efficacy of food materials or components by examining the expression of pathogenic/toxic genes, stress-responsive genes and disease/metabolism-associated genes.

Fluorescent dyes, such as cyanines (Cy3 and Cy5), fluoresceins (including FITC and FAM) and Alexa Fluor dyes, have been used in DNA microarray assays. New fluorescent dyes, Fluolid dyes, which have characteristics of higher light/temperature resistance and longer Stokes shifts, have been developed and applied for DNA microarray assays [26]. These fluorescent dyes are used to label target DNA either by direct labeling, where fluorescent dyes directly attached to nucleotides (e.g., deoxyuridine 5'-triphosphate or dUTP) are used to label DNA by nick translation or primer extension, or by indirect labeling, where small nucleotides, such as aminoallyl nucleotides, are used to label DNA first, and the primary amino group attached to DNA is subjected to a reaction with the N-hydroxysuccinimide ester group attached to a fluorescent dye. Alternatively, small nucleotides, such as biotinylated or digoxigenin-labeled ones, are used to label DNA first, and the labeled DNA is then detected by secondary molecules, such as fluorescently labeled streptavidin or anti-digoxigenin antibodies, respectively. Biotinylated or digoxigenin-labeled DNA can alternatively be detected by non-fluorescent assays, such as colorimetric and chemiluminescent ones by using chromogens, such as Seramun Green, Silverquant and True Blue, or chemiluminescent substrates, such as chloro-5-substituted adamantyl-1,2-dioxetane phosphate (CSPD) and luminol (see below).

3.2. Application of DNA Microarray Assay for Food Study

DNA microarrays used for food study are summarized in Table 3. DNA microarrays have been used to examine the following subjects: allergies such as latex and/or vegetable food allergy; poisoning by microbes, such as *Bacillus cereus, Clostridium botulinum, Campylobacter* spp., *Clostridium perfringens, Escherichia coli, Salmonella enterica* and *Staphylococcus aureus*; toxic effects of cadmium, mycotoxins, silver-nanoparticles and tetrodotoxin; contamination of microbes, such as *Alicyclobacillus* spp., *Arcobacter butzleri, Bacillus anthracis, Lactobacillus* spp., *Listeria monocytogenes, Yersinia enterocolitica* and *Yersinia pestis*; the efficacy of food and food materials, such as that in the absorption of phenolic acids, suppressing cancer, the plasma triglyceride-lowering effect, the lipid consumption in skeletal muscle and the improvement of diabetic symptoms and osteoporosis; mechanisms such as those involved in the response to drought stress, immune stress, inflammation, mucosal IgA antibodies and oxidative stress/DNA damage; and quality control and safety of food, such as the authenticity of food, food safety assessment and the identification of genetically modified organisms.

The food/food materials analyzed by DNA microarray assays include the following: bovine milk, cheese, fish, horseradish, meat (pork and chicken), pancake with chicken, pufferfish, rice and vegetables for the study of allergy, poisoning or toxicity; alfalfa, bread (whole-grain and fiber-rich), cantaloupe, cilantro, egg, fish, juice, maize, meat (beef, pork and poultry meat), milk, mung bean, potato, rice, sausage (Thai Nham) and water for the study of food contamination; beverage, cassava, chitooligosaccharide, dairy, herbs (e.g., licorice and those used in Hochuekkito), high-cholesterol/fat diet, imbibed soybean, phenolic preservatives, pineapple, polyunsaturated fatty acids, psyllium, quercetin, skim milk, sweet corn, tea and xanthan gum for study of their efficacy and mechanisms; and canola, cereal (e.g., barley, oat, rice and wheat), citrinin, cotton, crop, food additives, ginseng, *Kothala himbutu* (a medicinal plant), maize, olive, potato, royal jelly and soybean for the study of food quality control and safety. Other materials besides food are composts, digestates and waste for the study of food contamination.

The types of DNA microarray used can be classified into those for genotyping and gene expression analyses (Table 3). The sources of microarrays are either custom arrays or microarrays supplied by companies, such as Affymetrix (USA), Agilent Technologies (USA), Alere Technologies (Clondiag Chip Technologies, Germany), Amersham/GE Healthcare (USA), GeneSystems (France), Mitsubishi Rayon (Japan) and Pathogen Functional Genomic Research Center (USA).

The types of dye or substrate used to detect the signal are: fluorophors, such as Alexa Fluor 555/647, Cy3/Cy5, fluorescein/6-FAM, phycoerythrin, TAMRA and 3,3',5,5',-tetramethylbenzidine (TMB), or chromogens/chemiluminescent or colorimetric substrates for non-fluorescent assays, such as CSPD, luminol, Seramun Green, Silverquant and True Blue.

Table 3. Application of DNA microarray assay for detection and evaluation of food materials.

Food Source or Material	Material Detected or Subject Examined	Type of Microarray Used (Source [a]/Dye)	Reference
Allergy/Poisoning/Toxicity			
Bovine milk/Pork	Staphylococcal food poisoning	Genotyping (Clondiag/TMB)	Johler et al., 2011 [141]
Cheese	Staphylococcus aureus poisoning	Genotyping (Alere/TMB)	Johler et al., 2015 [142]
Cheese/Fish/Meat, etc.	Staphylococcus aureus poisoning	Genotyping (Clondiag/TMB)	Baumgartner et al., 2014 [143]
Citrinin	Mycotoxin toxicity	Gene expression (Custom/Cy3, Cy5)	Iwahashi et al., 2007 [144]
Food	Bacillus cereus poisoning	Genotyping (Custom, E)	Liu et al., 2007 [145]
Food	Coagulase-negative staphylococci	Genotyping (Custom/Cy5)	Seitter et al., 2011 [146]
Food	69 Salmonella virulence genes	Genotyping (Custom/Cy3)	Zou et al., 2011 [147]
Food	Salmonella serogroups	Genotyping (Custom, C/SG)	Braun et al., 2012 [148]
Food	Clostridium perfringens poisoning	Genotyping (Custom/Cy3, Cy5)	Lahti et al., 2012 [149]
Food	Allergen-specific response	Gene expression (Affymetrix/PE)	Martino et al., 2012 [150]
Food	Staphylococcal food poisoning	Genotyping (Clondiag/TMB)	Wattinger et al., 2012 [151]
Food	Silver-nanoparticle-induced genotoxicity	Gene expression (Agilent/Cy3, Cy5)	Xu et al., 2012 [152]
Food	46 Salmonella O serogroups	Genotyping (Custom/Cy3)	Guo et al., 2013 [153]
Food	Campylobacter pathotypes	Genotyping (Custom/Cy3)	Marotta et al., 2013 [154]
Food	Botulinum neurotoxin poisoning	Genotyping (Custom/PE)	Vanhomwegen et al., 2013 [155]
Food	117 antibiotic resistance genes	Genotyping (Custom, C/True Blue)	Strauss et al., 2015 [156]
Food additive	Toxicity in liver	Gene expression (Custom/Cy3, Cy5)	Stierum et al., 2008 [157]
Horseradish	Quorum sensing inhibitors	Gene expression (Custom/PE)	Jakobsen et al., 2012 [158]
Meat	Shiga toxin-producing Escherichia coli	Genotyping (GeneSystems/6-FAM)	Miko et al., 2014 [159]
Meat	Cephalosporin-resistant Escherichia coli	Genotyping (Alere/TMB)	Vogt et al., 2014 [160]
Meat/Milk	Coagulase-negative staphylococci	Genotyping (Custom/Cy3, Cy5)	Even et al., 2010 [161]
Pancake with chicken	Staphylococcus aureus poisoning	Genotyping (Clondiag/TMB)	Johler et al., 2013 [162]
Pork	Salmonella enterica pathogenicity genes	Genotyping (Custom/Alexa Fluor 555/647)	Hauser et al., 2011 [163]
Pufferfish	Tetrodotoxin accumulation	Gene expression (Custom/Cy3)	Feroudj et al., 2014 [164]
Rice	Cadmium toxicity	Gene expression (Custom, C/CSPD)	Zhang et al., 2012 [165]
Vegetable	Latex and/or vegetable food allergy	Gene expression (Affymetrix/PE)	Saulnier et al., 2014 [166]
Contamination			
Alfalfa/Cilantro/Mung bean, etc.	Detection of Yersinia enterocolitica	Genotyping (Custom/Cy3, Cy5)	Siddique et al., 2009 [167]
Beef	Pathogenic Escherichia coli O157	Gene expression (Custom/Cy3, Cy5)	Fratamico et al., 2011 [168]
Beer	Beer spoilage bacterial contamination	Beer spoilage bacterial contamination	Weber et al., 2008 [169]
Beef/Egg/Fish/Milk	26 probes for pathogenic bacteria	Genotyping (Custom/Cy3)	Wang et al., 2007 [170]
Bread (Whole-grain/Fiber-rich)	Intestinal microbiota composition	Genotyping (Agilent/Cy3, Cy5)	Lappi et al., 2013 [171]
Cantaloupe	24 probes for Listeria monocytogenes	Genotyping (Affymetrix/PE)	Laksanalamai et al., 2012 [172]

Table 3. *Cont.*

Food Source or Material	Material Detected or Subject Examined	Type of Microarray Used (Source [a]/Dye)	Reference
Chicken	Rapid analysis of pathogenic bacteria	Genotyping (Custom/Cy3, Cy5)	Quiñones et al., 2007 [173]
Chicken/Pork	Salmonella enterica probes	Genotyping (Custom/Alexa Fluor 555/647)	Hauser et al., 2012 [174]
Compost/Digestate/Waste	Microbial community	Genotyping (Custom/Cy3, Cy5)	Franke-Whittle et al., 2014 [175]
Egg/Meat/Milk, etc.	Listeria spp. contamination	Genotyping (Custom/Cy5)	Hmaied et al., 2014 [176]
Egg/Meat/Milk/Rice, etc.	16S rRNA probes for pathogens	Genotyping (Custom/Alexa Fluor 647)	Hwang et al., 2012 [177]
Food	250 probes for pathogenic bacteria	Genotyping (Custom/Cy3)	Kim et al., 2008 [178]
Food	Rapid analysis of pathogenic bacteria	Genotyping (Custom/Cy3)	Kim et al., 2010 [179]
Food	Rapid analysis of pathogenic bacteria	Genotyping (Custom, C/Luminol)	Donhauser et al., 2011 [180]
Food	Yersinia pestis/Bacillus anthracis	Genotyping (Custom/Alexa Fluor 555/647)	Goji et al., 2012 [181]
Food	50 probes for pathogenic bacteria	Genotyping (Custom/Cy3)	Lee et al., 2011 [182]
Food	Diversity of Arcobacter butzleri	Genotyping (Custom/Cy3, Cy5)	Merga et al., 2013 [183]
Food	Pathogenic Escherichia coli/Salmonella	Genotyping (Custom/Cy3)	Fischer et al., 2014 [184]
Food/Water	63 probes for pathogenic bacteria	Genotyping (Alere/TMB)	Kostić et al., 2010 [185]
Juice	Alicyclobacillus spp. contamination	Genotyping (Custom/TAMRA)	Jang et al., 2011 [186]
Maize	96 probes for mycotoxigenic fungi	Genotyping (Custom/Cy3, Cy5)	Lezar & Barros, 2010 [187]
Meat	Rapid analysis of pathogenic bacteria	Genotyping (Custom/Cy3, Cy5)	Suo et al., 2010 [188]
Meat (Ready-to-eat)	Listeria monocytogenes contamination	Genotyping (Custom/Cy3)	Bae et al., 2011 [189]
Potato	DNA/RNA pathogens	Gene expression (PFRGC/Cy3, Cy5)	Dobnik et al., 2014 [190]
Poultry meat	102 pathogenicity genes	Genotyping (Custom, C/SG)	Toboldt et al., 2014 [191]
Sausage (Thai Nham)	164 probes for lactobacilli	Genotyping (Custom/Alexa Fluor 555/647)	Rungrassamee et al., 2012 [192]
Water	26 probes for pathogenic bacteria	Genotyping (Custom/Cy3, Cy5)	Zhou et al., 2011 [193]
Water	Pathogenic Legionella spp.	Genotyping (Custom/Cy3)	Cao et al., 2014 [194]
Efficacy/Mechanism			
Beverage/Dairy/Food	Interaction between yeast and bacteria	Gene expression (Affymetrix/PE)	Mendes et al., 2013 [195]
Cassava	Drought stress response	Gene expression (Custom/Cy3)	Utsumi et al., 2012 [196]
Chitooligosaccharide	Immune responses in adipocytes	Gene expression (Illumina/NS)	Choi et al., 2012 [197]
Food	Metabolic change in white blood cells	Gene expression (Affymetrix/PE)	Kawakami et al., 2013 [198]
Food (High-cholesterol diet)	Osteoporosis risk	Gene expression (Affymetrix/PE)	You et al., 2011 [199]
Food (High-fat diet)	Inflammation-associated genes	Gene expression (Illumina/NS)	Ding et al., 2014 [200]
Herb (Hochuekkito)	Mucosal IgA antibody response	Gene expression (Custom/Cy3)	Matsumoto et al., 2010 [201]
Herb (Licorice)	Estrogen-like effect	Gene expression (Custom/ Cy3, Cy5)	Dong et al., 2007 [202]
Imbibed soybean	New protein food item	Gene expression (Custom/Cy3)	Tamura et al., 2014 [203]
Phenolic preservative	Oxidative stress/DNA damage	Gene expression (Custom/Cy3, Cy5)	Martin et al., 2014 [204]

Table 3. *Cont.*

Food Source or Material	Material Detected or Subject Examined	Type of Microarray Used (Source [a]/Dye)	Reference
Pineapple (*Ananas comosus*)	Absorption of phenolic acid	Gene expression (Custom/Cy3, Cy5)	Dang & Zhu, 2015 [205]
Polyunsaturated fatty acid, *etc.*	Growth and metabolic status of rats	Gene expression (Illumina/Cy3)	Castañeda-Gutiérrez *et al.*, 2014 [206]
Psyllium	Lipid consumption in skeletal muscle	Gene expression (Mitsubishi/Cy5)	Togawa *et al.*, 2013 [207]
Quercetin	Improvement of diabetic symptoms	Gene expression (Affymetrix/PE)	Kobori *et al.*, 2009 [208]
Skim milk	Survival of *L. monocytogenes*	Gene expression (Custom/Alexa Fluor 555/647)	Liu & Ream, 2008 [209]
Sweet corn	Effect of suppressing cancer	Gene expression (GE Healthcare/Cy5)	Tokuji *et al.*, 2009 [210]
Tea (*Eucommia ulmoides*)	Plasma triglyceride-lowering effect	Gene expression (Agilent/Cy3)	Kobayashi *et al.*, 2012 [211]
Xanthan gum	*Xanthomonas arboricola* metabolism	Genotyping (Custom/Cy3, Cy5)	Mayer *et al.*, 2011 [212]
Quality Control/Safety			
Canola	Genetically modified organism	Genotyping (Custom/Cy3)	Schmidt *et al.*, 2008 [213]
Canola/Cotton/Maize/Soybean	Genetically modified organism	Genotyping (Custom/Cy5)	Kim *et al.*, 2010 [214]
Cereal (Barley/Oat/Rice/Wheat)	Authenticity of plant	Genotyping (Custom/Fluorescein)	Rønning *et al.*, 2005 [215]
Crop	Authenticity of food	Genotyping (Custom/Cy3)	Voorhuijzen *et al.*, 2012 [216]
Ginseng	Food adulteration	Genotyping (Custom/Cy3)	Niu *et al.*, 2011 [217]
Kothala himbutu (Medicinal plant)	Food safety assessment	Genotyping (Affymetrix/PE)	Im *et al.*, 2008 [218]
Maize/Potato	Food safety assessment	Gene expression	van Dijk *et al.*, 2010 (review) [219]
Maize/Soybean, *etc.*	Genetically modified organism	Genotyping (Custom, C/Silverquant)	Leimanis *et al.*, 2006 [220]
Olive	Authenticity of plant	Genotyping (Custom/Cy3, Cy5)	Consolandi *et al.*, 2007 [221]
Potato	Food safety assessment	Gene expression (Custom/Cy3, Cy5)	van Dijk *et al.*, 2009 [222]
Royal jelly	Food safety assessment	Gene expression (Amersham/Cy5)	Kamakura *et al.*, 2005 [223]

[a] The sources of DNA microarrays are either custom arrays (Custom) or microarrays supplied by companies as follows: Agilent: Agilent Technologies, USA; Alere: Alere Technologies, Germany; Amersham/GE Healthcare: GE Healthcare, USA; Clondiag: Clondiag Chip Technologies (renamed as Alere Technologies), Germany; GeneSystems: GeneSystems, France; Mitsubishi: Mitsubishi Rayon, Japan; and PFRGC: Pathogen Functional Genomic Research Center, USA. The type of microarrays used includes fluorescent assays with indicated fluorescent dyes, and non-fluorescent assays (C: colorimetric/chemiluminescent assays; or E: assays with electric arrays) with indicated chromogenic dyes/chemiluminescent substrates. CSPD: chloro-5-substituted adamantyl-1,2-dioxetane phosphate; 6-FAM: 6-carboxyfluorescein; NS: not specified; PE: phycoerythrin; SG: Seramun Green; TAMRA: carboxytetramethylrhodamine; and TMB: 3,3',5,5'-tetramethylbenzidine.

3.3. Merits of DNA Microarray Assay

Applications and potentials of DNA microarray technologies (e.g., DNA, cDNA and oligonucleotide microarray assays) for the detection and identification of microbial pathogens, such as antibiotic resistance genes, virulence factors and strain subtypes, have been discussed by comparison with other DNA-based methods, including PCR [224–228]. While PCR-based methods are normally limited to the analysis of a single or a small number of pathogens, microarray technology can analyze a significant number of pathogens simultaneously, and thus it has potential for use in basic research and industrial applications, such as food safety assessment. Gene expression profiling by DNA microarray assay has an advantage of examining the expression of large numbers of genes in a single experiment, and thus has been widely used to analyze food samples and materials. DNA microarray technologies have also been applied to monitor genetically modified food [225,229] and traditional Chinese medicine [230,231], and to evaluate drug safety [232]. Degenkolbe *et al.* discussed how quality control was examined for the procedures in DNA microarray assay, such as mRNA preparation, cDNA synthesis, fluorescent dye-labeling, hybridization/imaging and data analysis, using plant leaf tissue as a source of mRNA [233].

While DNA microarray assay has been considered to be effective and sensitive for assaying microbial spoilage of food, it is expensive and requires technical expertise. Therefore, several alternative methods were developed to explore cost-effective but still high-throughput assay systems. Böhme *et al.* developed an efficient method for bacterial identification based on detection of the 16S rRNA gene by flow-through hybridization on membranes, coupled to ligation detection reaction, which may provide an alternative to a DNA microarray assay for the rapid, accurate and cost-effective identification of bacterial species in order to assess food quality and safety [234]. Atanasova and Druzhinina discussed the Phenotype MicroArray, which tests cell respiration as a reporter system to characterize the metabolism of food spoilage pathogens, including conidial fungi [235]. However, the number of probes in these alternative assays is generally up to 100, and DNA microarray assays would be more useful when the number is over 100.

One of the merits of a DNA microarray assay is that it provides information about pathway-based intracellular signaling, which is important to evaluate the efficacy and mechanism of action of food materials. For example, a variety of signaling pathways have been identified by DNA microarray assay for traditional herbal medicine, such as traditional Chinese medicine (TCM) and traditional Japanese medicine (Kampo), which are associated with effects on cell functions and diseases, such as anti-adipogenesis, anti-atherosclerosis, anti-carcinogenesis, anti-inflammation, apoptosis, chemoprevention, circulation disorder and neuroprotection [231]. For example, the mitogen-activated protein kinase (MAPK) signaling pathway was shown to be associated with the apoptotic effect of Inchin-ko-to (Kampo) [236] and the anti-carcinogenic effect of Juzen-taiho-to (Kampo) [237], while TGF-β1/Smad and IGF-1 signaling pathways were associated with the inhibitory effect of Kangxianling (TCM) on renal fibrosis [238] and the immune response against viral infection induced by VI-28 (TCM) [239], respectively. The pathways associated with environmental estrogens, such as phytoestrogens, which are also important food materials, include a variety of signaling pathways related to apoptosis, carcinogenesis, cell growth/proliferation, differentiation/development and inflammation [240]. Therefore, the information about pathway-based intracellular signaling provided by DNA microarray assays will add variability and sensitivity to the assay system.

4. Conclusions

We have here summarized recent progress in fluorescence-based bioassays developed and applied for the detection and evaluation of food materials. A comprehensive list of fluorescent dyes used in recent bioassays includes those in biosensing, cell assay, energy transfer-based assay, probing, protein/immunological assay and microarray/biochip assay. Among these technologies, fluorescence-based microarrays/biochips, such as antibody/protein microarrays, bead/suspension arrays, capillary/sensor arrays, DNA microarrays/PCR-based arrays, glycan/lectin arrays, immunoassay/ELISA-based arrays, microfluidic chips and tissue arrays, have been developed and used widely for food safety and quality as well as the search for effective components. Applications of DNA microarray assay were discussed for important issues, such as allergy/poisoning/toxicity, contamination, efficacy/mechanism and quality control/safety, based on a comprehensive list of references showing these cases. The merits of DNA microarray assays were discussed by pointing to their advantages over other technologies in terms of features such as the sensitivity and efficiency, the number of probes to be analyzed rapidly and simultaneously, and the quality and quantity of information about pathway-based intracellular signaling in response to food materials.

Acknowledgments: This research was supported partly by a Knowledge Cluster Initiative program and a Grant-in-aid for Basic Areas from the Ministry of Education, Culture, Sports, Science and Technology of Japan, and by a grant from Kyushu Sangyo University for supporting research and the development of new fluorescent dyes.

Author Contributions: Shin-Ichiro Isobe and Ryoiti Kiyama made the outline; Kentaro Nishi, Shin-Ichiro Isobe, Yun Zhu and Ryoiti Kiyama wrote the paper.

Conflicts of Interest: The authors declare no conflict of interest.

References

1. Valeur, B.; Berberan-Santos, M.N. *Molecular Fluorescence: Principles and Applications*, 2nd ed.; Wiley-VCH: Weinheim, Germany, 2013.

2. Grimm, J.B.; Heckman, L.M.; Lavis, L.D. The Chemistry of Small-Molecule Fluorogenic Probes. In *Fluorescence-Based Biosensors, Volume 113: From Concepts to Applications (Progress in Molecular Biology and Translational Science)*; Morris, M.C., Ed.; Academic Press: Oxford, UK, 2013; pp. 1–34.

3. Yu, W.; So, P.T.; French, T.; Gratton, E. Fluorescence Generalized Polarization of Cell Membranes: A Two-Photon Scanning Microscopy Approach. *Biophys. J.* **1996**, *70*, 626–636.

4. Parasassi, T.; Gratton, E.; Yu, W.M.; Wilson, P.; Levi, M. Two-photon fluorescence microscopy of laurdan generalized polarization domains in model and natural membranes. *Biophys. J.* **1997**, *72*, 2413–2429.

5. Bagatolli, L.A. To see or not to see: Lateral organization of biological membranes and fluorescence microscopy. *Biochim. Biophys. Acta* **2006**, *1758*, 1541–1556.

6. Johnson, I.; Spence, M.T.Z. *The Molecular Probes Handbook: A Guide to Fluorescent Probes and Labeling Technologies*, 11th ed.; Life Technologies, Inc.: Carlsbad, CA, USA, 2010.

7. Beppu, T.; Tomiguchi, K.; Masuhara, A.; Pu, Y.J.; Katagiri, H. Single Benzene Green Fluorophore: Solid-State Emissive, Water-Soluble, and Solvent- and pH-Independent Fluorescence with Large Stokes Shifts. *Angew Chem. Int. Ed. Engl.* **2015**, *54*, 7332–7335.

8. Alberts, B.; Johnson, A.; Lewis, J.; Raff, M.; Roberts, K.; Walter, P. *Molecular Biology of the Cell*, 5th ed.; Garland Science: New York, NY, USA, 2007.

9. Herzenberg, L.A.; Sweet, R.G.; Herzenberg, L.A. Fluorescence-activated cell sorting. *Sci. Am.* **1976**, *234*, 108–117.

10. Connally, R.E. Flow Cytometry. In *Fluorescence Applications in Biotechnology and Life Sciences*; Goldys, E.M., Ed.; Wiley-Blackwell: Hoboken, NJ, USA, 2009; pp. 245–268.

11. Volpi, E.V.; Bridger, J.M. FISH glossary: An overview of the fluorescence *in situ* hybridization technique. *Biotechniques* **2008**, *45*, 385–409.

12. Bailey, S.M.; Goodwin, E.H.; Cornforth, M.N. Strand-specific fluorescence *in situ* hybridization: the CO-FISH family. *Cytogenet. Genome Res.* **2004**, *107*, 14–17.

13. Kerppola, T.K. Design and implementation of bimolecular fluorescence complementation (BiFC) assays for the visualization of protein interactions in living cells. *Nat. Protoc.* **2006**, *1*, 1278–1286.

14. Kodama, Y.; Hu, C.D. Bimolecular fluorescence complementation (BiFC): A 5-year update and future perspectives. *Biotechniques* **2012**, *53*, 285–298.

15. Jaiswal, J.K.; Simon, S.M. Potentials and pitfalls of fluorescent quantum dots for biological imaging. *Trends Cell Biol.* **2004**, *14*, 497–504.

16. Yuan, J.; Wang, G. Lanthanide complex-based fluorescence label for time-resolved fluorescence bioassay. *J. Fluoresc.* **2005**, *15*, 559–568.

17. Duan, Y.; Liu, M.; Sun, W.; Wang, M.; Liu, S.; Li, Q.X. Recent Progress on Synthesis of Fluorescein Probes. *Mini-Rev. Org. Chem.* **2009**, *6*, 35–43.

18. DeBiasio, R.; Bright, G.R.; Ernst, L.A.; Waggoner, A.S.; Taylor, D.L. Five-parameter fluorescence imaging: Wound healing of living Swiss 3T3 cells. *J. Cell Biol.* **1987**, *105*, 1613–1622.

19. DeRisi, J.L.; Iyer, V.R.; Brown, P.O. Exploring the metabolic and genetic control of gene expression on a genomic scale. *Science* **1997**, *278*, 680–686.

20. Maurel, D.; Banala, S.; Laroche, T.; Johnsson, K. Photoactivatable and photoconvertible fluorescent probes for protein labeling. *ACS Chem. Biol.* **2010**, *5*, 507–516.

21. Panchuk-Voloshina, N.; Haugland, R.P.; Bishop-Stewart, J.; Bhalgat, M.K.; Millard, P.J.; Mao, F.; Leung, W.Y.; Haugland, R.P. Alexa dyes, a series of new fluorescent dyes that yield exceptionally bright, photostable conjugates. *J. Histochem. Cytochem.* **1999**, *47*, 1179–1188.

22. Ormö, M.; Cubitt, A.B.; Kallio, K.; Gross, L.A.; Tsien, R.Y.; Remington, S.J. Crystal structure of the Aequorea victoria green fluorescent protein. *Science* **1996**, *273*, 1392–1395.

23. Heim, R.; Tsien, R.Y. Engineering green fluorescent protein for improved brightness, longer wavelengths and fluorescence resonance energy transfer. *Curr. Biol.* **1996**, *6*, 178–182.

24. Pollok, B.A.; Heim, R. Using GFP in FRET-based applications. *Trends Cell. Biol.* **1999**, *9*, 57–60.

25. Kanemaru, T.; Hirata, K.; Takasu, S.; Isobe, S.; Mizuki, K.; Mataka, S.; Nakamura, K. A fluorescence scanning electron microscope. *Ultramicroscopy* **2009**, *109*, 344–349.

26. Zhu, Y.; Ogaeri, T.; Suzuki, J.; Dong, S.; Aoyagi, T.; Mizuki, K.; Takasugi, M.; Isobe, S.; Kiyama, R. Application of Fluolid-Orange-labeled probes for DNA microarray and immunological assays. *Biotechnol. Lett.* **2011**, *33*, 1759–1766.

27. Wuxiuer, D.; Zhu, Y.; Ogaeri, T.; Mizuki, K.; Kashiwa, Y.; Nishi, K.; Isobe, S.; Aoyagi, T.; Kiyama, R. Development of pathological diagnostics of human kidney cancer by multiple staining using new fluorescent Fluolid dyes. *Biomed. Res. Int.* **2014**, *2014*, 437871.

28. Schena, M. *Microarray Analysis*; Wiley-Liss: Hoboken, NJ, USA, 2002.

29. Jones, L.J.; Yue, S.T.; Cheung, C.Y.; Singer, V.L. RNA quantitation by fluorescence-based solution assay: RiboGreen reagent characterization. *Anal. Biochem.* **1998**, *265*, 368–374.

30. Titus, J.A.; Haugland, R.; Sharrow, S.O.; Segal, D.M. Texas Red, a hydrophilic, red-emitting fluorophore for use with fluorescein in dual parameter flow microfluorometric and fluorescence microscopic studies. *J. Immunol. Methods* **1982**, *50*, 193–204.

31. Jones, L.J.; Haugland, R.P.; Singer, V.L. Development and characterization of the NanoOrange protein quantitation assay: A fluorescence-based assay of proteins in solution. *Biotechniques* **2003**, *34*, 850–861.

32. Sobek, J.; Aquino, C.; Schlapbach, R. Analyzing Properties of Fluorescent Dyes Used for Labeling DNA in Microarray Experiments. *BioFiles* **2011**, *6*, 9–12.

33. Fare, T.L.; Coffey, E.M.; Dai, H.; He, Y.D.; Kessler, D.A.; Kilian, K.A.; Koch, J.E.; LeProust, E.; Marton, M.J.; Meyer, M.R.; *et al.* Effects of atmospheric ozone on microarray data quality. *Anal. Chem.* **2003**, *75*, 4672–4675.

34. Cox, W.G.; Beaudet, M.P.; Agnew, J.Y.; Ruth, J.L. Possible sources of dye-related signal correlation bias in two-color DNA microarray assays. *Anal. Biochem.* **2004**, *331*, 243–254.

35. Wang, L.; Lofton, C.; Popp, M.; Tan, W. Using luminescent nanoparticles as staining probes for Affymetrix GeneChips. *Bioconjug. Chem.* **2007**, *18*, 610–613.

36. Okuzaki, D.; Fukushima, T.; Tougan, T.; Ishii, T.; Kobayashi, S.; Yoshizaki, K.; Akita, T.; Nojima, H. Genopal™: A novel hollow fibre array for focused microarray analysis. *DNA Res.* **2010**, *17*, 369–379.

37. GE Healthcare. Amersham HyPer5 Dye. Data File 28-9299-03 AA. Available online: http://www.gelifesciences. co.jp/catalog/pdf/hyper5_df.pdf (accessed on 12 October 2015).

38. Harper, I.S. Labeling of Cells with Fluorescent Dyes. In *Fluorescence Applications in Biotechnology and Life Sciences*; Goldys, E.M., Ed.; Wiley-Blackwell: Hoboken, NJ, USA, 2009; pp. 27–45.

39. Ohba, Y.; Fujioka, Y.; Nakada, S.; Tsuda, M. Fluorescent Protein-Based Biosensors and Their Clinical Applications. In *Fluorescence-Based Biosensors, Volume 113: From Concepts to Applications (Progress in Molecular Biology and Translational Science)*; Morris, M.C., Ed.; Academic Press: Oxford, UK, 2013; pp. 313–348.

40. Altschuh, D.; Oncul, S.; Demchenko, A.P. Fluorescence sensing of intermolecular interactions and development of direct molecular biosensors. *J. Mol. Recognit.* **2006**, *19*, 459–477.

41. Jobgen, W.S.; Jobgen, S.C.; Li, H.; Meininger, C.J.; Wu, G. Analysis of nitrite and nitrate in biological samples using high-performance liquid chromatography. *J. Chromatogr. B Anal. Technol. Biomed. Life Sci.* **2007**, *851*, 71–82.

42. Danielli, A.; Porat, N.; Ehrlich, M.; Arie, A. Magnetic modulation biosensing for rapid and homogeneous detection of biological targets at low concentrations. *Curr. Pharm. Biotechnol.* **2010**, *11*, 128–137.

43. Vignali, D.A. Multiplexed particle-based flow cytometric assays. *J. Immunol. Methods* **2000**, *243*, 243–255.

44. Edwards, B.S.; Oprea, T.; Prossnitz, E.R.; Sklar, L.A. Flow cytometry for high-throughput, high-content screening. *Curr. Opin. Chem. Biol.* **2004**, *8*, 392–398.

45. White, N.S.; Errington, R.J. Fluorescence techniques for drug delivery research: theory and practice. *Adv. Drug Deliv. Rev.* **2005**, *57*, 17–42.

46. Wolff, M.; Kredel, S.; Wiedenmann, J.; Nienhaus, G.U.; Heilker, R. Cell-based assays in practice: Cell markers from autofluorescent proteins of the GFP-family. *Comb. Chem. High Throughput Screen.* **2008**, *11*, 602–609.

47. Hanson, G.T.; Hanson, B.J. Fluorescent probes for cellular assays. *Comb. Chem. High Throughput Screen.* **2008**, *11*, 505–513.

48. Ulrich, H.; Martins, A.H.; Pesquero, J.B. RNA and DNA aptamers in cytomics analysis. *Cytometry A* **2004**, *59*, 220–231.

49. Svobodová, K.; Cajthaml, T. New *in vitro* reporter gene bioassays for screening of hormonal active compounds in the environment. *Appl. Microbiol. Biotechnol.* **2010**, *88*, 839–847.

50. Mullassery, D.; Horton, C.A.; Wood, C.D.; White, M.R. Single live-cell imaging for systems biology. *Essays Biochem.* **2008**, *45*, 121–133.

51. Salipalli, S.; Singh, P.K.; Borlak, J. Recent advances in live cell imaging of hepatoma cells. *BMC Cell Biol.* **2014**, *15*, 26.

52. Szöllosi, J.; Damjanovich, S.; Mátyus, L. Application of fluorescence resonance energy transfer in the clinical laboratory: routine and research. *Cytometry* **1998**, *34*, 159–179.

53. Léonard, A.; Rueff, J.; Gerber, G.B.; Léonard, E.D. Usefulness and limits of biological dosimetry based on cytogenetic methods. *Radiat. Prot. Dosimetry* **2005**, *115*, 448–454.

54. Lokeshwar, V.B.; Selzer, M.G. Urinary bladder tumor markers. *Urol. Oncol.* **2006**, *24*, 528–537.

55. Chun, H.; Lee, D.S.; Kim, H.C. Bio-cell chip fabrication and applications. *Methods Mol. Biol.* **2009**, *509*, 145–158.

56. Falconer, E.; Lansdorp, P.M. Strand-seq: A unifying tool for studies of chromosome segregation. *Semin. Cell Dev. Biol.* **2013**, *24*, 643–652.

57. Apáti, Á.; Pászty, K.; Erdei, Z.; Szebényi, K.; Homolya, L.; Sarkadi, B. Calcium signaling in pluripotent stem cells. *Mol. Cell. Endocrinol.* **2012**, *353*, 57–67.

58. Brunelle, J.K.; Zhang, B. Apoptosis assays for quantifying the bioactivity of anticancer drug products. *Drug Resist. Updates* **2010**, *13*, 172–179.

59. Marras, S.A.; Tyagi, S.; Kramer, F.R. Real-time assays with molecular beacons and other fluorescent nucleic acid hybridization probes. *Clin. Chim. Acta* **2006**, *363*, 48–60.

60. Sokolova, V.; Epple, M. Synthetic pathways to make nanoparticles fluorescent. *Nanoscale* **2011**, *3*, 1957–1962.

61. Kricka, L.J.; Fortina, P. Analytical ancestry: "Firsts" in fluorescent labeling of nucleosides, nucleotides, and nucleic acids. *Clin. Chem.* **2009**, *55*, 670–683.

62. Musa-Aziz, R.; Boron, W.F.; Parker, M.D. Using fluorometry and ion-sensitive microelectrodes to study the functional expression of heterologously-expressed ion channels and transporters in Xenopus oocytes. *Methods* **2010**, *51*, 134–145.

63. Sánchez, J.G.; Kouznetsov, V.V. Antimycobacterial susceptibility testing methods for natural products research. *Braz. J. Microbiol.* **2010**, *41*, 270–277.

64. Jiang, T.; Xing, B.; Rao, J. Recent developments of biological reporter technology for detecting gene expression. *Biotechnol. Genet. Eng. Rev.* **2008**, *25*, 41–75.

65. Jahnz, M.; Schwille, P. Enzyme assays for confocal single molecule spectroscopy. *Curr. Pharm. Biotechnol.* **2004**, *5*, 221–229.

66. Liu, T.; Liu, B.; Zhang, H.; Wang, Y. The fluorescence bioassay platforms on quantum dots nanoparticles. *J. Fluoresc.* **2005**, *15*, 729–733.

67. Medintz, I.L.; Uyeda, H.T.; Goldman, E.R.; Mattoussi, H. Quantum dot bioconjugates for imaging, labelling and sensing. *Nat. Mater.* **2005**, *4*, 435–446.

68. Zhang, Y.; Wang, T.H. Quantum dot enabled molecular sensing and diagnostics. *Theranostics* **2012**, *2*, 631–654.

69. Akinfieva, O.; Nabiev, I.; Sukhanova, A. New directions in quantum dot-based cytometry detection of cancer serum markers and tumor cells. *Crit. Rev. Oncol. Hematol.* **2013**, *86*, 1–14.

70. Arterburn, J.B.; Oprea, T.I.; Prossnitz, E.R.; Edwards, B.S.; Sklar, L.A. Discovery of selective probes and antagonists for G-protein-coupled receptors FPR/FPRL1 and GPR30. *Curr. Top. Med. Chem.* **2009**, *9*, 1227–1236.

71. Maghzal, G.J.; Krause, K.H.; Stocker, R.; Jaquet, V. Detection of reactive oxygen species derived from the family of NOX NADPH oxidases. *Free Radic. Biol. Med.* **2012**, *53*, 1903–1918.

72. Horobin, R.W.; Stockert, J.C.; Rashid-Doubell, F. Uptake and localisation of small-molecule fluorescent probes in living cells: A critical appraisal of QSAR models and a case study concerning probes for DNA and RNA. *Histochem. Cell Biol.* **2013**, *139*, 623–637.

73. Vernall, A.J.; Hill, S.J.; Kellam, B. The evolving small-molecule fluorescent-conjugate toolbox for Class A GPCRs. *Br. J. Pharmacol.* **2014**, *171*, 1073–1084.

74. Kerppola, T.K. Visualization of molecular interactions using bimolecular fluorescence complementation analysis: Characteristics of protein fragment complementation. *Chem. Soc. Rev.* **2009**, *38*, 2876–2886.

75. Miller, K.E.; Kim, Y.; Huh, W.K.; Park, H.O. Bimolecular fluorescence complementation (BiFC) analysis: Advances and recent applications for genome-wide interaction studies. *J. Mol. Biol.* **2015**, *427*, 2039–2055.

76. Wu, D.; Sylvester, J.E.; Parker, L.L.; Zhou, G.; Kron, S.J. Peptide reporters of kinase activity in whole cell lysates. *Biopolymers* **2010**, *94*, 475–486.

77. Noble, J.E.; Bailey, M.J. Quantitation of protein. *Methods Enzymol.* **2009**, *463*, 73–95.

78. Freudenberg, J.A.; Bembas, K.; Greene, M.I.; Zhang, H. Non-invasive, ultra-sensitive, high-throughput assays to quantify rare biomarkers in the blood. *Methods* **2008**, *46*, 33–38.

79. Guo, H.; Sun, S. Lanthanide-doped upconverting phosphors for bioassay and therapy. *Nanoscale* **2012**, *4*, 6692–6706.

80. Sakudo, A.; Nakamura, I.; Ikuta, K.; Onodera, T. Recent developments in prion disease research: diagnostic tools and *in vitro* cell culture models. *J. Vet. Med. Sci.* **2007**, *69*, 329–337.

81. Barletta, J. Applications of real-time immuno-polymerase chain reaction (rt-IPCR) for the rapid diagnoses of viral antigens and pathologic proteins. *Mol. Aspects Med.* **2006**, *27*, 224–253.

82. Ligler, F.S.; Taitt, C.R.; Shriver-Lake, L.C.; Sapsford, K.E.; Shubin, Y.; Golden, J.P. Array biosensor for detection of toxins. *Anal. Bioanal. Chem.* **2003**, *377*, 469–477.

83. Bovee, T.F.H.; Heskamp, H.H.; Hamers, A.R.M.; Hoogenboom, L.A.P.; Nielen, M.W.F. Validation of a rapid yeast estrogen bioassay, based on the expression of green fluorescent protein, for the screening of estrogenic activity in calf urine. *Anal. Chim. Acta* **2005**, *529*, 57–64.

84. Bovee, T.F.; Bor, G.; Heskamp, H.H.; Hoogenboom, R.L.; Nielen, M.W. Validation and application of a robust yeast estrogen bioassay for the screening of estrogenic activity in animal feed. *Food Addit. Contam.* **2006**, *23*, 556–568.

85. Nielen, M.W.; Bovee, T.F.; van Engelen, M.C.; Rutgers, P.; Hamers, A.R.; van Rhijn, J.H.; Hoogenboom, L.R. Urine testing for designer steroids by liquid chromatography with androgen bioassay detection and electrospray quadrupole time-of-flight mass spectrometry identification. *Anal. Chem.* **2006**, *78*, 424–431.

86. Rijk, J.C.; Bovee, T.F.; Wang, S.; van Poucke, C.; van Peteghem, C.; Nielen, M.W. Detection of anabolic steroids in dietary supplements: The added value of an androgen yeast bioassay in parallel with a liquid chromatography-tandem mass spectrometry screening method. *Anal. Chim. Acta* **2009**, *637*, 305–314.

87. Toorians, A.W.; Bovee, T.F.; de Rooy, J.; Stolker, L.A.; Hoogenboom, R.L. Gynaecomastia linked to the intake of a herbal supplement fortified with diethylstilbestrol. *Food Addit. Contam. Part A* **2010**, *27*, 917–925.

88. Simons, R.; Vincken, J.P.; Roidos, N.; Bovee, T.F.; van Iersel, M.; Verbruggen, M.A.; Gruppen, H. Increasing soy isoflavonoid content and diversity by simultaneous malting and challenging by a fungus to modulate estrogenicity. *J. Agric. Food Chem.* **2011**, *59*, 6748–6758.

89. Rijk, J.C.; Ashwin, H.; van Kuijk, S.J.; Groot, M.J.; Heskamp, H.H.; Bovee, T.F.; Nielen, M.W. Bioassay based screening of steroid derivatives in animal feed and supplements. *Anal. Chim. Acta* **2011**, *700*, 183–188.

90. Hall, D.A.; Ptacek, J.; Snyder, M. Protein microarray technology. *Mech. Ageing Dev.* **2007**, *128*, 161–167.

91. Gehring, A.G.; Albin, D.M.; Reed, S.A.; Tu, S.I.; Brewster, J.D. An antibody microarray, in multiwell plate format, for multiplex screening of foodborne pathogenic bacteria and biomolecules. *Anal. Bioanal. Chem.* **2008**, *391*, 497–506.

92. Lian, W.; Wu, D.; Lim, D.V.; Jin, S. Sensitive detection of multiplex toxins using antibody microarray. *Anal. Biochem.* **2010**, *401*, 271–279.

93. Pauly, D.; Kirchner, S.; Stoermann, B.; Schreiber, T.; Kaulfuss, S.; Schade, R.; Zbinden, R.; Avondet, M.A.; Dorner, M.B.; Dorner, B.G. Simultaneous quantification of five bacterial and plant toxins from complex matrices using a multiplexed fluorescent magnetic suspension assay. *Analyst* **2009**, *134*, 2028–2039.

94. Sun, Y.; Xu, J.; Li, W.; Cao, B.; Wang, D.D.; Yang, Y.; Lin, Q.X.; Li, J.L.; Zheng, T.S. Simultaneous detection of ochratoxin A and fumonisin B1 in cereal samples using an aptamer-photonic crystal encoded suspension array. *Anal. Chem.* **2014**, *86*, 11797–11802.

95. Wang, X.; Mu, Z.; Shangguan, F.; Liu, R.; Pu, Y.; Yin, L. Rapid and sensitive suspension array for multiplex detection of organophosphorus pesticides and carbamate pesticides based on silica-hydrogel hybrid microbeads. *J. Hazard. Mater.* **2014**, *273*, 287–292.

96. Nolan, J.P.; Sklar, L.A. Suspension array technology: Evolution of the flat-array paradigm. *Trends Biotechnol.* **2002**, *20*, 9–12.

97. Niamnont, N.; Mungkarndee, R.; Techakriengkrai, I.; Rashatasakhon, P.; Sukwattanasinitt, M. Protein discrimination by fluorescent sensor array constituted of variously charged dendritic phenylene-ethynylene fluorophores. *Biosens. Bioelectron.* **2010**, *26*, 863–867.

98. Tan, J.; Li, R.; Jiang, Z.T. Discrimination of fresh fruit juices by a fluorescent sensor array for carboxylic acids based on molecularly imprinted titania. *Food Chem.* **2014**, *165*, 35–41.

99. Yu, Y.; Mishra, S.; Song, X.; Lasanajak, Y.; Bradley, K.C.; Tappert, M.M.; Air, G.M.; Steinhauer, D.A.; Halder, S.; Cotmore, S.; *et al.* Functional glycomic analysis of human milk glycans reveals the presence of virus receptors and embryonic stem cell biomarkers. *J. Biol. Chem.* **2012**, *287*, 44784–44799.

100. Wang, H.; Li, H.; Zhang, W.; Wei, L.; Yu, H.; Yang, P. Multiplex profiling of glycoproteins using a novel bead-based lectin array. *Proteomics* **2014**, *14*, 78–86.

101. Hu, W.; Li, X.; He, G.; Zhang, Z.; Zheng, X.; Li, P.; Li, C.M. Sensitive competitive immunoassay of multiple mycotoxins with non-fouling antigen microarray. *Biosens. Bioelectron.* **2013**, *50*, 338–344.

102. Peters, J.; Cardall, A.; Haasnoot, W.; Nielen, M.W. 6-Plex microsphere immunoassay with imaging planar array detection for mycotoxins in barley. *Analyst* **2014**, *139*, 3968–3976.

103. Priest, C. Surface patterning of bonded microfluidic channels. *Biomicrofluidics* **2010**, *30*, 32206.

104. Sakamoto, C.; Yamaguchi, N.; Nasu, M. Rapid and simple quantification of bacterial cells by using a microfluidic device. *Appl. Environ. Microbiol.* **2005**, *71*, 1117–1121.

105. Ikeda, M.; Yamaguchi, N.; Tani, K.; Nasu, M. Rapid and simple detection of food poisoning bacteria by bead assay with a microfluidic chip-based system. *J. Microbiol. Methods* **2006**, *67*, 241–247.

106. Le, N.C.; Gel, M.; Zhu, Y.; Dacres, H.; Anderson, A.; Trowell, S.C. Real-time, continuous detection of maltose using bioluminescence resonance energy transfer (BRET) on a microfluidic system. *Biosens. Bioelectron.* **2014**, *62*, 177–181.

107. Chau, H.W.; Goh, Y.K.; Si, B.C.; Vujanovic, V. An innovative brilliant blue FCF method for fluorescent staining of fungi and bacteria. *Biotechnol. Histochem.* **2011**, *86*, 280–287.

108. Stroot, J.M.; Leach, K.M.; Stroot, P.G.; Lim, D.V. Capture antibody targeted fluorescence *in situ* hybridization (CAT-FISH): dual labeling allows for increased specificity in complex samples. *J. Microbiol. Methods* **2012**, *88*, 275–284.

109. Han, X.; Wang, H.; Chen, H.; Mei, L.; Wu, S.; Jia, G.; Cheng, T.; Zhu, S.; Lin, X. Development and primary application of a fluorescent liquid bead array for the simultaneous identification of multiple genetically modified maize. *Biosens. Bioelectron.* **2013**, *49*, 360–366.

110. Deng, G.; Xu, K.; Sun, Y.; Chen, Y.; Zheng, T.; Li, J. High sensitive immunoassay for multiplex mycotoxin detection with photonic crystal microsphere suspension array. *Anal. Chem.* **2013**, *85*, 2833–2840.

111. Khandurina, J.; Anderson, A.A.; Olson, N.A.; Stege, J.T.; Guttman, A. Large-scale carbohydrate analysis by capillary array electrophoresis: Part 2. Data normalization and quantification. *Electrophoresis* **2004**, *25*, 3122–3127.

112. Kong, H.; Volokhov, D.V.; George, J.; Ikonomi, P.; Chandler, D.; Anderson, C.; Chizhikov, V. Application of cell culture enrichment for improving the sensitivity of mycoplasma detection methods based on nucleic acid amplification technology (NAT). *Appl. Microbiol. Biotechnol.* **2007**, *77*, 223–232.

113. Zhu, S.; Fushimi, H.; Komatsu, K. Development of a DNA microarray for authentication of ginseng drugs based on 18S rRNA gene sequence. *J. Agric Food Chem.* **2008**, *56*, 3953–3959.

114. Otsuka, C.; Minami, I.; Oda, K. Hypoxia-inducible genes encoding small EF-hand proteins in rice and tomato. *Biosci. Biotechnol. Biochem.* **2010**, *74*, 2463–2469.

115. Reverter, A.; Henshall, J.M.; McCulloch, R.; Sasazaki, S.; Hawken, R.; Lehnert, S.A. Numerical analysis of intensity signals resulting from genotyping pooled DNA samples in beef cattle and broiler chicken. *J. Anim. Sci.* **2014**, *92*, 1874–1885.

116. Brunner, C.; Hoffmann, K.; Thiele, T.; Schedler, U.; Jehle, H.; Resch-Genger, U. Novel calibration tools and validation concepts for microarray-based platforms used in molecular diagnostics and food safety control. *Anal. Bioanal. Chem.* **2015**, *407*, 3181–3191.

117. Tang, W.; Coughlan, S.; Crane, E.; Beatty, M.; Duvick, J. The application of laser microdissection to in planta gene expression profiling of the maize anthracnose stalk rot fungus *Colletotrichum graminicola*. *Mol. Plant Microbe Interact.* **2006**, *19*, 1240–1250.

118. Bidzhieva, B.; Laassri, M.; Chumakov, K. MAPREC assay for quantitation of mutants in a recombinant flavivirus vaccine strain using near-infrared fluorescent dyes. *J. Virol. Methods* **2011**, *175*, 14–19.

119. Morisset, D.; Dobnik, D.; Hamels, S.; Zel, J.; Gruden, K. NAIMA: target amplification strategy allowing quantitative on-chip detection of GMOs. *Nucleic Acids Res.* **2008**, *36*, e118.

120. Huang, A.; Qiu, Z.; Jin, M.; Shen, Z.; Chen, Z.; Wang, X.; Li, J.W. High-throughput detection of food-borne pathogenic bacteria using oligonucleotide microarray with quantum dots as fluorescent labels. *Int. J. Food Microbiol.* **2014**, *185*, 27–32.

121. Abdullahi, I.; Gryshan, Y.; Rott, M. Amplification-free detection of grapevine viruses using an oligonucleotide microarray. *J. Virol. Methods* **2011**, *178*, 1–15.

122. Choi, S.H. Hexaplex PCR assay and liquid bead array for detection of stacked genetically modified cotton event 281-24-236 × 3006-210-23. *Anal. Bioanal. Chem.* **2011**, *401*, 647–655.

123. Panicker, G.; Call, D.R.; Krug, M.J.; Bej, A.K. Detection of pathogenic *Vibrio* spp. in shellfish by using multiplex PCR and DNA microarrays. *Appl. Environ. Microbiol.* **2004**, *70*, 7436–7444.

124. Chen, W.; Yu, S.; Zhang, C.; Zhang, J.; Shi, C.; Hu, Y.; Suo, B.; Cao, H.; Shi, X. Development of a single base extension-tag microarray for the detection of pathogenic *Vibrio* species in seafood. *Appl. Microbiol. Biotechnol.* **2011**, *89*, 1979–1990.

125. Germini, A.; Mezzelani, A.; Lesignoli, F.; Corradini, R.; Marchelli, R.; Bordoni, R.; Consolandi, C.; de Bellis, G. Detection of genetically modified soybean using peptide nucleic acids (PNAs) and microarray technology. *J. Agric. Food Chem.* **2004**, *52*, 4535–4540.

126. Ngundi, M.M.; Shriver-Lake, L.C.; Moore, M.H.; Lassman, M.E.; Ligler, F.S.; Taitt, C.R. Array biosensor for detection of ochratoxin A in cereals and beverages. *Anal. Chem.* **2005**, *77*, 148–154.

127. Herrmann, M.; Veres, T.; Tabrizian, M. Enzymatically-generated fluorescent detection in micro-channels with internal magnetic mixing for the development of parallel microfluidic ELISA. *Lab Chip* **2006**, *6*, 555–560.

128. Han, J.H.; Kim, H.J.; Sudheendra, L.; Gee, S.J.; Hammock, B.D.; Kennedy, I.M. Photonic crystal lab-on-a-chip for detecting staphylococcal enterotoxin B at low attomolar concentration. *Anal. Chem.* **2013**, *85*, 3104–3109.

129. Shriver-Lake, L.C.; Taitt, C.R.; Ligler, F.S. Applications of array biosensor for detection of food allergens. *J. AOAC Int.* **2004**, *87*, 1498–1502.

130. Ngundi, M.M.; Shriver-Lake, L.C.; Moore, M.H.; Ligler, F.S.; Taitt, C.R. Multiplexed detection of mycotoxins in foods with a regenerable array. *J. Food Prot.* **2006**, *69*, 3047–3051.

131. Weingart, O.G.; Gao, H.; Crevoisier, F.; Heitger, F.; Avondet, M.A.; Sigrist, H. A bioanalytical platform for simultaneous detection and quantification of biological toxins. *Sensors* **2012**, *12*, 2324–2339.

132. Ngundi, M.M.; Taitt, C.R. An array biosensor for detection of bacterial and toxic contaminants of foods. *Methods Mol. Biol.* **2006**, *345*, 53–68.

133. Zhang, X.; Liu, F.; Yan, R.; Xue, P.; Li, Y.; Chen, L.; Song, C.; Liu, C.; Jin, B.; Zhang, Z.; Yang, K. An ultrasensitive immunosensor array for determination of staphylococcal enterotoxin B. *Talanta* **2011**, *85*, 1070–1074.

134. Wang, Z.; Fan, Y.; Chen, J.; Guo, Y.; Wu, W.; He, Y.; Xu, L.; Fu, F. A microfluidic chip-based fluorescent biosensor for the sensitive and specific detection of label-free single-base mismatch via magnetic beads-based "sandwich" hybridization strategy. *Electrophoresis* **2013**, *34*, 2177–2184.

135. Law, J.W.; Ab Mutalib, N.S.; Chan, K.G.; Lee, L.H. Rapid methods for the detection of foodborne bacterial pathogens: principles, applications, advantages and limitations. *Front. Microbiol.* **2015**, *5*, 770.

136. Josefsen, M.H.; Bhunia, A.K.; Engvall, E.O.; Fachmann, M.S.; Hoorfar, J. Monitoring Campylobacter in the poultry production chain—From culture to genes and beyond. *J. Microbiol. Methods* **2015**, *112*, 118–125.

137. Gui, J.; Patel, I.R. Recent advances in molecular technologies and their application in pathogen detection in foods with particular reference to yersinia. *J. Pathog.* **2011**, *2011*, 310135.

138. Lauri, A.; Mariani, P.O. Potentials and limitations of molecular diagnostic methods in food safety. *Genes Nutr.* **2009**, *4*, 1–12.

139. Vassioukovitch, O.; Orsini, M.; Paparini, A.; Gianfranceschi, G.; Cattarini, O.; di Michele, P.; Montuori, E.; Vanini, G.C.; Romano Spica, V. Detection of metazoan species as a public health issue: simple methods for the validation of food safety and quality. *Biotechnol. Annu. Rev.* **2005**, *11*, 335–354.

140. Kiyama, R.; Zhu, Y. DNA microarray-based gene expression profiling of estrogenic chemicals. *Cell. Mol. Life Sci.* **2014**, *71*, 2065–2082.

141. Johler, S.; Layer, F.; Stephan, R. Comparison of virulence and antibiotic resistance genes of food poisoning outbreak isolates of *Staphylococcus aureus* with isolates obtained from bovine mastitis milk and pig carcasses. *J. Food Prot.* **2011**, *74*, 1852–1859.

142. Johler, S.; Weder, D.; Bridy, C.; Huguenin, M.C.; Robert, L.; Hummerjohann, J.; Stephan, R. Outbreak of staphylococcal food poisoning among children and staff at a Swiss boarding school due to soft cheese made from raw milk. *J. Dairy Sci.* **2015**, *9*, 2944–2948.

143. Baumgartner, A.; Niederhauser, I.; Johler, S. Virulence and resistance gene profiles of *Staphylococcus aureus* strains isolated from ready-to-eat foods. *J. Food Prot.* **2014**, *77*, 1232–1236.

144. Iwahashi, H.; Kitagawa, E.; Suzuki, Y.; Ueda, Y.; Ishizawa, Y.H.; Nobumasa, H.; Kuboki, Y.; Hosoda, H.; Iwahashi, Y. Evaluation of toxicity of the mycotoxin citrinin using yeast ORF DNA microarray and Oligo DNA microarray. *BMC Genomics* **2007**, *8*, 95.

145. Liu, Y.; Elsholz, B.; Enfors, S.O.; Gabig-Ciminska, M. Confirmative electric DNA array-based test for food poisoning *Bacillus cereus*. *J. Microbiol. Methods* **2007**, *70*, 55–64.

146. Seitter, M.; Nerz, C.; Rosenstein, R.; Götz, F.; Hertel, C. DNA microarray based detection of genes involved in safety and technologically relevant properties of food associated coagulase-negative staphylococci. *Int. J. Food Microbiol.* **2011**, *145*, 449–458.

147. Zou, W.; Al-Khaldi, S.F.; Branham, W.S.; Han, T.; Fuscoe, J.C.; Han, J.; Foley, S.L.; Xu, J.; Fang, H.; Cerniglia, C.E.; Nayak, R. Microarray analysis of virulence gene profiles in *Salmonella* serovars from food/food animal environment. *J. Infect. Dev. Ctries.* **2011**, *5*, 94–105.

148. Braun, S.D.; Ziegler, A.; Methner, U.; Slickers, P.; Keiling, S.; Monecke, S.; Ehricht, R. Fast DNA serotyping and antimicrobial resistance gene determination of *Salmonella enterica* with an oligonucleotide microarray-based assay. *PLoS ONE* **2012**, *7*, e46489.

149. Lahti, P.; Lindström, M.; Somervuo, P.; Heikinheimo, A.; Korkeala, H. Comparative genomic hybridization analysis shows different epidemiology of chromosomal and plasmid-borne cpe-carrying *Clostridium perfringens* type A. *PLoS ONE* **2012**, *7*, e46162.

150. Martino, D.J.; Bosco, A.; McKenna, K.L.; Hollams, E.; Mok, D.; Holt, P.G.; Prescott, S.L. T-cell activation genes differentially expressed at birth in CD^{4+} T-cells from children who develop IgE food allergy. *Allergy* **2012**, *67*, 191–200.

151. Wattinger, L.; Stephan, R.; Layer, F.; Johler, S. Comparison of *Staphylococcus aureus* isolates associated with food intoxication with isolates from human nasal carriers and human infections. *Eur. J. Clin. Microbiol. Infect. Dis.* **2012**, *31*, 455–464.

152. Xu, L.; Li, X.; Takemura, T.; Hanagata, N.; Wu, G.; Chou, L.L. Genotoxicity and molecular response of silver nanoparticle (NP)-based hydrogel. *J. Nanobiotechnology* **2012**, *10*, 16.

153. Guo, D.; Liu, B.; Liu, F.; Cao, B.; Chen, M.; Hao, X.; Feng, L.; Wang, L. Development of a DNA microarray for molecular identification of all 46 *Salmonella* O serogroups. *Appl. Environ. Microbiol.* **2013**, *79*, 3392–3399.

154. Marotta, F.; Zilli, K.; Tonelli, A.; Sacchini, L.; Alessiani, A.; Migliorati, G.; di Giannatale, E. Detection and genotyping of *Campylobacter jejuni* and *Campylobacter coli* by use of DNA oligonucleotide arrays. *Mol. Biotechnol.* **2013**, *53*, 182–188.

155. Vanhomwegen, J.; Berthet, N.; Mazuet, C.; Guigon, G.; Vallaeys, T.; Stamboliyska, R.; Dubois, P.; Kennedy, G.C.; Cole, S.T.; Caro, V.; *et al.* Application of high-density DNA resequencing microarray for detection and characterization of botulinum neurotoxin-producing clostridia. *PLoS ONE* **2013**, *8*, e67510.

156. Strauss, C.; Endimiani, A.; Perreten, V. A novel universal DNA labeling and amplification system for rapid microarray-based detection of 117 antibiotic resistance genes in Gram-positive bacteria. *J. Microbiol. Methods* **2015**, *108*, 25–30.

157. Stierum, R.; Conesa, A.; Heijne, W.; van Ommen, B.; Junker, K.; Scott, M.P.; Price, R.J.; Meredith, C.; Lake, B.G.; Groten, J. Transcriptome analysis provides new insights into liver changes induced in the rat upon dietary administration of the food additives butylated hydroxytoluene, curcumin, propyl gallate and thiabendazole. *Food Chem. Toxicol.* **2008**, *46*, 2616–2628.

158. Jakobsen, T.H.; Bragason, S.K.; Phipps, R.K.; Christensen, L.D.; van Gennip, M.; Alhede, M.; Skindersoe, M.; Larsen, T.O.; Høiby, N.; Bjarnsholt, T.; *et al.* Food as a source for quorum sensing inhibitors: iberin from horseradish revealed as a quorum sensing inhibitor of Pseudomonas aeruginosa. *Appl. Environ. Microbiol.* **2012**, *78*, 2410–2421.

159. Miko, A.; Rivas, M.; Bentancor, A.; Delannoy, S.; Fach, P.; Beutin, L. Emerging types of Shiga toxin-producing *E. coli* (STEC) O178 present in cattle, deer, and humans from Argentina and Germany. *Front. Cell. Infect. Microbiol.* **2014**, *4*, 78.

160. Vogt, D.; Overesch, G.; Endimiani, A.; Collaud, A.; Thomann, A.; Perreten, V. Occurrence and genetic characteristics of third-generation cephalosporin-resistant *Escherichia coli* in Swiss retail meat. *Microb. Drug Resist.* **2014**, *20*, 485–494.

161. Even, S.; Leroy, S.; Charlier, C.; Zakour, N.B.; Chacornac, J.P.; Lebert, I.; Jamet, E.; Desmonts, M.H.; Coton, E.; Pochet, S.; *et al.* Low occurrence of safety hazards in coagulase negative staphylococci isolated from fermented foodstuffs. *Int. J. Food Microbiol.* **2010**, *139*, 87–95.

162. Johler, S.; Tichaczek-Dischinger, P.S.; Rau, J.; Sihto, H.M.; Lehner, A.; Adam, M.; Stephan, R. Outbreak of Staphylococcal food poisoning due to SEA-producing *Staphylococcus aureus*. *Foodborne Pathog. Dis.* **2013**, *10*, 777–781.

163. Hauser, E.; Hebner, F.; Tietze, E.; Helmuth, R.; Junker, E.; Prager, R.; Schroeter, A.; Rabsch, W.; Fruth, A.; Malorny, B. Diversity of *Salmonella enterica* serovar Derby isolated from pig, pork and humans in Germany. *Int. J. Food Microbiol.* **2011**, *151*, 141–149.

164. Feroudj, H.; Matsumoto, T.; Kurosu, Y.; Kaneko, G.; Ushio, H.; Suzuki, K.; Kondo, H.; Hirono, I.; Nagashima, Y.; Akimoto, S.; *et al.* DNA microarray analysis on gene candidates possibly related to tetrodotoxin accumulation in pufferfish. *Toxicon* **2014**, *77*, 68–72.

165. Zhang, M.; Liu, X.; Yuan, L.; Wu, K.; Duan, J.; Wang, X.; Yang, L. Transcriptional profiling in cadmium-treated rice seedling roots using suppressive subtractive hybridization. *Plant Physiol. Biochem.* **2012**, *50*, 79–86.

166. Saulnier, N.; Nucera, E.; Altomonte, G.; Rizzi, A.; Pecora, V.; Aruanno, A.; Buonomo, A.; Gasbarrini, A.; Patriarca, G.; Schiavino, D. Gene expression profiling of patients with latex and/or vegetable food allergy. *Eur. Rev. Med. Pharmacol. Sci.* **2012**, *16*, 1197–1210.

167. Siddique, N.; Sharma, D.; Al-Khaldi, S.F. Detection of *Yersinia enterocolitica* in alfalfa, mung bean, cilantro, and mamey sapote (*Pouteria sapota*) food matrices using DNA microarray chip hybridization. *Curr. Microbiol.* **2009**, *59*, 233–239.

168. Fratamico, P.M.; Wang, S.; Yan, X.; Zhang, W.; Li, Y. Differential gene expression of *E. coli* O157:H7 in ground beef extract compared to tryptic soy broth. *J. Food Sci.* **2011**, *76*, 79–87.

169. Weber, D.G.; Sahm, K.; Polen, T.; Wendisch, V.F.; Antranikian, G. Oligonucleotide microarrays for the detection and identification of viable beer spoilage bacteria. *J. Appl. Microbiol.* **2008**, *105*, 951–962.

170. Wang, X.W.; Zhang, L.; Jin, L.Q.; Jin, M.; Shen, Z.Q.; An, S.; Chao, F.H.; Li, J.W. Development and application of an oligonucleotide microarray for the detection of food-borne bacterial pathogens. *Appl. Microbiol. Biotechnol.* **2007**, *76*, 225–233.

171. Lappi, J.; Salojärvi, J.; Kolehmainen, M.; Mykkänen, H.; Poutanen, K.; de Vos, W.M.; Salonen, A. Intake of whole-grain and fiber-rich rye bread *versus* refined wheat bread does not differentiate intestinal microbiota composition in Finnish adults with metabolic syndrome. *J. Nutr.* **2013**, *143*, 648–655.

172. Laksanalamai, P.; Joseph, L.A.; Silk, B.J.; Burall, L.S.; Tarr, C.L.; Gerner-Smidt, P.; Datta, A.R. Genomic characterization of *Listeria* monocytogenes strains involved in a multistate listeriosis outbreak associated with cantaloupe in US. *PLoS ONE* **2012**, *7*, e42448.

173. Quiñones, B.; Parker, C.T.; Janda, J.M., Jr.; Miller, W.G.; Mandrell, R.E. Detection and genotyping of *Arcobacter* and *Campylobacter* isolates from retail chicken samples by use of DNA oligonucleotide arrays. *Appl. Environ. Microbiol.* **2007**, *73*, 3645–3655.

174. Hauser, E.; Tietze, E.; Helmuth, R.; Junker, E.; Prager, R.; Schroeter, A.; Rabsch, W.; Fruth, A.; Toboldt, A.; Malorny, B. Clonal dissemination of *Salmonella enterica* serovar Infantis in Germany. *Foodborne Pathog. Dis.* **2012**, *9*, 352–360.

175. Franke-Whittle, I.H.; Confalonieri, A.; Insam, H.; Schlegelmilch, M.; Körner, I. Changes in the microbial communities during co-composting of digestates. *Waste Manag.* **2014**, *34*, 632–641.

176. Hmaïed, F.; Helel, S.; le Berre, V.; François, J.M.; Leclercq, A.; Lecuit, M.; Smaoui, H.; Kechrid, A.; Boudabous, A.; Barkallah, I. Prevalence, identification by a DNA microarray-based assay of human and food isolates *Listeria* spp. from Tunisia. *Pathol. Biol.* **2014**, *62*, 24–29.

177. Hwang, B.H.; Shin, H.H.; Seo, J.H.; Cha, H.J. Specific multiplex analysis of pathogens using a direct 16S rRNA hybridization in microarray system. *Anal. Chem.* **2012**, *84*, 4873–4879.

178. Kim, H.J.; Park, S.H.; Lee, T.H.; Nahm, B.H.; Kim, Y.R.; Kim, H.Y. Microarray detection of food-borne pathogens using specific probes prepared by comparative genomics. *Biosens. Bioelectron.* **2008**, *24*, 238–246.

179. Kim, D.H.; Lee, B.K.; Kim, Y.D.; Rhee, S.K.; Kim, Y.C. Detection of representative enteropathogenic bacteria, *Vibrio* spp., pathogenic *Escherichia coli*, *Salmonella* spp., *Shigella* spp., and *Yersinia enterocolitica*, using a virulence factor gene-based oligonucleotide microarray. *J. Microbiol.* **2010**, *48*, 682–688.

180. Donhauser, S.C.; Niessner, R.; Seidel, M. Sensitive quantification of *Escherichia coli* O157:H7, *Salmonella enterica*, and *Campylobacter jejuni* by combining stopped polymerase chain reaction with chemiluminescence flow-through DNA microarray analysis. *Anal. Chem.* **2011**, *83*, 3153–3160.

181. Goji, N.; Macmillan, T.; Amoako, K.K. A New Generation Microarray for the Simultaneous Detection and Identification of *Yersinia pestis* and *Bacillus anthracis* in Food. *J. Pathog.* **2012**, *2012*, 627036.

182. Lee, J.Y.; Kim, B.C.; Chang, K.J.; Ahn, J.M.; Ryu, J.H.; Chang, H.I.; Gu, M.B. A subtractively optimized DNA microarray using non-sequenced genomic probes for the detection of food-borne pathogens. *Appl. Biochem. Biotechnol.* **2011**, *164*, 183–193.

183. Merga, J.Y.; Williams, N.J.; Miller, W.G.; Leatherbarrow, A.J.; Bennett, M.; Hall, N.; Ashelford, K.E.; Winstanley, C. Exploring the diversity of *Arcobacter butzleri* from cattle in the UK using MLST and whole genome sequencing. *PLoS ONE* **2013**, *8*, e55240.

184. Fischer, J.; Rodríguez, I.; Baumann, B.; Guiral, E.; Beutin, L.; Schroeter, A.; Kaesbohrer, A.; Pfeifer, Y.; Helmuth, R.; Guerra, B. blaCTX-M-15-carrying *Escherichia coli* and *Salmonella* isolates from livestock and food in Germany. *J. Antimicrob. Chemother.* **2014**, *69*, 2951–2958.

185. Kostić, T.; Stessl, B.; Wagner, M.; Sessitsch, A.; Bodrossy, L. Microbial diagnostic microarray for food- and water-borne pathogens. *Microb. Biotechnol.* **2010**, *3*, 444–454.

186. Jang, J.H.; Kim, S.J.; Yoon, B.H.; Ryu, J.H.; Gu, M.B.; Chang, H.I. Detection of *Alicyclobacillus* species in fruit juice using a random genomic DNA microarray chip. *J. Food Prot.* **2011**, *74*, 933–938.

187. Lezar, S.; Barros, E. Oligonucleotide microarray for the identification of potential mycotoxigenic fungi. *BMC Microbiol.* **2010**, *10*, 87.

188. Suo, B.; He, Y.; Paoli, G.; Gehring, A.; Tu, S.I.; Shi, X. Development of an oligonucleotide-based microarray to detect multiple foodborne pathogens. *Mol. Cell. Probes* **2010**, *24*, 77–86.

189. Bae, D.; Crowley, M.R.; Wang, C. Transcriptome analysis of Listeria monocytogenes grown on a ready-to-eat meat matrix. *J. Food Prot.* **2011**, *74*, 1104–1111.

190. Dobnik, D.; Morisset, D.; Lenarčič, R.; Ravnikar, M. Simultaneous detection of RNA and DNA targets based on multiplex isothermal amplification. *J. Agric. Food Chem.* **2014**, *62*, 2989–2996.

191. Toboldt, A.; Tietze, E.; Helmuth, R.; Junker, E.; Fruth, A.; Malorny, B. Molecular epidemiology of *Salmonella enterica* serovar Kottbus isolated in Germany from humans, food and animals. *Vet. Microbiol.* **2014**, *170*, 97–108.

192. Rungrassamee, W.; Tosukhowong, A.; Klanchui, A.; Maibunkaew, S.; Plengvidhya, V.; Karoonuthaisiri, N. Development of bacteria identification array to detect lactobacilli in Thai fermented sausage. *J. Microbiol. Methods* **2012**, *91*, 341–353.

193. Zhou, G.; Wen, S.; Liu, Y.; Li, R.; Zhong, X.; Feng, L.; Wang, L.; Cao, B. Development of a DNA microarray for detection and identification of *Legionella pneumophila* and ten other pathogens in drinking water. *Int. J. Food Microbiol.* **2011**, *145*, 293–300.

194. Cao, B.; Liu, X.; Yu, X.; Chen, M.; Feng, L.; Wang, L. A new oligonucleotide microarray for detection of pathogenic and non-pathogenic *Legionella* spp. *PLoS ONE* **2014**, *9*, e113863.

195. Mendes, F.; Sieuwerts, S.; de Hulster, E.; Almering, M.J.; Luttik, M.A.; Pronk, J.T.; Smid, E.J.; Bron, P.A.; Daran-Lapujade, P. Transcriptome-based characterization of interactions between *Saccharomyces cerevisiae* and *Lactobacillus delbrueckii* subsp. *bulgaricus* in lactose-grown chemostat cocultures. *Appl. Environ. Microbiol.* **2013**, *79*, 5949–5961.

196. Utsumi, Y.; Tanaka, M.; Morosawa, T.; Kurotani, A.; Yoshida, T.; Mochida, K.; Matsui, A.; Umemura, Y.; Ishitani, M.; Shinozaki, K.; *et al.* Transcriptome analysis using a high-density oligomicroarray under drought stress in various genotypes of cassava: An important tropical crop. *DNA Res.* **2012**, *19*, 335–345.

197. Choi, E.H.; Yang, H.P.; Chun, H.S. Chitooligosaccharide ameliorates diet-induced obesity in mice and affects adipose gene expression involved in adipogenesis and inflammation. *Nutr. Res.* **2012**, *32*, 218–228.

198. Kawakami, Y.; Yamanaka-Okumura, H.; Sakuma, M.; Mori, Y.; Adachi, C.; Matsumoto, Y.; Sato, T.; Yamamoto, H.; Taketani, Y.; Katayama, T.; *et al.* Gene expression profiling in peripheral white blood cells in response to the intake of food with different glycemic index using a DNA microarray. *J. Nutrigenet. Nutrigenomics* **2013**, *6*, 154–168.

199. You, L.; Sheng, Z.Y.; Tang, C.L.; Chen, L.; Pan, L.; Chen, J.Y. High cholesterol diet increases osteoporosis risk via inhibiting bone formation in rats. *Acta Pharmacol. Sin.* **2011**, *32*, 1498–1504.

200. Ding, Y.; Li, J.; Liu, S.; Zhang, L.; Xiao, H.; Li, J.; Chen, H.; Petersen, R.B.; Huang, K.; Zheng, L. DNA hypomethylation of inflammation-associated genes in adipose tissue of female mice after multigenerational high fat diet feeding. *Int. J. Obes.* **2014**, *38*, 198–204.

201. Matsumoto, T.; Noguchi, M.; Hayashi, O.; Makino, K.; Yamada, H. Hochuekkito, a Kampo (traditional Japanese herbal) Medicine, Enhances Mucosal IgA Antibody Response in Mice Immunized with Antigen-entrapped Biodegradable Microparticles. *Evid. Based Complement. Altern. Med.* **2010**, *7*, 69–77.

202. Dong, S.; Inoue, A.; Zhu, Y.; Tanji, M.; Kiyama, R. Activation of rapid signaling pathways and the subsequent transcriptional regulation for the proliferation of breast cancer MCF-7 cells by the treatment with an extract of *Glycyrrhiza glabra* root. *Food Chem. Toxicol.* **2007**, *45*, 2470–2478.

203. Tamura, T.; Kamei, A.; Ueda, R.; Arai, S.; Mura, K. Characterization of the quality of imbibed soybean at an early stage of pre-germination for the development of a new protein food item. *Biosci. Biotechnol. Biochem.* **2014**, *78*, 115–123.

204. Martín, J.M.; Freire, P.F.; Daimiel, L.; Martínez-Botas, J.; Sánchez, C.M.; Lasunción, M.Á.; Peropadre, A.; Hazen, M.J. The antioxidant butylated hydroxyanisole potentiates the toxic effects of propylparaben in cultured mammalian cells. *Food Chem. Toxicol.* **2014**, *72*, 195–203.

205. Dang, Y.J.; Zhu, C.Y. Genomic Study of the Absorption Mechanism of p-Coumaric Acid and Caffeic Acid of Extract of Ananas Comosus L. Leaves. *J. Food Sci.* **2015**, *80*, 504–509.

206. Castañeda-Gutiérrez, E.; Moser, M.; García-Ródenas, C.; Raymond, F.; Mansourian, R.; Rubio-Aliaga, I.; Viguet-Carrin, S.; Metairon, S.; Ammon-Zufferey, C.; Avanti-Nigro, O.; *et al.* Effect of a mixture of bovine milk oligosaccharides, Lactobacillus rhamnosus NCC4007 and long-chain polyunsaturated fatty acids on catch-up growth of intra-uterine growth-restricted rats. *Acta Physiol.* **2014**, *210*, 161–173.

207. Togawa, N.; Takahashi, R.; Hirai, S.; Fukushima, T.; Egashira, Y. Gene expression analysis of the liver and skeletal muscle of psyllium-treated mice. *Br. J. Nutr.* **2013**, *109*, 383–393.

208. Kobori, M.; Masumoto, S.; Akimoto, Y.; Takahashi, Y. Dietary quercetin alleviates diabetic symptoms and reduces streptozotocin-induced disturbance of hepatic gene expression in mice. *Mol. Nutr. Food Res.* **2009**, *53*, 859–868.

209. Liu, Y.; Ream, A. Gene expression profiling of Listeria monocytogenes strain F2365 during growth in ultrahigh-temperature-processed skim milk. *Appl. Environ. Microbiol.* **2008**, *74*, 6859–6866.

210. Tokuji, Y.; Akiyama, K.; Yunoki, K.; Kinoshita, M.; Sasaki, K.; Kobayashi, H.; Wada, M.; Ohnishi, M. Screening for beneficial effects of oral intake of sweet corn by DNA microarray analysis. *J. Food Sci.* **2009**, *74*, 197–203.

211. Kobayashi, Y.; Hiroi, T.; Araki, M.; Hirokawa, T.; Miyazawa, M.; Aoki, N.; Kojima, T.; Ohsawa, T. Facilitative effects of *Eucommia ulmoides* on fatty acid oxidation in hypertriglyceridaemic rats. *J. Sci. Food Agric.* **2012**, *92*, 358–365.

212. Mayer, L.; Vendruscolo, C.T.; Silva, W.P.; Vorhölter, F.J.; Becker, A.; Pühler, A. Insights into the genome of the xanthan-producing phytopathogen *Xanthomonas arboricola* pv. pruni 109 by comparative genomic hybridization. *J. Biotechnol.* **2011**, *155*, 40–49.

213. Schmidt, A.M.; Sahota, R.; Pope, D.S.; Lawrence, T.S.; Belton, M.P.; Rott, M.E. Detection of genetically modified canola using multiplex PCR coupled with oligonucleotide microarray hybridization. *J. Agric. Food Chem.* **2008**, *56*, 6791–6800.

214. Kim, J.H.; Kim, S.Y.; Lee, H.; Kim, Y.R.; Kim, H.Y. An event-specific DNA microarray to identify genetically modified organisms in processed foods. *J. Agric. Food Chem.* **2010**, *58*, 6018–6026.

215. Rønning, S.B.; Rudi, K.; Berdal, K.G.; Holst-Jensen, A. Differentiation of important and closely related cereal plant species (Poaceae) in food by hybridization to an oligonucleotide array. *J. Agric. Food Chem.* **2005**, *53*, 8874–8880.

216. Voorhuijzen, M.M.; van Dijk, J.P.; Prins, T.W.; van Hoef, A.M.; Seyfarth, R.; Kok, E.J. Development of a multiplex DNA-based traceability tool for crop plant materials. *Anal. Bioanal. Chem.* **2012**, *402*, 693–701.

217. Niu, L.; Mantri, N.; Li, C.G.; Xue, C.; Wohlmuth, H.; Pang, E.C. Detection of Panax quinquefolius in Panax ginseng using "subtracted diversity array". *J. Sci. Food Agric.* **2011**, *91*, 1310–1315.

218. Im, R.; Mano, H.; Nakatani, S.; Shimizu, J.; Wada, M. Safety evaluation of the aqueous extract Kothala himbutu (*Salacia reticulata*) stem in the hepatic gene expression profile of normal mice using DNA microarrays. *Biosci. Biotechnol. Biochem.* **2008**, *72*, 3075–3083.

219. Van Dijk, J.P.; Leifert, C.; Barros, E.; Kok, E.J. Gene expression profiling for food safety assessment: examples in potato and maize. *Regul. Toxicol. Pharmacol.* **2010**, *58*, 21–25.

220. Leimanis, S.; Hernández, M.; Fernández, S.; Boyer, F.; Burns, M.; Bruderer, S.; Glouden, T.; Harris, N.; Kaeppeli, O.; Philipp, P.; *et al.* A microarray-based detection system for genetically modified (GM) food ingredients. *Plant Mol. Biol.* **2006**, *61*, 123–139.

221. Consoland, C.; Palmieri, L.; Doveri, S.; Maestri, E.; Marmiroli, N.; Reale, S.; Lee, D.; Baldoni, L.; Tosti, N.; Severgnini, M.; *et al.* Olive variety identification by ligation detection reaction in a universal array format. *J. Biotechnol.* **2007**, *129*, 565–574.

222. Van Dijk, J.P.; Cankar, K.; Scheffer, S.J.; Beenen, H.G.; Shepherd, L.V.; Stewart, D.; Davies, H.V.; Wilkockson, S.J.; Leifert, C.; Gruden, K.; *et al.* Transcriptome analysis of potato tubers—Effects of different agricultural practices. *J. Agric. Food Chem.* **2009**, *57*, 1612–1623.

223. Kamakura, M.; Maebuchi, M.; Ozasa, S.; Komori, M.; Ogawa, T.; Sakaki, T.; Moriyama, T. Influence of royal jelly on mouse hepatic gene expression and safety assessment with a DNA microarray. *J. Nutr. Sci. Vitaminol.* **2005**, *51*, 148–155.

224. Al-Khaldi, S.F.; Martin, S.A.; Rasooly, A.; Evans, J.D. DNA microarray technology used for studying foodborne pathogens and microbial habitats: Mini review. *J. AOAC Int.* **2002**, *85*, 906–910.

225. Liu-Stratton, Y.; Roy, S.; Sen, C.K. DNA microarray technology in nutraceutical and food safety. *Toxicol. Lett.* **2004**, *150*, 29–42.

226. Kostrzynska, M.; Bachand, A. Application of DNA microarray technology for detection, identification, and characterization of food-borne pathogens. *Can. J. Microbiol.* **2006**, *52*, 1–8.

227. Roy, S.; Sen, C.K. cDNA microarray screening in food safety. *Toxicology* **2006**, *221*, 128–133.

228. Rasooly, A.; Herold, K.E. Food microbial pathogen detection and analysis using DNA microarray technologies. *Foodborne Pathog. Dis.* **2008**, *5*, 531–550.

229. Kuiper, H.A.; Kok, E.J.; Engel, K.H. Exploitation of molecular profiling techniques for GM food safety assessment. *Curr. Opin. Biotechnol.* **2003**, *14*, 238–243.

230. Li, W.F.; Jiang, J.G.; Chen, J. Chinese medicine and its modernization demands. *Arch. Med. Res.* **2008**, *39*, 246–251.

231. Kiyama, R. DNA microarray assay (DMA) for screening and characterization of traditional herbal medicine. In *Applications of DNA Microarray to Drug Discovery and Development*; Cheng, F., Ed.; CRC Press/Taylor and Francis: Boca Raton, FL, USA, in press.

232. Afshari, C.A.; Nuwaysir, E.F.; Barrett, J.C. Application of complementary DNA microarray technology to carcinogen identification, toxicology, and drug safety evaluation. *Cancer Res.* **1999**, *59*, 4759–4760.

233. Degenkolbe, T.; Hannah, M.A.; Freund, S.; Hincha, D.K.; Heyer, A.G.; Köhl, K.I. A quality-controlled microarray method for gene expression profiling. *Anal. Biochem.* **2005**, *346*, 217–224.

234. Böhme, K.; Cremonesi, P.; Severgnini, M.; Villa, T.G.; Fernández-No, I.C.; Barros-Velázquez, J.; Castiglioni, B.; Calo-Mata, P. Detection of food spoilage and pathogenic bacteria based on ligation detection reaction coupled to flow-through hybridization on membranes. *Biomed. Res. Int.* **2014**, *2014*, 156323.

235. Atanasova, L.; Druzhinina, I.S. Review: Global nutrient profiling by Phenotype MicroArrays: A tool complementing genomic and proteomic studies in conidial fungi. *J. Zhejiang Univ. Sci. B* **2010**, *11*, 151–168.

236. Sakaida, I.; Tsuchiya, M.; Kawaguchi, K.; Kimura, T.; Terai, S.; Okita, K. Herbal medicine Inchin-ko-to (TJ-135) prevents liver fibrosis and enzyme-altered lesions in rat liver cirrhosis induced by a choline-deficient L-amino acid-defined diet. *J. Hepatol.* **2003**, *38*, 762–769.

237. Zheng, H.C.; Noguchi, A.; Kikuchi, K.; Ando, T.; Nakamura, T.; Takano, Y. Gene expression profiling of lens tumors, liver and spleen in α-crystallin/SV40 T antigen transgenic mice treated with Juzen-taiho-to. *Mol. Med. Rep.* **2014**, *9*, 547–552.

238. Dong, F.X.; Zhang, X.Z.; Wu, F.; He, L.Q. The effects of kangxianling on renal fibrosis as assessed with a customized gene chip. *J. Tradit. Chin. Med.* **2012**, *32*, 229–233.

239. Pan-Hammarström, Q.; Wen, S.; Hammarström, L. Cytokine gene expression profiles in human lymphocytes induced by a formula of traditional Chinese medicine, vigconic VI-28. *J. Interferon Cytokine Res.* **2006**, *26*, 628–636.

240. Kiyama, R.; Wada-Kiyama, Y. Estrogenic endocrine disruptors: molecular mechanisms of action. *Environ. Int.* **2015**, *83*, 11–40.

Antibody Microarray for *E. coli* O157:H7 and Shiga Toxin in Microtiter Plates

Andrew G. Gehring , Jeffrey D. Brewster, Yiping He, Peter L. Irwin,
George C. Paoli, Tawana Simons, Shu-I Tu and Joseph Uknalis

Abstract: Antibody microarray is a powerful analytical technique because of its inherent ability to simultaneously discriminate and measure numerous analytes, therefore making the technique conducive to both the multiplexed detection and identification of bacterial analytes (*i.e.*, whole cells, as well as associated metabolites and/or toxins). We developed a sandwich fluorescent immunoassay combined with a high-throughput, multiwell plate microarray detection format. Inexpensive polystyrene plates were employed containing passively adsorbed, array-printed capture antibodies. During sample reaction, centrifugation was the only strategy found to significantly improve capture, and hence detection, of bacteria (pathogenic *Escherichia coli* O157:H7) to planar capture surfaces containing printed antibodies. Whereas several other sample incubation techniques (e.g., static *vs.* agitation) had minimal effect. Immobilized bacteria were labeled with a red-orange-fluorescent dye (Alexa Fluor 555) conjugated antibody to allow for quantitative detection of the captured bacteria with a laser scanner. Shiga toxin 1 (Stx1) could be simultaneously detected along with the cells, but none of the agitation techniques employed during incubation improved detection of the relatively small biomolecule. Under optimal conditions, the assay had demonstrated limits of detection of ~5.8 × 10^5 cells/mL and 110 ng/mL for *E. coli* O157:H7 and Stx1, respectively, in a ~75 min total assay time.

Reprinted from *Sensors*. Cite as: Gehring, A.G.; Brewster, J.D.; He, Y.; Irwin, P.L.; Paoli, G.C.; Simons, T.; Tu, S.-I.; Uknalis, J. Antibody Microarray for *E. coli* O157:H7 and Shiga Toxin in Microtiter Plates. *Sensors* **2015**, *15*, 30429–30442.

1. Introduction

According to the U.S. Centers for Disease Control and Prevention, approximately 48 million illnesses; 128,000 hospitalizations; and 3000 deaths per year in the United States alone are attributed to ingestion of contaminated foods [1]. Traditional microbial culture methods can detect and identify a single, specific bacterium contaminant in foods, but the approach may require days or weeks to complete and typically, quantitative data is not generated. Such specific detection of very small numbers (e.g., 1 cell/mL) of pathogenic bacteria in complex food matrices necessitates methods with extremely high sensitivity. The quest for faster assay times (of minutes to hours) combined with quantitative, low level detection results has stimulated the development of rapid microbial methods, many of which are biosensor based [2–4].

A notorious bacterial pathogen, *Escherichia coli* O157:H7, can cause severe sickness (e.g., hemorrhagic colitis and hemolytic uremic syndrome) and death for some infected by the microorganism [5]. Sickness associated with foodborne *E. coli* O157:H7 is an important problem in the United States where past multistate outbreaks have been associated with meat [6] and produce [7]. *E. coli* O157:H7 is classified as a "zero tolerance" adulterant and is therefore perceived as a major concern due to the threat of incidental contamination of foods with the pathogen. Therefore, considerable effort has been undertaken to develop specific, rapid methods for the detection of pathogens associated with foodborne outbreaks [2,8,9].

Rapid methods with the capacity to screen for analytes of differing size (e.g., ranging from biomolecular toxins to whole bacterial cells) can be useful for multivariate analysis [10]. In addition, the desire to screen large numbers of samples for reliable food safety monitoring necessitates high-throughput analytical processing. Nucleic acid microarrays have exhibited enormous potential for pathogen screening [11,12]. Similarly, protein microarrays comprised of antibodies as biorecognition elements orthogonally arrayed in spots or parallel printed stripes have also been generated for the detection and typing of pathogens. Several examples of antibody arrays that show promise for the multiplex detection of bacterial cells and/or toxins in complex sample matrices (e.g., foods) have been developed [13–17], as well as commercialized [18]. The evolution, application, and merits of antibody, or protein, microarrays have been reviewed elsewhere [19–24].

Past research in this group has demonstrated the high-throughput and multiplex capability of antibody microarray in multiwell format [15]. This study presents a streamlined and improved version of that system with an optimized assay that considerably reduces the overall assay time with a concomitantly better limit of detection (LOD) for bacterial cells. A "bottleneck" in improvement of LOD has been that bacteria suspended in aqueous medium are relatively immobile in part due to their density being essentially that of water. Hence, under static incubation conditions, non-flagellated and/or dead bacteria (essentially large "particles") that may exhibit Brownian motion travel an insignificant distance when suspended in aqueous medium. Even the metabolic-dependent motion of flagellated bacteria is quite slow [25]. Therefore, under their own accord, most bacteria suspended in bulk solvent do not come in close contact with planar binding surfaces, which, in this study, was passively adsorbed with capture antibodies to relatively inexpensive polystyrene plate well bottoms that served as microarray substrates. At low concentrations ($\leqslant 10^6$/mL), the cells are relatively dispersed so that binding events are rare. Increased assay sensitivity necessitates improved antibody-based immobilization of bacteria to solid supports. Dielectrophoresis [26] and direct radiation force combined with ultrasound acoustic streaming [27] have been employed as means to improve immobilization via active partitioning of bacteria from liquid phase to static, antibody-coated, solid substrates. Other groups, such as Ball *et al.* [28], have employed centrifugation to mechanically force bacteria to a capture surface. Our efforts focused on the latter technique given its simplicity, rapidity, and particularly its availability for immediate application with our transparent, polystyrene array substrates. A major part of this investigation compared the efficacy (*i.e.*, increased fluorescence responses associated with bacterial capture) of binding for bacteria (*E. coli* O157:H7) versus the biomolecule, Shiga toxin 1 (Stx1; a protein synthesis inhibitor that is produced by Shigatoxigenic strains of *E. coli*), at the capture surface of a microarray substrate as influenced by various incubation conditions (static, mixing, and centrifugation).

2. Experimental Section

2.1. Materials

Reagents used in this research were: glycerol, tablets of phosphate-buffered saline (PBS; 10 mM phosphate, 2.7 mM KCl, 137 mM NaCl, pH 7.4), fraction V bovine serum albumin (BSA) from Sigma (St. Louis, MO, USA) and NeutrAvidin from Thermo Scientific (Waltham, MA, USA). Plates used were MicroAmp® 384-well reaction/microarray source plates (polypropylene, conical wells) from PE Biosystems (Carlsbad, CA, USA) and antibodies were printed on black-walled, clear/transparent and flat-bottomed, polystyrene 96-multiwell microtiter/microarray destination plates (well dimensions—6.6 mm diameter, ~11 mm height) with (FLUOTRAC 600) surfaces from Greiner Bio-One North America Inc. (Monroe, NC, USA). Anti-*E. coli* O157:H7 antibody (unmodified or biotinylated; polyclonal IgG affinity purified for target, exclusivity purified against non-target *E. coli* strains) raised in goats was obtained from Kirkegaard and Perry Laboratories, Inc. (Gaithersburg, MD, USA). Alexa Fluor 555 (AF555) dye labeling kit (from Invitrogen, Carlsbad, CA USA) was used to prepare fluorescent BSA and antibody conjugates. Stx1 and anti-Stx1 antibody solution comprised of equal parts of 9C9 (IgG$_1$; A, A$_1$, B neutralizing), 10D11 (IgG$_2$b; A, A$_1$, B neutralizing), and 13C4 (IgG$_1$κ; B neutralizing) murine monoclonal antibodies initially constituted in 50% glycerol in nH$_2$O (employed for analyte capture) and 3C10 (IgG$_1$; A, A$_1$, B neutralizing) monoclonal antibody, also reconstituted in 50% glycerol (employed for analyte labeling after conjugation with AF555 fluorescent dye) were from Toxin Technology (Sarasota, FL, USA). Strain B1409 of *E. coli* O157:H7 became available to our research center via a route of multiple destinations that last passed through the Centers for Disease Control and Prevention (Atlanta, GA, USA).

Modified Brain Heart Infusion broth was from Becton Dickinson (Sparks, MD, USA). Any chemicals not mentioned were at least of reagent grade.

2.2. Apparatus

Solutions of biorecognition elements (antibodies in this manuscript) were orthogonally array printed into 96-well microtiter plate wells using an Omnigrid Accent Pro from Bucher (Basel, Switzerland) outfitted with a single, Stealth printing pin (model SMP3; TeleChem International, Inc., Sunnyvale, CA, USA). (Laser-induced fluorescence images were obtained with an LS400 scanner from Tecan Group Ltd. (Männedorf, Switzerland). Shaking of mictrotiter plates were conducted on a Titer Plate Shaker (Lab-Line Instruments, Inc.; Melrose Park, IL, USA) at slow-moderate speed setting. Microtiter plates were centrifuged in an Eppendorf refrigerated centrifuge (model 5810R) using an A-4-62 rotor (Eppendorf AG, Hamburg, Germany). Ultraviolet-Visible spectrophotometric readings were taken with a Cary 50 UV-Vis spectrophotometer (Varian, Inc., Palo Alto, CA, USA). Enumeration of intact bacterial cells was achieved with the aid of a Petroff-Hausser counting chamber obtained from Thomas Scientific (Swedesboro, NJ, USA).

2.3. Growth and Enumeration of Bacteria

Immediately prior to use, a frozen culture of stationary phase E. coli O157:H7 was thawed and added to modified Brain Heart Infusion broth (10 mL). This was incubated at 37 °C for 18 h with shaking at 160 rpm. Serial dilutions of cultures were enumerated using a Petroff-Hausser counting chamber as described by Gehring, et al. [29].

2.4. Conjugation of Antibodies with Fluorescent Dye

An AF555 dye labeling kit was used to prepare fluorescent BSA and antibody conjugates following the manufacturer's instructions, briefly: BSA or antibody was diluted to ~1 mg/mL in 0.1 M carbonate buffer (pH 8.3), dye was added to ~0.5 mL of protein solution and incubated for 1 h at room temperature (RT) with stirring, the mix was eluted (using 10 mM PBS, pH 7.2 containing azide) through a gel filtration column to separate labeled protein from unbound dye and fractions of the first of two resolved colored bands were collected and pooled. The absorbance of pooled fractions was measured at 280 and 555 nm using a UV-Vis spectrophotometer in order to determine dye incorporation stoichiometry and antibody conjugate concentration.

2.5. Antibody Preparation and Microarray Printing

Biotinylated and non-biotinylated anti-E. coli O157:H7 capture antibodies (that were obtained as lyophilized reagents) were rehydrated in 50% glycerol to a concentration of 1 mg/mL that was further diluted to 1:30 in PBS containing 5% glycerol working solutions for microarray printing. (Glycerol was employed to prevent evaporation of the printed spots as well as to maintain hydration of the capture antibodies [30]). Anti-Stx antibodies were similarly reconstituted to 0.25 mg/mL in nH$_2$O as directed by the supplier and further diluted 1:4 with PBS (containing 5% glycerol) for array printing.

Approximately 25 μL of capture antibody solution was pipetted into separate wells of a MicroAmp source plate on the microarray printer (positioned atop a 4 °C cooled thermal block during printing). Immediately prior to printing, source plates were centrifuged at 200× g for 2 min to remove any air bubbles. Array contact printing was performed with the following parameters—preprints/blots = 20; contact time = 0; dip and print acceleration = 10 cm/s^2, and print velocity = 2 cm/s using an SMP3 (spot diameter of ~100 μm) pin that delivered ~0.7 nL per contact stroke. In each well, 2 columns of 8 spots per antibody were printed with a horizontal and vertical separation of 150 μm. After printing, all wells were visually examined, often with the assistance of a stereo light microscope (~10–20 × magnification) to ensure that spots were uniformly printed. Following array printing, spotted destination plates sat for 1 h at RT before use.

2.6. Antibody Microarray Detection of Bacteria and Shiga Toxin 1 in Multiwell Plates

A schematic for the fluorescence, sandwich immunoassay as applied to the multiwell antibody microarray-based detection *E. coli* O157 bacteria and Stx1 is depicted in Figure 1. The assay generally followed the one previously described for microarray slides [31] with several modifications. All immunoassay procedures and reagents were at RT. Wells of the destination plate, preprinted with capture antibody, were washed with 200 μL PBST (PBS containing 0.05% Tween 20), immediately emptied via rapid inversion of the plate, and any remaining liquid was removed by striking the plate (upside down) onto an absorbent towel laid flat on a laboratory bench. This wash procedure was repeated once with PBST. The plate wells were blocked with 50 μL of 1% BSA in PBS for 30 min. The plates were washed (as above) following removal of this BSA solution. Analyte (100 μL, or as indicated otherwise, of samples containing bacterial stock or Stx serially diluted in PBS) was then added, and each array was subjected to incubation (static, unless otherwise indicated) for 1 h (or time as otherwise indicated) to allow analyte capture. During the incubation for capture, the reporter antibody solutions were prepared (1:50 for AF555-labeled antibody conjugates) with PBST. The reporter antibodies were shielded from light during all experiments. The wells were washed twice with PBST and excess liquid was removed as above. Next, 50 μL reporter antibody solution was added to each well, which was subjected to static incubation for 1 h (unless otherwise indicated) at RT.

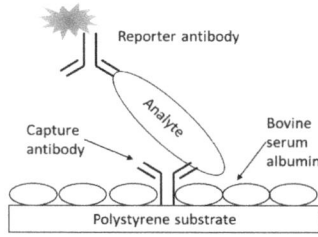

Figure 1. Fluorescence sandwich immunoassay schematic. Represented are analyte (e.g., bacterial cell or proteinaceous toxin) bound at a microarrayed spot of capture antibodies printing on the bottom of a well of a multiwell, black-walled, clear-bottomed, polystyrene microtiter plate. The bare polystyrene was further blocked with BSA and following analyte capture, reporter antibodies, conjugated with fluorescent molecules, were used to sandwich and label the analyte prior to detection with laser-induced fluorescence scanning.

2.7. Scanning Electron Microscopy

After centrifugation concentration ($3220 \times g$ for 5 min) of live *E. coli* O157:H7 cells in capture antibody (non-biotinylated) microarray-printed polystyrene microtiter plates, the cells fixed with 200 μL of 2.5% glutaraldehyde (Electron Microscopy Sciences, Hatfield, PA USA) for 30 min. The plates were then rinsed twice for 30 min each with ~200 μL per well of 0.1 M imidazole, (Electron Microscopy Sciences). The cell-associated bottoms (samples) of the wells were removed with a cork borer. The samples were then sequentially washed for 30 min intervals each with 2 mL of 50%, 80%, 90%, and finally 95% ethanol (The Warner-Graham Company, Cockeysville, MD, USA). The samples were momentarily held in and then washed with ~2 mL of 100% ethanol three times before being critically point dried. The samples were then stacked in a wire basket, separated by cloth, and placed in a Critical Point Drying Apparatus, (Denton Vacuum, Inc., Cherry Hill, NJ, USA), using liquid carbon dioxide (Welco Co, Allentown, PA, USA) for approximately 20 min. The samples were then removed and mounted on stubs and sputter gold coated for 30 s (EMS 150R ES, Electron Microscopy Sciences). They were then observed with Scanning Electron Microscope, FEI Quanta 200 F, (Hillsboro, OR, USA) with an accelerating voltage of 5–10 KV in high vacuum mode. It was observed that the critical point drying process shrunk the polystyrene discs to about 1/3 their original size.

2.8. Microarray Scanning and Data Analysis

Wells were washed twice with PBST and then were scanned at the appropriate fluorescence setting (AF555-excitation: 543 nm, emission filter: 590 nm) on the

LS400 scanner using single channel scanning mode. During conjugate incubation, 543 and 643 nm lasers were turned on to warm up and stabilize for at least 30 min. Typical LS400 instrument scanning parameters, set and controlled via Array-Pro Analyzer software (ver. 4.5.1.73; Tecan Group Ltd., Männedorf, Switzerland) interface included: autofocusing in well mode, PMT gain that ranged from 100 to 150, 20 μm resolution, small pinhole setting, and optimization of integration time = 1. The multiwell plates had to be inverted during scanning.

Each well, which contained 8 printed spots per antibody, was considered an experimental unit. Fluorescence intensities (in fluorescence units or (arbitrary fluorescence units) AFUs) of sample spots and local/proximal (adjacent to sample spots in the same well) background pixels were obtained using the ArrayPro Software. Net spot intensities (sample responses minus individual, corresponding concentric and median background responses) were compared, and the 2 highest and 2 lowest values from each set of 8 were discarded. At least 3 technical replicates (single columns of 8 printed antibody spots in 3 individual microtiter plate wells) were generated for each concentration level of analyte tested. Some, but not all experiments were replicated over multiple days of experimentation. (No significant discrepancies between day-to-day replication were observed, data not shown.) Net intensities were then averaged and standard deviations, presented as error bars in plots, were computed for the means.

3. Results and Discussion

Maximizing binding of target is a critical factor in microarray detection. NeutrAvidin (NAv) is a deglycosylated form of avidin with a higher biotin binding specificity and lower non-specific binding. In a simple biotinylated globular protein (employing fluorescently labeled and biotinylated BSA that simulated biotinylated antibody) binding study, NAv, streptavidin (SAv), or biotinylated BSA (subsequently reacted with SAv) in different buffers were compared for capture efficacy of the dye-labeled protein following passive adsorption to polystyrene. Cursory results indicated that highest capture was with the NAv system, but the improvement was only marginal (~2×). Remarkably, direct adsorption of dye-labeled BSA elicited the same level of fluorescence as the SAv systems (data not shown). Direct adsorption of capture antibody to the well bottoms presented itself as an attractive and reasonably effective (with respect to fluorescence response) alternative, especially if the SAv/biotin binding system could be avoided altogether. Such passive adsorption of capture antibody serving as a foundation for fluorescence sandwich immunoassays was used throughout this study in conjunction with microarray detection (Figure 1).

The magnitude of microarray response was a function of the amount of time the analyte (bacteria or proteinaceous toxin) was in contact with the antibody arrayed plate well bottoms and that of the dwell time of the fluorescent antibody conjugate with captured analyte. The combination of 60 min analyte incubation with 60 min conjugate incubation, respectively, (or 60', 60') had the greatest response over all cell dilutions (Figure 2A). The next lowest plot is the 60', 5' (60 min sample incubation/capture, 5 min labeling antibody conjugate incubation; statistical difference P = 0.147, between these two curves was not significant) where the similar number of targets do not have the time to be detected by fluorescent antibody and the response curve saturates (particularly evident in the inset log-log plot in Figure 2A). For these and all subsequent results reported herein, a Student's t-based statistical analysis was employed to test the homogeneity between regression coefficients of selected data sets [32]. Mass transport of reporter conjugate to the well surface tethered analyte appears to be a diffusion controlled, rate-limiting step in the assay. An even more interesting observation is between the 5', 60' and 60', 5' response curves (P = 0.00229). With the 5', 60' incubation, there were presumably fewer captured targets (bacterial analyte) for the fluorescent antibody to interact with as compared to 60', 5' where a concentration dependent curve became more evident. Since the 60', 5' curve exhibited greater sensitivity; this result indicated that the greater "bottleneck" was the analyte incubation time. The 5', 5' incubation conditions yielded the lowest response (P = 0.0408 *versus* 60', 60' and P = 0.000691 *vs.* 60', 5'). Time of bacterial analyte contact was the determining factor for such static incubations. Using the same reaction conditions, similar, but much less profound trends were observed when the analyte was a proteinaceous toxin (Figure 2B). There was only a slight difference between the 60', 60' *vs.* 5', 5' curve (P = 0.0693) whereas there were no statistically significance differences between 60', 60' *vs.* 60', 5' (P = 0.433), 60', 60' *vs.* 5', 60' (P = 0.261), 60', 5' *vs.* 5', 60' (P = 0.623), 60', 5' *vs.* 5', 5' (P = 0.107), and 5', 60' *vs.* 5', 5' (P = 0.0753).

Since bacterial analyte incubation time appeared to be the primary rate-limiting step during the immunoassay portion of plate-based microarray detection, various incubation treatments during analyte capture were tested to determine if detection could be improved relative to static incubation conditions (Figure 3). Three additional conditions were compared and they included: (1) shaking-moderate mixing speed on a platform shaker; (2) aspirating/dispensing-analyte mixtures were repeatedly (~3×) mixed manually via aspirating and dispensing with a multi-channel pipettor once every 5 min during the total analyte capture reaction time; and (3) centrifugation-analyte mixtures were added to the microtiter plate that was subsequently placed into a centrifuge outfitted with a swinging bucket rotor, the plate was spun for 5 min, and the mixture was aspirated/dispensed with a pipettor prior to additional incubation or a washing step.

It was not too surprising that centrifugation by far elicited the highest microarray response and best limit of detection for the live cells (Figure 3A,B) since bacterial cells are slightly denser than water. All of the curves for Figure 3A were statistically different except for "static" *vs.* "aspirated/dispensed." (P = 0.0396 for 4× centrifuged *vs.* shaken; P = 0.0329 for 4× centrifuged *vs.* aspirated/dispensed; P = 0.0333 for 4× centrifuged *vs.* static; P = 0.00533 for shaken *vs.* aspirated/dispensed; P = 0.00704 for shaken *vs.* static; P = 0.317 for aspirated/dispensed *vs.* static.) Response levels were marginally (~2×) higher when non-biotinylated antibodies were used (Figure 3B). For Figure 3B, all of the curves were significantly different except again for "aspirated/dispensed" *vs.* "static." (P = 0.000862, 4× centrifuged *vs.* aspirated/dispensed; P = 0.000942, 4× centrifuged *vs.* static; P = 0.174, aspirated/dispensed *vs.* static.) This result was reproducible and suggested that, as otherwise might be anticipated, a sub-population of the employed polyclonal antibodies had their antigen-binding sites deactivated (via steric hindrance and/or disruption of potential electrostatic interaction by amino acid functional groups) as a result of biotinylation. In other words, random, undirected biotinylation of antibodies may lower antibody specificity if epitope-binding site amines are blocked with biotin moieties.

Centrifugation had significantly less effect on heat-killed cells (Figure 3C; none of the curves in Figure 3C were significantly different with P ranging from 0.499 to 0.850) and, as would be expected with a relatively small biomolecular analyte, no considerable influence on the capture of the proteinaceous toxin Stx1 (Figure 3D) with P ranging from 0.130 to 0.587 expect for an unexpected difference between "aspirated/dispensed" *vs.* "static" (P = 0.0460). Upon visualization with light microscopy (data not shown), heat-killed *E. coli* O157:H7 cells have the appearance of disrupted cells fragmented into multiple pieces of various shapes, sizes, and density that has considerably more surface area available for binding by the polyclonal antibodies employed. Such fragments apparently do not share the same fluidic transport behavior observed with live (intact) cells.

Response after shaking was unexpectedly low for bacterial cells and even more surprisingly low with the multiple aspirating/dispensing technique (Figure 3A,B). It was hypothesized that continuous aspirating/dispensing would be analogous to improved capture typically observed in flow systems. However, it is possible that such action, and to a lesser extent, shaking, caused sheering of the bacteria from the surfaced of the antibody-coated well bottom/substrate. In addition, shaking possibly caused cells to be forced to the sides of the wells and therefore they did not interact with the printed antibodies on the bottom of the wells.

Repeated centrifugation of live cells did not appear to significantly improve binding to (non-biotinylated) capture antibody as observed by the marginal increase in microarray response (Figure 3E). Yet, upon statistical analysis, significant improvements were observed with 4× *vs.* 2× (P = 0.0477), 4× *vs.* 1× (P = 0.00339), 3× *vs.* 2× (P = 0.00673), 3× *vs.* 1× (P = 0.000453), and 2× *vs.* 1× (P = 0.0197), but not with $4\times$ *vs.* $3\times$ (P = 0.399). Together with the results observed for Stx1 dose-response (Figure 3D), this was evidence that the 60 min analyte incubation time could be considerably reduced upon substitution with a 5 min centrifugation step. Only slight improvement in microarray response for intact cells would be expected with additional centrifugation steps. Unfortunately, additional centrifugation would detrimentally add to the total assay time.

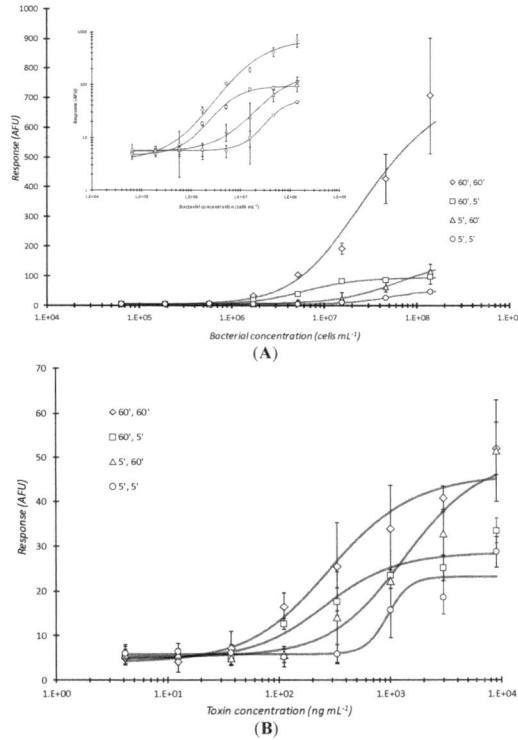

Figure 2. Effects of varying analyte and conjugation reaction times on microarray detection of bacterial cells or toxin. Serially diluted analyte, live *E. coli* O157:H7 (**A**) or Stx1 (**B**) were statically incubated in microtiter plates microarray printed with passively adsorbed biotinylated capture antibodies for the first time (5 or 60 min) indicated and then further reacted with fluorescent dye labeled antibody conjugates for the second time (5 or 60 min) also indicated. As with all subsequent plots in this report, the above curves show the microarray response in arbitrary fluorescence units (AFU) versus bacterial or toxin concentration. Each data point represented the mean ± standard deviation for 4 of 8 daily technical replicates (with 2 highest and 2 lowest values dropped) from serial dilution series combined with other similarly treated replicates from experiments repeated no less than 2 days and no more than 4 days.

257

The assay conditions used for the generation of the 1x curve in Figure 3E were considered optimized and the final assay conditions for this investigation. A limit of detection of ~5.8×10^5 cells/mL for live cells could be inferred from the 1x data set being a prospective lower detectable limit response value, minus its standard deviation, that was distinguishable from the baseline value at the lowest concentration plus $3\times$ the standard deviation for that value.

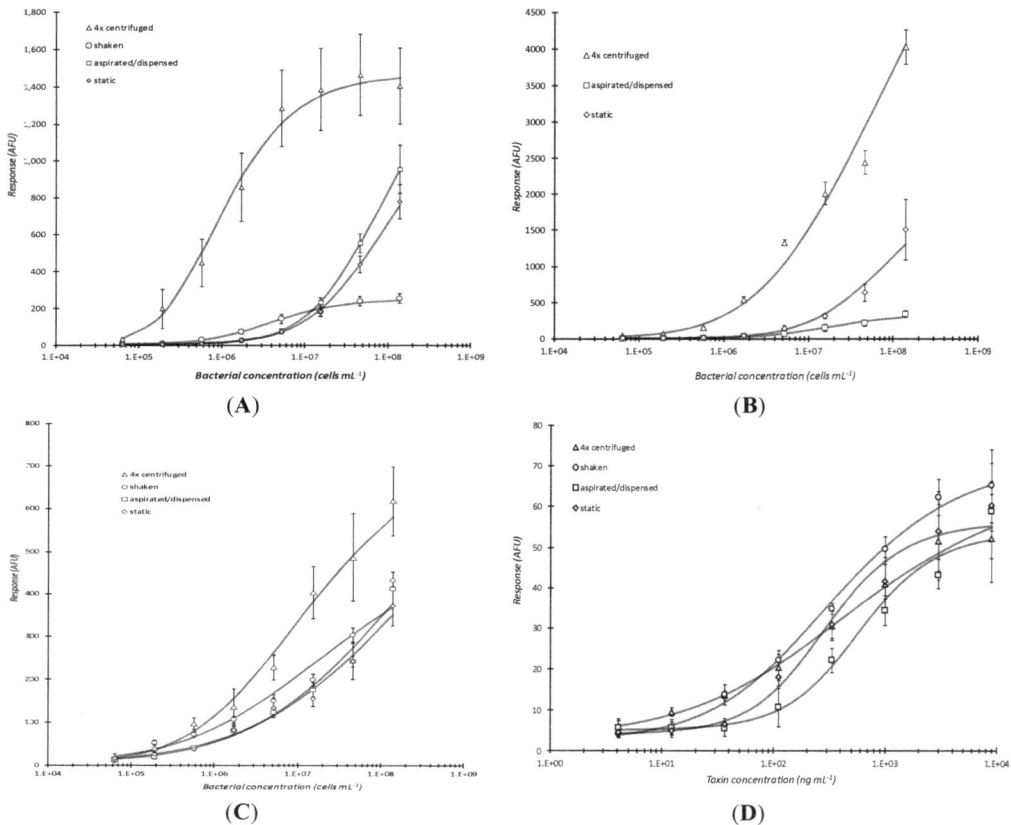

Figure 3. *Cont.*

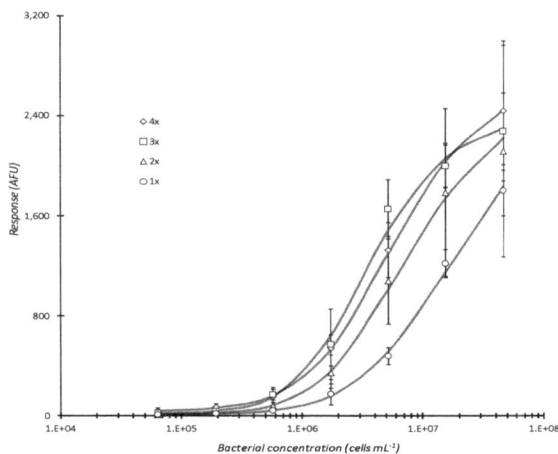

(**E**)

Figure 3. Comparison of incubation conditions in microarray detection of bacteria. Serially diluted analyte, live *E. coli* O157:H7, captured by passively adsorbed biotinylated (**A**) or non-biotinylated anti-*E. coli* O157:H7 antibody (**B**), heat-killed *E. coli* O157:H7 captured by biotinylated anti-O157:H7 antibody (**C**), or Stx1 captured with biotinylated anti-Stx1 antibody (**D**) were subjected to different incubation conditions during analyte incubation. Response *vs.* concentration curves are displayed above for static, aspirating/dispensing, shaking, and centrifugation incubation for 60 min at RT. The effect of multiple (1–4×), 5 min centrifugations (with resuspension of analyte mixture following each centrifugation) followed by static incubation for the remainder of 1 h total incubation time was also assessed for live *E. coli* O157:H7 that were reacted with non-biotinylated capture antibody (**E**). Each data point represented the mean ± standard deviation for 4 of 8 daily technical replicates (with 2 highest and 2 lowest values dropped) from serial dilution series combined with other similarly treated replicates from experiments repeated no less than 2 days and no more than 4 days.

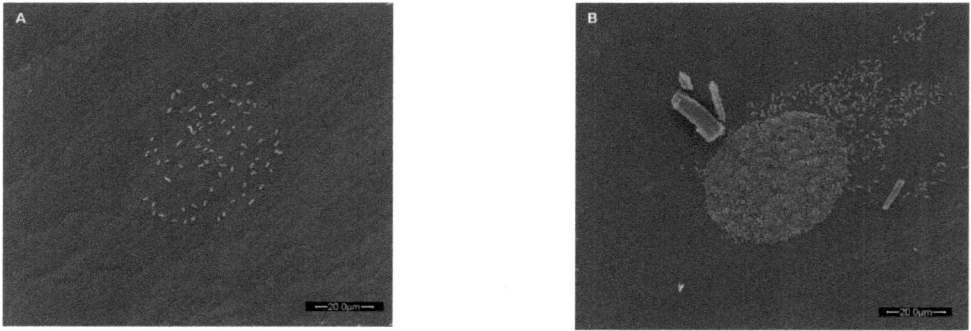

Figure 4. Scanning electron micrograph of bacteria associated with a single microarrayed spot of capture antibody. Live *E. coli* O157:H7, at two initial concentrations of 10^5 CFU/mL (**A**) and 10^8 CFU/mL (**B**) captured by array-printed antibodies (non-biotinylated anti-*E. coli* O157:H7 antibody) passively adsorbed to polystyrene well bottoms of a microtiter plate. Prior to SEM analysis, the BSA blocked plates were centrifuged ($3220 \times$ g for 5 min) to promote capture of bacteria (The scale bar is 20.0 μm in length).

Figure 4 displays scanning electron micrographs of fixed and dried *E. coli* O157:H7 cells captured by antibodies passively adsorbed to the well bottoms of polystyrene microtiter plates. Centrifugation was employed to enhance capture of the cells. Such an investigation may be used to correlate the fluorescence response versus the actual total number of cells associated with the microarrayed capture antibody spot. However, a very interesting observation was made that was particularly evident in Figure 4B. Whether array printing nucleic acids or antibodies, any excess unbound biorecognition element may bind outside of the intended printing area. Further binding of target (and subsequent label to the target) in these regions results in a smear often referred to as a "comet tailing". Such comet tailing has been observed throughout years of our array-based research using various antibody systems at different concentration levels and washing techniques (data not shown). In Figure 4B, the comet tailing appears to represent excess capture antibody not thoroughly removed via washing and since washes were very rapid (<1 min) it thus provides evidence that adsorption of functionally active antibody onto "virgin" polystyrene was almost instantaneous.

The established conditions (5 min centrifugation of analyte with array-printed antibodies and 60 min fluorescent antibody conjugate reaction) were considered an optimal compromise between immunoassay response and total assay time. The combination of these conditions was applied to the co-detection of *E. coli* O157:H7 bacterial cells and Stx1 toxin (Figure 5). Heat-killed cells were specifically selected for testing not only because the effect of centrifugation on bacterial analyte had already been assessed, but more so since it was of concern to task the assay (*i.e.*, test for potential antibody cross-reactivity) with disrupted cells to insure that internal cell components would not affect the assay response when both analytes were combined for detection. Note, the selected Shigatoxigenic strain, B1409, of *E. coli* O157:H7 employed in this study produces only Stx2, and not Stx1. Figure 5A clearly shows that detection of the bacteria was essentially no different (P = 0.104) in the presence or absence of the toxin. The limit of detection, based on the criteria employed with Figure 3E (above), for the heat-killed cells (Figure 5A) was the same for the live cells being ~5.8×10^5 cells/mL. Conversely, detection of Stx1, in the presence of the bacteria, exhibited essentially the same dose-response curve as observed for detection of toxin alone under the same conditions (refer to Figure 3D). The detection limit, also determined as above, for Stx1 was ~110 ng/mL as derived from the data presented in Figure 5B.

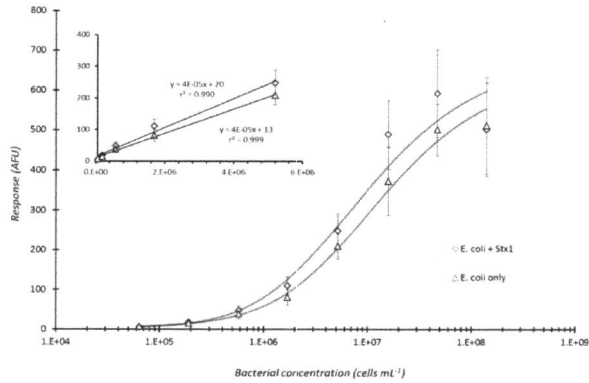

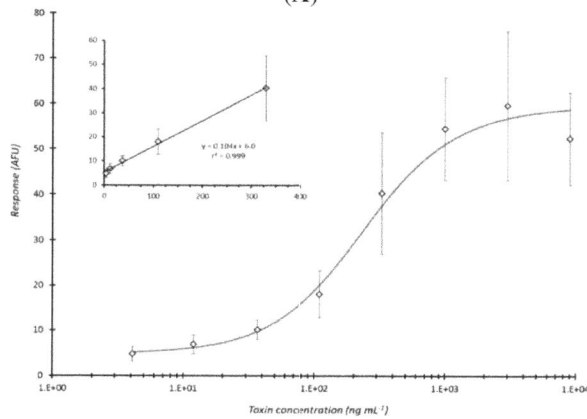

Figure 5. Microarray detection of bacteria and/or toxin under optimized immunoassay conditions. Heat-killed *E. coli* O157:H7 and Stx1 toxin were combined and 3-fold serially diluted in PBS to yield concentration ranges of 1.4×10^8 to 6.4×10^4 cells·mL^{-1} and 9×10^3 to 4.1 ng·mL^{-1} for the bacteria and toxin, respectively. The samples were subjected to microarray detection using immunoassay conditions (sample centrifuged for 5 min with capture antibody (non-biotinylated for *E. coli* and biotinylated for toxin, microarray printed to the bottoms of microtiter plate wells) and then reacted with fluorescent antibody conjugate for 60 min before laser-induced fluorescence scanning) optimized in this investigation. The dose-response curves in (**A**) exhibit the microarray response versus concentration for *E. coli* O157:H7 in the presence or absence of Stx1 whereas (**B**) displays the dose-response curve for Stx1 in the presence of serially diluted *E. coli* O157:H7. Each data point represented the mean ± standard deviation for 4 of 8 daily technical replicates (with 2 highest and 2 lowest values dropped) from serial dilution series combined with other similarly treated replicates from experiments repeated once.

4. Conclusions

This investigation demonstrated that elimination of the costly streptavidin/biotin binding system with passive adsorption of microarray printed antibodies in 96-well, relatively inexpensive polystyrene microtiter plates can be a useful and cost-reduced method for high-throughput, multiplexed detection of analytes. This work further revealed that a 60 min static incubation may be replaced with a much shorter (5 min) centrifugation step that significantly increased detection response for intact bacteria (*E. coli* O157:H7) but not for a considerably smaller proteinaceous toxin (Stx1, a biomolecule of approx. molecular weight of 68 kD). Agitation of aqueous mixtures of analyte by shaking or manual aspiration/dispensing only marginally enhanced detection of live bacteria but had no impact on the detection of Stx1 indicating that intact cells, and not fragments or relatively small biomolecules, were solely influenced by the applied centrifugal force. A prospective alternative to centrifugation may be employing multiwell plates that incorporate biorecognition element arrayed filter membranes, a combination recently exhibited by [33]. Any process that forces analyte and antibody probe and/or subsequent reporter probe into close association will be advantageous to detection as exhibited by the method herein.

With the introduction of centrifugation during exposure of sample to capture antibody, a significant reduction in total assay time was afforded thus representing a major milestone towards the future development of an array-based assay that may be employed for typing mixed cultures within an 8 h workshift. This timeframe will allow for sample preparation (e.g., pre-filtration of extraneous matrix) and a brief growth enrichment culture that may be conducted in an MPN fashion if quantitation is desired. Therefore, intent was to limit the total immunoassay time to ~2.5 h, hence conditions that enhanced assay performance were primarily judged from signal amplitude, and secondarily for absolute error associated with individual data points. Though, as is often observed with immunoassay response curves, absolute error levels increased with analyte concentration, however, relative error generally remained constant. As described above, LOD determination only involved using background response and error as compared to near LOD response and error. Overall, this optimized assay yields conservatively determined limits of detection of 5.8×10^5 cells/mL for both live and heat-killed *E. coli* O157:H7 and 110 ng/mL for Stx1 in a total assay time of ~75 min. These results represent an ~40% improvement in bacterial detection limit for *E. coli* O157:H7 with a corresponding 50% reduction in total assay time as compared with an analogous assay previously developed by this group [15]. Though the results were promising, there is always room for improvement for this as well as other rapid methods since detection of "zero tolerance" pathogens ultimately requiring the need to detect a single cell in approx. 100 g or more of food.

In the future, these assays may incorporate automated plate handling, washing, and pipetting systems, as well as automated sample preparation for enhancement of sensitivity via target concentration achieved with cross-flow microfiltration [34] or antibody-coated paramagnetic particle-based immunomagnetic separation. This multiplex protein microarray format, performed in individual wells of 96-multiwell plates, may be used for high-throughput screening in clinical diagnostics and food testing as well as the characterization of biorecognition elements.

Acknowledgments: Acknowledgments: We gratefully thank Alan Lightfield from the Eastern Regional Research Center, Agricultural Research Service, U.S. Department of Agriculture for mass spectral analysis of select reagents. Mention of brand or firm names does not constitute an endorsement by the USDA over others of a similar nature not mentioned. The USDA is an equal opportunity employer.

Author Contributions: Author Contributions: Andrew G. Gehring, Jeffrey D. Brewster; Shu-I Tu, and Joseph Uknalis conceived and designed the experiments; Andrew G. Gehring, Tawana Simons, and Joseph Uknalis performed the experiments; Andrew G. Gehring, Jeffrey D. Brewster, Yiping He, Peter L. Irwin, George C. Paoli, and Joseph Uknalis analyzed the data: Andrew G. Gehring, George C. Paoli, Yiping He, and Joseph Uknalis provided background information; Andrew G. Gehring, Tawana Simons, and Joseph Uknalis compiled figures/tables and wrote the manuscript; Andrew G. Gehring, Jeffrey D. Brewster, Yiping He, Peter Irwin, George C. Paoli, Tawana Simons, Shu-I Tu, and Joseph Uknalis edited the manuscript.

Conflicts of Interest: Conflicts of Interest: The authors declare no conflict of interest.

References

1. Scallan, E.; Hoekstra, R.M.; Angulo, F.J.; Tauxe, R.V.; Widdowson, M.-A.; Roy, S.L.; Jones, J.L.; Griffin, P.M. Foodborne illness acquired in the United States—Major pathogens. *Emerg. Infect. Dis.* **2011**, *17*, 7–15.

2. Fratamico, P.; Gehring, A.; Karns, J.; van Kessel, J. Chapter 2: Detecting Pathogens in Cattle and Meat. In *Improving the Safety of Fresh Meat*; Sofos, J., Ed.; Woodhead Publishing, Ltd.: Cambridge, UK, 2005; pp. 24–55.

3. Lazcka, O.; Campo, F.J.D.; Muñoz, F.X. Pathogen detection: A perspective of traditional methods and biosensors. *Biosens. Bioelectron.* **2007**, *22*, 1205–1217.

4. Bhunia, A.K. Biosensors and Bio-Based Methods for the Separation and Detection of Foodborne Pathogens. In *Advances in Food and Nutrition Research*; Taylor, S.L., Ed.; Academic Press: Waltham, MA, USA, 2008; pp. 1–44.

5. Rangel, J.M.; Sparling, P.H.; Crowe, C.; Griffin, P.M.; Swerdlow, D.L. Epidemiology of *Escherichia coli* O157:H7 outbreaks, United States, 1982–2002. *Emerg. Infect. Dis.* **2005**, *11*, 603–609.

6. Anonymous. Surveillance for Foodborne Disease Outbreaks—United States, 2007. *Morb. Mortal. Wkly. Rep.* **2010**, *59*, 973–979.

7.   Anonymous.   Ongoing multistate outbreak of *Escherichia coli* serotype O157:H7 infections associated with consumption of fresh spinach—United States, September 2006. *MMWR Morb. Mortal. Wkly. Rep.* **2006**, *55*, 1045–1046.

8.   Velusamy, V.; Arshak, K.; Korostynska, O.; Oliwa, K.; Adley, C. An overview of foodborne pathogen detection: In the perspective of biosensors. *Biotechnol. Adv.* **2010**, *28*, 232–254.

9.   Jasson, V.; Jacxsens, L.; Luning, P.; Rajkovic, A.; Uyttendaele, M. Alternative microbial methods: An overview and selection criteria. *Food Microbiol.* **2010**, *27*, 710–730.

10.   Sapsford, K.E.; Ngundi, M.M.; Moore, M.H.; Lassman, M.E.; Shriver-Lake, L.C.; Taitt, C.R.; Ligler, F.S. Rapid Detection of Foodborne Contaminants Using an Array Biosensor. *Sens. Actuators B Chem.* **2006**, *113*, 599–607.

11.   Liu, Y.; Fratamico, P. *Escherichia coli* O antigen typing using DNA microarrays. *Mol. Cell. Probes* **2006**, *20*, 239–244.

12.   Suo, B.; He, Y.; Paoli, G.; Gehring, A.; Tu, S.I.; Shi, X. Development of an oligonucleotide-based microarray to detect multiple foodborne pathogens. *Mol. Cell. Probes* **2010**, *24*, 77–86.

13.   Shriver-Lake, L.C.; Turner, S.; Taitt, C.R. Rapid detection of *Escherichia coli* O157:H7 spiked into food matrices. *Anal. Chim. Acta* **2007**, *584*, 66–71.

14.   Chen, C.-S.; Durst, R.A. Simultaneous detection of *Escherichia coli* O157:H7, Salmonella spp. and Listeria monocytogenes with an array-based immunosorbent assay using universal protein G-liposomal nanovesicles. *Talanta* **2006**, *69*, 232–238.

15.   Gehring, A.G.; Albin, D.M.; Reed, S.A.; Tu, S.-I.; Brewster, J.D. An antibody microarray, in multiwell plate format, for multiplex screening of foodborne pathogenic bacteria and biomolecules. *Anal. Bioanal. Chem.* **2008**, *391*, 497–506.

16.   Anjum, M.F.; Tucker, J.D.; Sprigings, K.A.; Woodward, M.J.; Ehricht, R. Use of Miniaturized Protein Arrays for *Escherichia coli* O Serotyping. *Clin. Vaccine Immunol.* **2006**, *13*, 561–567.

17.   Desmet, C.; Blum, L.J.; Marquette, C.A. Multiplex microarray ELISA versus classical ELISA, a comparison study of pollutant sensing for environmental analysis. *Environ. Sci. Processes Impacts* **2013**, *15*, 1876–1882.

18.   Ligler, F.S.; Sapsford, K.E.; Golden, J.P.; Shriver-Lake, L.C.; Taitt, C.R.; Dyer, M.A.; Barone, S.; Myatt, C.J. The array biosensor: Portable, automated systems. *Anal. Sci.* **2007**, *23*, 5–10.

19.   Kingsmore, S.F. Multiplexed protein measurement: Technologies and applications of protein and antibody arrays. *Nat. Rev. Drug Discov.* **2006**, *5*, 310–320.

20.   Seidel, M.; Niessner, R. Automated analytical microarrays: A critical review. *Anal. Bioanal. Chem.* **2008**, *391*, 1521–1544.

21.   Zhang, Z.; Li, P.; Hu, X.; Zhang, Q.; Ding, X.; Zhang, W. Microarray technology for major chemical contaminants analysis in food: Current status and prospects. *Sensors* **2012**, *12*, 9234–9252.

22.   Gehring, A.G.; Tu, S.-I. High-Throughput Biosensors for Multiplexed Food-Borne Pathogen Detection. *Annl. Rev. Anal. Chem.* **2011**, *4*, 151–172.

23. Borrebaeck, C.A.K.; Wingren, C. Design of high-density antibody microarrays for disease proteomics: Key technological issues. *J. Proteom.* **2009**, *72*, 928–935.

24. Haab, B.B. Applications of antibody array platforms. *Curr. Opin. Biotechnol.* **2006**, *17*, 415–421.

25. Schaefer, D.W.; Berne, B.J. Number fluctuation spectroscopy of motile microorganisms. *Biophys. J.* **1975**, *15*, 785–794.

26. Bao, Q.; Tian, Y.; Li, W.; Xu, Z.; Xuan, Z.; Hu, S.; Dong, W.; Yang, J.; Chen, Y.; Xue, Y.; *et al.* A Complete Sequence of the *T. tengcongensis* Genome. *Genome Res.* **2002**, *12*, 689–700.

27. Kuznetsova, L.A.; Coakley, W.T. Applications of ultrasound streaming and radiation force in biosensors. *Biosens. Bioelectron.* **2007**, *22*, 1567–1577.

28. Ball, H.J.; Mackie, D.P.; Finlay, D.; McNair, J.; Pollock, D.A. An antigen capture ELISA test using monoclonal antibodies for the detection of Mycoplasma californicum in milk. *Vet. Immunol. Immunopathol.* **1990**, *25*, 269–278.

29. Gehring, A.G.; Patterson, D.L.; Tu, S.I. Use of a light-addressable potentiometric sensor for the detection of *Escherichia coli* O157:H7. *Anal. Biochem.* **1998**, *258*, 293–298.

30. MacBeath, G.; Schreiber, S.L. Printing proteins as microarrays for high-throughput function determination. *Science* **2000**, *289*, 1760–1763.

31. Gehring, A.G.; Albin, D.M.; Bhunia, A.K.; Reed, S.A.; Tu, S.-I.; Uknalis, J. Antibody Microarray Detection of *Escherichia coli* O157:H7: Quantification, Assay Limitations, and Capture Efficiency. *Anal. Chem.* **2006**, *78*, 6601–6607.

32. Steele, R.G.D.; Torrie, J.H. *Principles and Procedures of Statistics, with Special Reference to the Biological Sciences*; McGraw-Hill Book Company: New York, NY, USA, 1960.

33. Le Goff, G.C.; Desmet, C.; Brès, J.-C.; Rigal, D.; Blum, L.J.; Marquette, C.A. Multipurpose high-throughput filtering microarrays (HiFi) for DNA and protein assays. *Biosens. Bioelectron.* **2010**, *26*, 1142–1151.

34. Li, X.; Ximenes, E.; Amalaradjou, M.A.R.; Vibbert, H.B.; Foster, K.; Jones, J.; Liu, X.; Bhunia, A.K.; Ladisch, M.R. Rapid sample processing for detection of food-borne pathogens via cross-flow microfiltration. *Appl. Environ. Microbiol.* **2013**, *79*, 7048–7054.

Portable Nanoparticle-Based Sensors for Food Safety Assessment

Gonca Bülbül and Silvana Andreescu

Abstract: The use of nanotechnology-derived products in the development of sensors and analytical measurement methodologies has increased significantly over the past decade. Nano-based sensing approaches include the use of nanoparticles (NPs) and nanostructures to enhance sensitivity and selectivity, design new detection schemes, improve sample preparation and increase portability. This review summarizes recent advancements in the design and development of NP-based sensors for assessing food safety. The most common types of NPs used to fabricate sensors for detection of food contaminants are discussed. Selected examples of NP-based detection schemes with colorimetric and electrochemical detection are provided with focus on sensors for the detection of chemical and biological contaminants including pesticides, heavy metals, bacterial pathogens and natural toxins. Current trends in the development of low-cost portable NP-based technology for rapid assessment of food safety as well as challenges for practical implementation and future research directions are discussed.

Reprinted from *Sensors*.   Cite as: Bülbül, G.; Hayat, A.; Andreescu, S. Portable Nanoparticle-Based Sensors for Food Safety Assessment. *Sensors* **2015**, *15*, 30736–30758.

1. Introduction

Food safety remains a major concern worldwide. The presence of unsafe levels of chemical and biological toxins in food represents a serious threat to the safety of the food supply and public health. According to the World Health Organization (WHO), foodborne illnesses predominantly affect the economy of underdeveloped nations. Food safety issues in developing countries are widely recognized; estimates indicate around 1500 annually diarrheal episodes occurring globally, 75% of which are attributed to biological contamination of food, resulting in ~3 million deaths [1]. The WHO has placed food safety among its top 11 priorities. The U.S. Centers for Disease Control and Prevention have estimated that 48 million Americans get sick because of contaminated food, 128,000 are hospitalized and 3000 die due to foodborne diseases [2].

In order to manage and overcome the problems related to foodborne illnesses, it is important to develop easy-to-use tests that can rapidly measure the presence of toxic contaminants in food so that remedial actions can be taken. Most analysis of chemical and biological contaminants is performed in centralized laboratories and only a limited number of samples can be tested. More effective methods are

267

needed to facilitate the screening of potentially contaminated samples remotely at food production and handling locations where resources or specialized equipment are not available. Several types of enzyme-based and bioaffinity assays have been reported as alternatives to conventional analytical instrumentation [3–7], albeit with few examples of food safety applications [8–11]. While progress has been made, these assays are still complex and costly, involve multiple analysis steps, addition of sensitive reagents and expensive instrumentation; and are not portable. Easy-to-use inexpensive methods that could be deployed remotely to site locations would significantly improve management and control of food quality and safety.

Nanotechnology-derived products have provided a wide range of material candidates that can be used to increase portability and enhance stability, selectivity and sensitivity of sensors and analytical measurement technologies. Nanotechnology is most widely used in electronics, sensing, biomaterials and catalysis [12–14] and more recently has made its way into the food industry [15]. Current applications comprise: nanomaterial-based encapsulation and delivery systems, antibacterial nanoparticles, NP additives for increasing the flavour and shelf-life of food products and for, tracking, tracing and brand protection [15,16]. Developments in the control of size, surface properties and assembly of NP systems provide opportunities for the development of advanced sensing systems and portable instrumentation that incorporate nanotechnology enabled solutions. Colorimetric [17] and electrochemical [18] detection systems have already been integrated with low-cost platforms such as patterned paper enabling on-site analysis. These portable, low cost and user-friendly sensors have been developed as alternative to conventional analytical methods for point of care medical diagnosis [19], environmental monitoring and food quality control [20]. Application of screen printed carbon electrodes (SPCE) as a portable platform in electrochemical sensors for environmental monitoring and food quality control have been extensively reported [21]. This review summarizes recent advancements in the design and development of NP-based sensors for assessing food safety. Selected examples the from literature on NP-based detection schemes, operational parameters and applications for measurement of food contaminants, as well as challenges for practical implementation and future research directions are discussed.

2. Common Types of Nanostructures in Nanotechnology-Based Sensing Approaches

2.1. Gold NPs

Most common nanotechnology-based sensing approaches utilize noble metal NPs such as gold [22] and silver [23–26]. Such applications are enabled by the useful optical properties of these NPs which can be tuned by changing the size, shape, local environment, and the synthesis method [27,28]. AuNPs have been used as carriers [29] for the biorecognition element such as antibodies or aptamers and as labels for signal transduction and amplification [30]. The basis for the absorption-based colorimetric sensing involves aggregation-induced interparticle surface plasmon coupling of AuNPs which results in a visible color change from red to blue. This concept has provided a practical platform for detection of any target analyte that triggers the AuNPs aggregation or re-dispersion [31]. AuNPs have been widely used to increase surface area and conductivity in electrochemical sensors. A variety of colorimetric and electrochemical assays based on AuNPs have been reported for the detection of chemical contaminants such as alkali and alkaline earth metal ions [32–34], heavy metal ions [35–38] and for assessment of microbiological food contamination like bacteria [39,40]. An example of colorimetric AuNP-based sensing for detection of pathogenic bacteria is shown in Figure 1. A detailed review on AuNP-based sensing has been published [31].

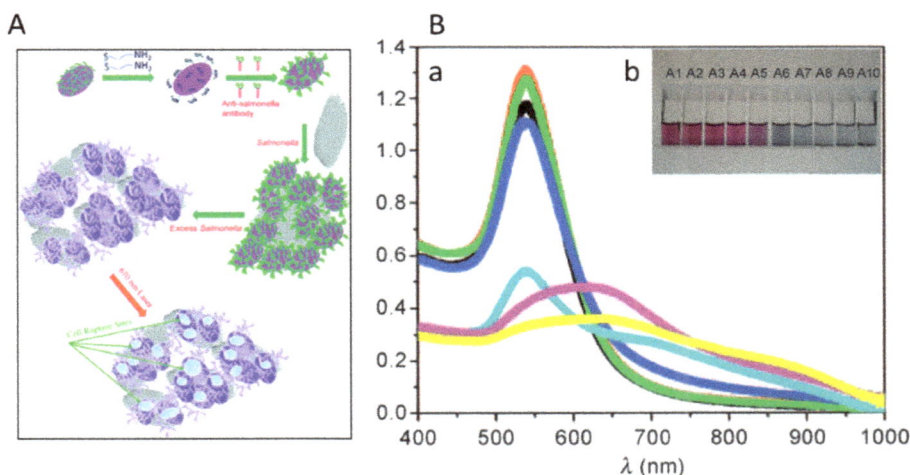

Figure 1. (**A**) Schematic of a colorimetric AuNPs based approach to selectively target and destroy pathogenic bacteria with antibody-conjugated oval-shaped AuNPs; (**B-a**) Absorption-profile of anti-salmonella- antibody-conjugated oval-shaped AuNPs in the absence of bacteria (**red**) due to the addition of 10^6 CFU/mL *E. coli* bacteria (**green**) or different concentrations of *Salmonella bacteria* (10 (**black**), 50 (**dark blue**), 10^3 (**light blue**), 10^4 (**pink**), and 10^5 bacteria (**yellow**)). The new band at around 680 nm following the addition of *Salmonella*, indicates the aggregation of the AuNPs (**B-b**) Photograph shows colorimetric change upon the addition of (A1) 10, (A2) 50, (A3) 100, (A4) 500, (A5) 1000, (A6) 5000, (A7) 10,000, (A8) 50,000, (A9) 100,000, and (A10) 500,000 *Salmonella* (reprinted with permission from [41]).

2.2. Silver NPs

Several sensing systems based on the optical properties of silver NPs (AgNPs) have been reported [24,42]. The color change between dispersed and aggregated AgNPs from yellow to brown can be associated with the concentration change of a target molecule [43]. Based on this principle, various AgNP-based assays have been developed for detection of metal ions [44], proteins [45], melamine [46] and pesticides [47].

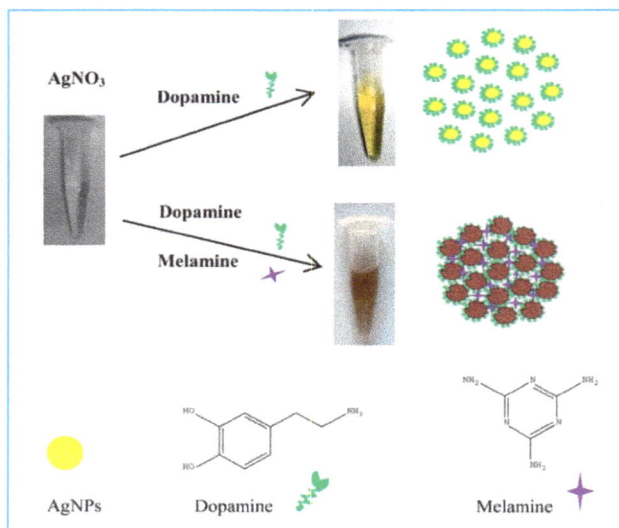

Figure 2. Schematic illustration of colorimetric method with AgNPs/dopamine system for the detection of melamine (reproduced from [48] with permission of The Royal Society of Chemistry).

Figure 2 shows an example of sensing of melamine based on the optical properties of AgNPs [48]. As compared to AuNPs, AgNPs retain higher extinction coefficients and have lower cost [49]. However, less focus has been placed on AgNPs based sensing due to the following limitations: (1) the functionalization of AgNPs can cause chemical degradation of NPs to silver ions and (2) the surface of AgNPs can be easily oxidized [26].

2.3. Cerium Oxide NPs

The use of cerium oxide NPs, or nanoceria as an active sensing component in portable assays is rapidly emerging [50–53]. Nanoceria has the ability to change redox states and surface properties due to the presence of dual reversible oxidation state of cerium Ce(III)/Ce(IV) on the NP surface. Nanoceria particles have been found to possess peroxidase-, superoxide- and oxidase-like activity [54] suggesting that they could potentially replace these enzymes in the development of analytical assays [51]. Nanoceria is optically active and can develop unique color patterns depending upon specific interactions at their surface. These changes can be associated with a target analyte which make these particles an attractive choice from colorimetric sensing [55]. We have used the optical changes of nanoceria upon interaction with phenolics and H_2O_2 to fabricate portable colorimetric sensors for the detection of food antioxidants and glucose in which nanoceria acts as a colorimetric probe [51, 55,56]. When used in conjunction with other metal oxides it is possible to establish a multi-array sensory panel in which each sensor provides a unique signature that could be used for cross-validation and increased accuracy [55,57]. Recently, we have utilized the enzyme mimetic properties of nanoceria to develop a portable enzyme-less electrochemical aptasensor for detection of food contaminants such as mycotoxins [50]. The general sensing design is shown in Figure 3.

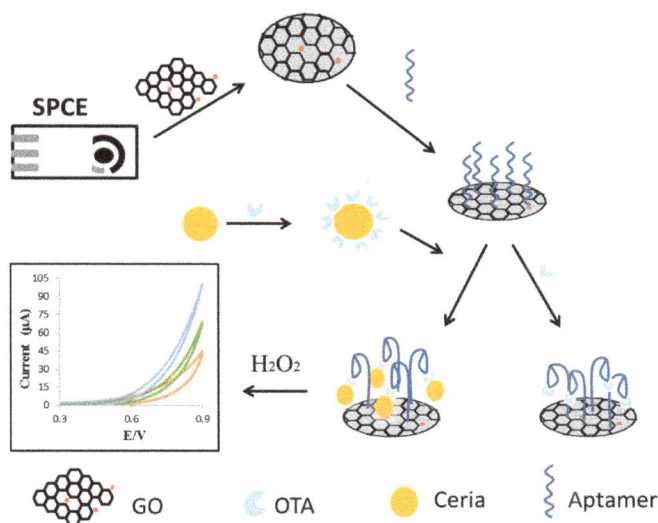

Figure 3. Non-enzymatic electrochemical aptasensor based on the use of a nanoceria tag and graphene oxide (GO) on a screen-printed electrode (SPCE) for the detection of ochratoxin A (OTA) (reproduced from [50] with permission of The Royal Society of Chemistry).

2.4. Carbon Nanotubes and Graphene

Graphene is an isolated single atomic layer of graphite which is currently utilized in electrochemical analysis due to its exceptional conductivity. Since the investigation of exfoliation and characterization of graphene by Geim and Novoselov in 2004 [58], the remarkable electronic transport properties of individual graphene sheets have been demonstrated in many studies [59,60]. The large surface area, high conductivity and ease of modification with biomolecules have been applied to a variety of biosensing systems [61–63], particularly those with electrochemical detection [64,65]. Graphene-based composite materials have been fabricated for electrochemical detection of food contaminants such as bisphenol A [66,67], hydrazine and nitrite [68], organophosphorus pesticides [69] and bacteria [70]. In addition, colorimetric detection of melamine [71], ochratoxin A [72], and mercury (II) (Hg^{2+}) and silver (I) (Ag^+) [73] ions using graphene-based composite materials has also been reported. Applications of graphene-based nanomaterials have been reviewed [74].

2.5. Magnetic Nanoparticles

Magnetic nanoparticles (MNPs) have been mostly utilized as immobilization supports in sensing assays and in the development of immunomagnetic separations and magnetically loaded and controlled sensoring platforms [75]. Functionalized MNPs with a variety of surface groups are readily available, permitting development of different strategies for detection of a variety of analytes [76]. Their large surface area and the increased possibilities for enhancing the assay kinetics, control the loading and improve the immobilization efficiency are some of the advantages of MNPs, which make them one of the most widely used NPs for detection and removal of food contaminants [77]. A multiplexed magnetic microsphere immunoassay for detection of food pathogens developed by Kim *et al.* showed good operational performance in spiked foodstuff such as apple juice, green pepper, tomato, ground beef, alfalfa sprouts, milk, lettuce, spinach, and chicken washes [78]. Magnetic beads conjugated with bacteriophage were utilized for the detection of *E. coli* in drinking water with a detection limit of 10 cfu/mL after pre-enrichment [79]. A sensor design by integrating magnetic nanobeads for detection of organophosphate insecticides using acetylcholinesterase was demonstrated on screen printed carbon electrode surface [80].

2.6. Low-Cost Platforms for Portable NP-Based Detection

Two types of transducer platforms are preferred for the development of inexpensive portable NP-based sensors: (1) screen printed electrodes and (2) paper. The screen printing technology is a low cost process that has been extensively used in artistic applications and for design electronic circuits. In the 80s, the screen printing technology was extended to the fabrication of portable electrochemical sensors, making them more suitable for commercialization [81]. Biosensors based on screen printed electrodes offer the advantages of reduced cost, ease in automation and good reproducibility and sensitivity characteristics. Various nanomaterials including but

272

not limited to carbon nanomaterials, CeO_2, Au, Ag and ZnO NPs have been added as active sensing components to working electrodes to increase surface area, add catalytic properties and amplify electrochemical signals. Some approaches involve the addition of NPs or nanotubes in the composition of screen-printed inks [82]. In other procedures, nanomaterials are drop-casted in DMF/water or electrodeposited on the working electrode surface [83]. The use of nanomaterials in the design of SPCE electrodes provides the following benefits [76]: (a) immobilization support for biomolecules increasing stability and bioactivity; (b) mediator to promote electron transfer reactions, lower the working potential and prevent interferences problems, improving sensitivity and selectivity; (c) electroactive label for electrochemical striping techniques to generate an electrochemical signal; (d) catalyst to amplify the electrochemical signal, enhancing sensitivity. Several examples of disposable nanomaterials-based SPCE electrochemical biosensors have been reported for the detection of food contaminants including pesticides, bacterial toxins and mycotoxins as well as for the detection of food antioxidants [84]. Specific examples will be discussed in the following section.

Another type of material that has received significant attention as a sensing platform in the last few years is paper. Starting with Whitesides' report in 2007, there has been a tremendous effort to develop paper-based low-cost sensors as alternatives to conventional methods for field analysis [17,85]. Paper is the simplest, most affordable and abundant material. Examples of paper bioassays include patterned paper fabricated by photolithography for detection of glucose and bovine serum albumin [17], inkjet-printed paperfluidic immuno-chemical sensing device [86], aptamer—NP-based lateral flow devices for detection of DNA sequences [87], inkjet-printed enzyme sensors for the detection of bisphenol A in field samples [88,89]. Paper based sensors are miniaturized, disposable and can be used for on-site analysis. Conductive materials can be added to modify the paper surface and enable electrochemical detection. These platforms have been integrated with colorimetric [90] and electrochemical [85,91] detection methods. The use of paper-based electrochemical sensors has been demonstrated for detection of analytes of interest in environmental monitoring, health care and food quality control [92]. Baxter *et al.* have proposed a simple and economical process to fabricate gold electrodes on paper using a camera flach sintering step [93]. Nie *et al.* [94] have integrated an electrochemical paper sensor with a commercial glucometer and have demonstrated applicability for on-site analysis of ethanol in food. The detection involved the enzymatic conversion of ethanol with alcohol dehydrogenase in the presence of β-NAD+. Ferricyanide was used as a mediator to enhance electron transfer [94]. Most paper based biosensors are still in their early developmental stage, especially in the field of food quality control. We have recently developed a portable and reagentless NP-based paper platform to detect oxidase enzyme substrates (e.g., glucose) [51] and polyphenols in food samples (e.g., wild mushrooms, wine, juice and green tea) [55] using the redox and surface chelating properties of nanoceria. Detection of the analyte was performed by quantifying the color change of the particles after reaction with the product of the enzymatic reaction (H_2O_2) or the polyphenol. All reagents needed for detection were immobilized onto paper. There was no need for addition of external reagents or the use of a power supply to perform analysis; the only step needed for analysis was the addition of the analyte. The enzyme sensor was very robust and stable over several months. The nanoceria-based antioxidant assay operates similar to a small sensor patch that changes color after contact with antioxidants (Figure 4). The sensor provided an optical signature of the antioxidant power of the sample [57]. When used in conjunction with other metal oxides it is possible to establish a multisensor panel in which each sensor provides a unique signature that could be used for cross-validation and increased accuracy [55,57]. This sensor has demonstrated applicability and excellent performances of field analysis of the brewing conditions for a large number of green tea samples [95].

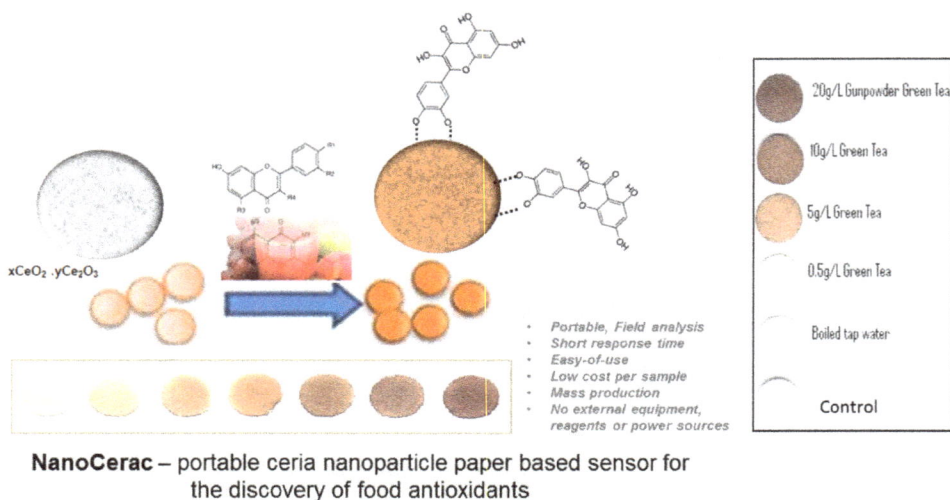

NanoCerac – portable ceria nanoparticle paper based sensor for the discovery of food antioxidants

Figure 4. Operational concept of the NanoCerac assay based on surface-immobilized nanoceria particles for field analysis of antioxidants (reproduced from [55] with permission of The Royal Society of Chemistry).

3. NP-Based Technologies for the Detection of Biological and Chemical Contaminants

3.1. Current Status of Foodborne-Related Illnesses

The food industry is the largest manufacturing sector in the world [96]. Unwashed foods, inadequate processing or cooking, poor quality of water can be a direct source of foodborne diseases. Excessive use of agrochemicals (pesticides, plant growth regulators, veterinary drugs) and presence of environmental contaminants and harmful microorganisms such as pathogenic bacteria, viruses, or parasites are the common causes of food contamination. According to the U.S. Food and Drug Administration (US-FDA) the primary contaminants leading to foodborne illness are pathogenic microorganisms including: *Bacillus cereus*, *Clostridium botulinum*, *Escherichia coli* (*E. coli*), hepatitis A, *Listeria monocytogenes*, *Noroviruses*, *Salmonella* and *Staphylococus aureus*, among others. Other contaminants, such as the toxic fungal metabolites known as mycotoxins are rapidly emerging. Currently, there are limited methods for field detection of toxins and foodborne pathogens, making early identification of a possible contamination difficult. To prevent and manage contamination, there is a need to develop portable and inexpensive detection tools that can provide effective screening and facilitate the analysis at food production and handling locations where specialized equipment is not available. Being able to assess the safety of food from the production process until it reaches the consumer is important for assessing food related health risks. The integration of NPs in the analysis methods for detection of food contaminants has improved the detection sensitivity and increased portability. However, the implementation of these methods for the analysis of real samples still remains as a challenge due to the uncontrollable properties of NPs in complex environments such as aggregation, non-specific signals due to interferences and chemical reactions with food constituents. In the following sections we discuss recently reported NP-based portable assays for the detection of chemical and biological contamination with their analytical performance characteristics for the analysis of food.

3.2. Detection of Microbial Contamination

The presence of pathogens in food or water can cause foodborne infections. They comprise bacteria, viruses, fungi and parasites [97]. The most common foodborne infections are caused by *Campylobacter* spp., *Salmonella* spp. and *E. coli* O157:H7 [98,99]. The conventional method for food pathogen detection is colony counting (CFU) on an agar plate which takes 2–3 days for initial results, and up to 1 week for confirming pathogen specificity [98]. This method is time consuming and laborious. Polymerase chain reaction (PCR) and enzyme-linked immunosorbent assay-based (ELISA) can be used as alternative to traditional CFU methods [100].

While these methods have high sensitivity and selectivity, both PCR and ELISA are still slow, labour-intensive and costly to implement in resource-limited settings. Current research activities target development of methods that can be adapted on portable platforms to enable rapid testing of a range of pathogens with potential for on-site analysis [101]. This section provides examples of recently developed NP-based assays for the detection of pathogens, with specific examples for *E. coli* and *Salmonella* spp.

E. coli O157:H7 [102], which can contaminate ground beef, raw milk, poultry products, cold sandwiches, vegetables, and drinking water supplies [103–105], is recognized as one of the most dangerous pathogens. Most methods designed for on-site detection of *E. coli* involve competitive displacement assays [39] or immuno-chromatographic test strips [106]. Miranda *et al.* developed a hybrid colorimetric enzymatic nanocomposite biosensor for the detection of *E. coli* in aqueous solutions based on enzyme amplification. The efficiency of the method was demonstrated in both solution and test strip format [40]. In this design, cationic AuNPs featuring quaternary amine head groups are electrostatically bound to an anionic enzyme, β-galactosidase, leading to inhibition of the enzyme activity. Upon binding of bacteria to the AuNPs, β-galactosidase is released restoring its activity. The binding event was quantified by colorimetric means. Using this strategy, bacteria at concentrations of 1×10^2 bacteria/mL in solution and 1×10^4 bacteria/mL in a field-friendly test strip format have been quantified (Figure 5) [40].

An immuno-chromatographic (IC) test strip against *E. coli* O157 in enriched samples (raw beef, pork, bovine feces and swine feces) was developed by Jung *et al.* [107]. The test, fabricated in a sandwich format, utilizes a murine monoclonal antibody against *E. coli* O157:H7 conjugated to colloidal AuNPs. Sensitivity of the IC strip was assessed with a10 fold diluted *E. coli* O157:H7 sample with a range of 1.8×10^7 to 1.8 colony-forming units (CFU)/mL in enriched raw beef. The detection limit was 1.8×10^5 CFU/mL without enrichment and 1.8 CFU/mL after enrichment. 48 of pure bacteria cultures (32 *E. coli* strains and 16 non-*E. coli* strains) were tested to determine the specificity.

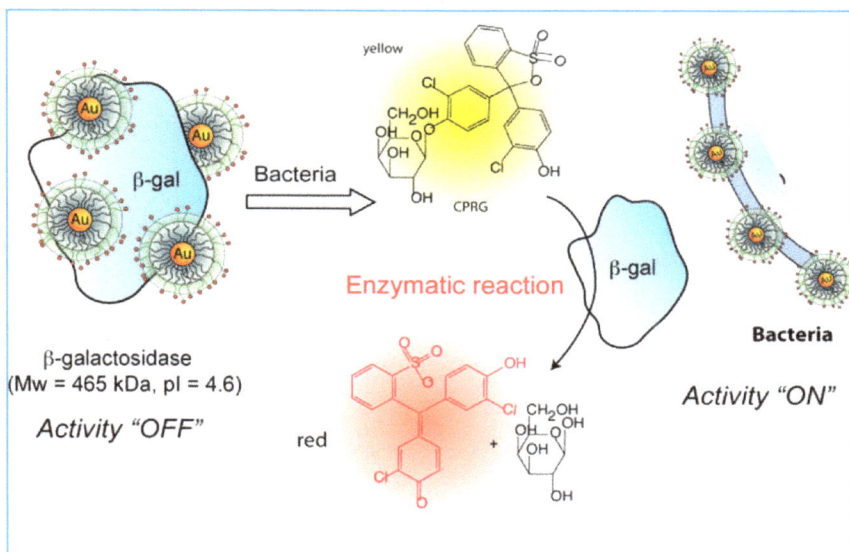

Figure 5. The schematic representation of the sensing mechanism of the colorimetric enzymatic nanocomposite biosensor for the detection of *E. coli* (reprinted with permission from [40]; copyright 2015 American Chemical Society).

Three strains showed positive signal to *E. coli* O157:H7 by the IC strip while the other 29 *E. coli* serotypes were negative. Among the non-*E. coli* strains, only *Citrobacter amalonaticus* yielded a positive signal. The specificity of the strip was higher with pork samples (98.8%) than with bovine (87.9%) and swine (93.4%) feces samples. The assay results were in agreement with the traditional culture procedure for the analysis of samples enriched with *E. coli* O157:H7. In another work, Hossain *et al.* reported a paper test strip for detection of *E. coli* [101]. The strip was based on the measurement of intracellular enzymes (β-galactosidase or β-glucuronidase) activity. The test was fabricated on paper strips (0.5 × 8 cm), onto which either 5-bromo-4-chloro-3-indolyl-β-D-glucuronide sodium salt (XG), chlorophenol red β-galactopyranoside (CPRG) or both and $FeCl_3$ were entrapped using sol– gel-derived silica inks in different zones via an ink-jet printing technique. The XG is broken down by β-glucuronidase into D-glucuronic acid and ClBrindoxyl followed by oxidation of the latter into ClBr-indigo dye, a blue product. On the other hand, CPRG (yellow colour) is broken down by β-galactosidase into chlorophenol red, a red-magenta product. The enzyme β-galactosidase has been widely used for counting total coliforms because the coliforms are generally β-galactosidase positive [108]. In this design [105] formation of red magenta on paper indicated the presence of coliforms. Most *E. coli* strains possess β-glucuronidase activity [109] but the pathogenic *E. coli* O157:H7 doesn't [110]. Therefore, the formation of blue color on the paper strips indicated the presence of non-pathogenic *E. coli*, while absence of blue color (and presence of red-magenta) was used as an indicator for pathogenic *E. coli*. Using immunomagnetic NPs for selective pre-concentration, the limit of detection was 5 CFU/mL for *E. coli* O157:H7 and 20 CFU/mL for *E. coli* BL21, within 30 min without cell culturing. Jokerst *et al.* developed a paper-based sensor for colorimetric detection of foodborne pathogens by measuring the color change when an enzyme associated with the pathogen of interest reacts with a chromogenic substrate (Figure 6) [99]. When combined with an enrichment procedure step of 12 h or less, the paper-based device was capable of detecting bacteria at a concentration of 10^1 CFU/cm^2.

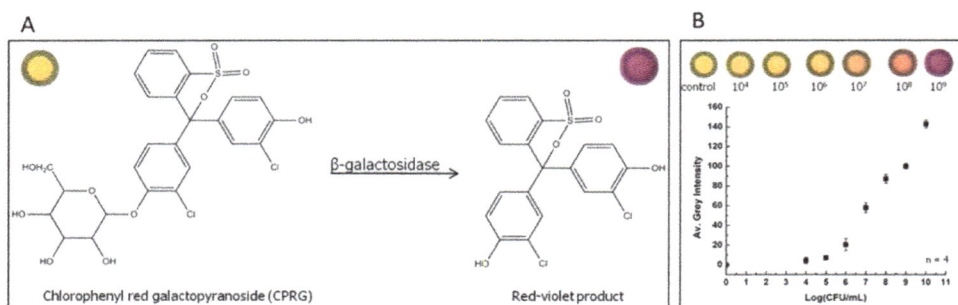

Figure 6. (**A**) Schematic shows the enzymatic reaction between galactosidase and chlorophenyl red galactopyranoside; (**B**) Calibration curve for detection of *E. coli* has been drawn for determination of the limit of detection for live bacterial assay (reprinted with permission from [99]; copyright 2015 American Chemical Society).

Several lateral flow strip test kits which enable low-cost detection of *E. coli* O157 in food are commercially available: MaxSignal®, RapidChek®, Gen-Probe®, IQuum®, Watersafe®, and others. The limit of detection of bacteria using such strips ranges from 10^4 to 10^7 CFU/mL without an enrichment step. The detection limit for *E. coli* O157:H7 is ~10^5 CFU/mL. Yet, culturing steps of at least 8 h are mostly required in order to reach a detection limit of 1 CFU/mL [101].

Another foodborne pathogen, *Salmonella* spp. causes one of the most common and widely distributed bacterial diseases, salmonellosis [111] with thousands of cases reported annually [111–113]. Most reported NP-based assays for *Salmonella* involve the use of immunomagnetic separation with immuno-modified magnetic NPs (MNPs). Joo et al detected *Salmonella* in milk by using MNPs and TiO$_2$ nanocrystals [114]. In this design, *Salmonella* was selectively captured, concentrated and separated from solution by antibody-immobilized magnetic NPs (Figure 7). Subsequent binding of antibody-conjugated nanocrystals to the MNP–*Salmonella*

278

complexes was monitored by absorbance measurement. The method enabled detection of 100 CFU/mL for *Salmonella* in milk.

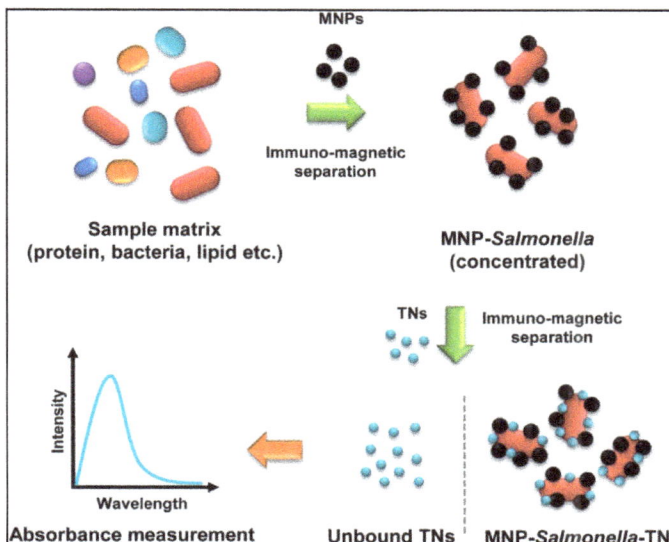

Figure 7. A schematic representation of the pathogenic bacteria detection method by using magnetic nanoparticles and optical nanoprobes (reproduced from [114] with permission of The Royal Society of Chemistry).

Huang *et al.* demonstrated the use of amine-functionalized (AF) MNPs for rapid capture and removal of bacterial pathogens from water, food matrixes and urine samples [115]. The positive charges on the surface of AF-MNPs induce strong electrostatic interactions with the negatively charged sites on the surface of bacterial pathogens resulting in efficient adsorption of bacteria on the particle surface.

3.3. Detection of Pesticides

The increased use of pesticides in agriculture raises public concern regarding the safety of food products. Among pesticides, organophosphorus (OP) and carbamates (C) are the most widely employed, representing ~40% of the world market of this class of compounds [116,117]. Their mode of action involves inhibition of acetylcholinesterase (AChE) enzyme which catalyzes the hydrolysis of neurotransmitter acetylcholine [118,119]. Reference methods for these compounds include chromatographic techniques (GC and HPLC) and coupled chromatographic-spectrometric procedures such as GC-MS and HPLC-MS. These techniques are expensive, time-consuming, and are not easily adaptable for *in situ* monitoring. Screening of multiple samples for pesticide contamination at

279

site locations requires sensitive, selective and robust methods that can be used on site [120]. To increase portability and improve detection capabilities several examples of NP-based assays have been developed as alternative to chromatographic techniques [121]. Despite many reports on sensors for pesticides, the application of these devices for the analysis of real food sample has been scarcely demonstrated.

The most popular sensing configurations involve measurement of AChE inhibition with detection of the enzyme activity before and after exposure to pesticides. AChE activity is typically measured by the colorimetric Ellman assay [122]. Hossain *et al.* developed a paper-based solid phase sensor fabricated by inkjet printing the AChE enzyme within sol-gel derived silica layers onto paper [123]. Pesticides were detected by measuring the residual AChE activity on paper, by using the Ellman's colorimetric assay. The detection was demonstrated on lateral flow and dipstick formats with detection limits of ~100 nM for paraoxon and 30 nM for aflatoxin B1, and a rapid response time (<5 min). In follow up work, Hossain *et al.* improved the design by depositing all the required reagents together with the enzyme onto paper [124]. Figure 8 shows an example of colorimetric "dipstick" bioassay based on AChE-catalyzed enlargement of AuNPs (3 nm) co-entrapped with the enzyme on paper [125]. Both the acetylthiocholine substrate and Au(III) salt were spotted on paper. Hydrolysis of the enzyme substrate generated thiocholine, which further reduced the Au(III) to AuNPs, inducing particle growth and resulting in an increase in color intensity. The color produced was correlated with enzyme inhibition by pesticides. A linear range from 500 nM to 1 mM was reported for paraoxon.

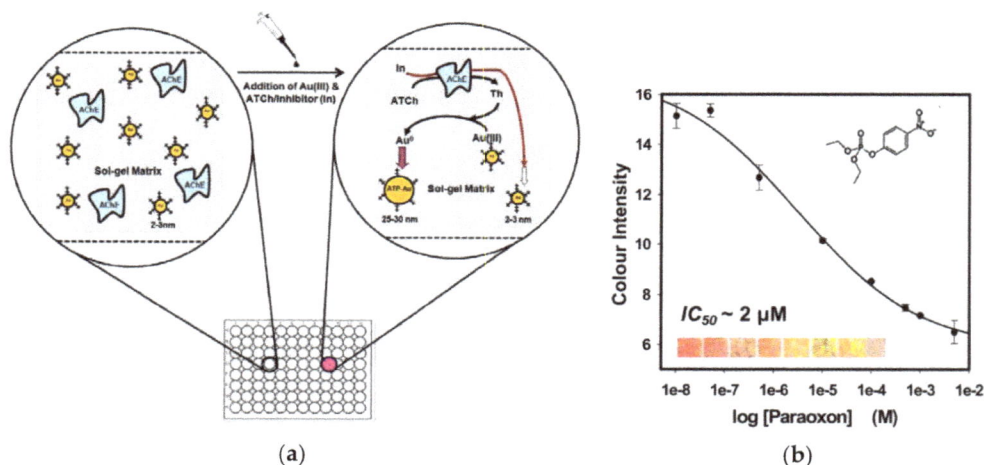

(a)   (b)

Figure 8. (a) Principle of a AChE colorimetric dipstick based on catalytic enlargement of AuNPs and (b) calibration curve for the detection of paraoxon (reproduced from [125] with permission of The Royal Society of Chemistry).

Liang *et al.* reported a colorimetric method using a bi-enzyme system, AChE and choline oxidase (ChO), and Fe_3O_4 MNPs with peroxidase mimetic activity to detect of OPs and nerve agents [126]. ChO catalyzed the conversion of the product of the AChE reaction, choline to hydrogen peroxide (H_2O_2). The produced H_2O_2 was then detected colorimetrically by a color change generated by the catalytic action of the MNPs on the oxidation of 3,5,3′,5′-tetramethylbenzidine (TMB). In the presence of pesticides, the enzymatic activity of AChE was inhibited and less H_2O_2 was produced. The decrease in the color intensity was used to quality AChE inhibition by pesticides (Figure 9).

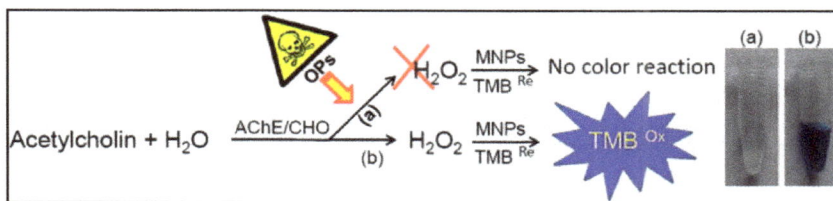

Figure 9. Sensing principle of the assay based on Fe_3O_4 magnetic nanoparticle for OPs detection (reprinted with permission from [126]; copyright 2015 American Chemical Society).

Detection of pesticides has also been realized with NP-based electrochemical immunosensing on low cost SPCE [127]. Figure 10 shows an example of a SPCE electrode functionalized with ZrO_2 NPs for the detection of phosphorylated AChE. ZrO_2NPs were used as sorbents for enzyme capture while quantum dots (QDs) were used as tags to label anti-AChE antibody and form a sandwich-like immunoreaction. The immunocaptured QD were determined by electrochemical stripping analysis of Cd ions after an acid-dissolution step of the QDs. The assay was used to detect AChE activity and paraoxon as an example of OP target and could be potentially extended to analysis in food.

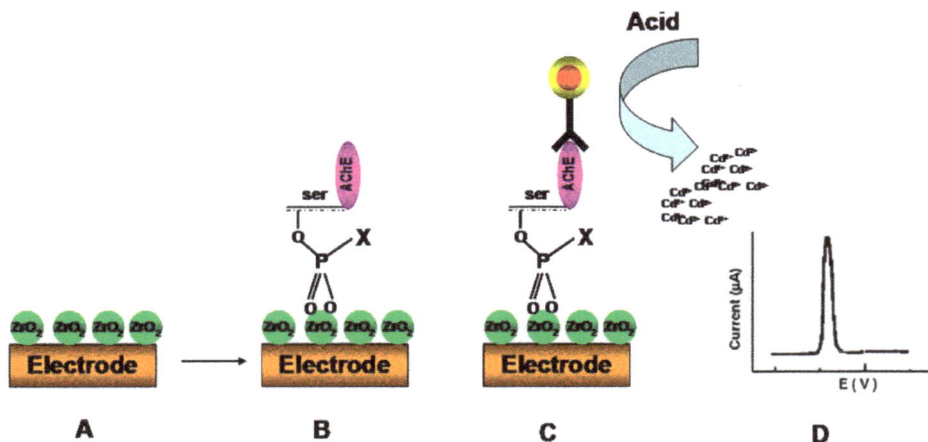

Figure 10. Electrochemical immunosensing of phosphorylated AChE using a ZrO_2 NP-modified SPE (**A**) with selective capturing of phosphorylated AChE adducts (**B**) and detection via (**C**) Immunoreaction between bound phosphorylated AChE adducts and QD-labeled anti-AChE antibody (**D**) Representative voltammogram (reprinted with permission from [127]).

In addition to systems based on optical and electrochemical transduction, functionalized NPs can also be used to develop assays with chemiluminescent (CL) detection. A recent report demonstrates the use of luminol-functionalized AgNPs (Lum-AgNPs) as functional nanomaterials (Figure 11) for the detection of OP and C pesticides, including dimethoate, dipterex, carbaryl, chlorpyrifos, and carbofuran, at a concentration of 24 µg/mL [128]. The Lum-AgNPs were used in conjunction with a H_2O_2 based CL detection to generate a CL "fingerpring" related to each specific pesticide.

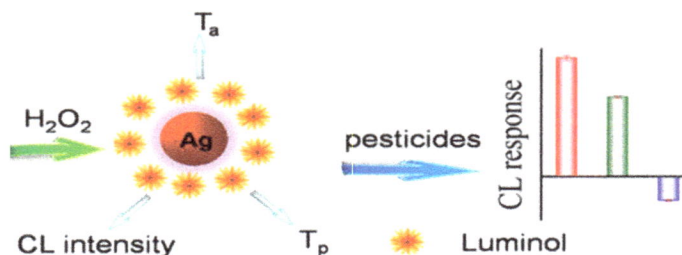

Figure 11. Chemiluminescence-based detection of pesticides using Lum-functionalized AgNPs (reprinted with permission from [128]; copyright 2015 American Chemical Society).

3.4. Detection of Metal Contaminants

Several metal ions such as arsenic, cadmium, lead and mercury are present into the environment and can be found as residues in food. Excess exposure to toxic metal ions can cause neurological, reproductive, cardiovascular, and developmental disorders [129]. Several NP-based sensors have been developed for detection of metal ions [129]. Most of these assays are based on Au and Ag NPs aggregation [34,42]. A portable lab-on-chip system for colorimetric detection of metal ions in water based on AuNP aggregation was developed by Zhao *et al.* [130]. The system provided detection limits of 30 ppb for Pb^{2+} and 89 ppb for Al^{3+}. AuNPs based colorimetric detection was also demonstrated for Hg^{2+} [131–135], Cu^{2+} and Ag^+ [136], Mn^{2+} [137], Cd^{2+} [138], Fe^{3+}, Pb^{2+}, Al^{3+}, Cu^{2+}, and Cr^{3+} [135]. Baxter *et al.* proposed a simple and economical process for fabricating gold electrodes by applying sintered AuNPs stabilised with 4-(dimethylamino)pyridine to filter paper and used this platform to detect Cu ions [93].

3.5. Detection of Mycotoxins

Toxin fungal metabolites known as mycotoxins can contaminate a wide range of agricultural commodities and are high priority targets for the development of new bioassays. It is estimated that at least 25% of the grain produced worldwide is contaminated with mycotoxins. Even small concentrations of mycotoxins can induce significant health problems including vomiting, kidney disease, liver disease, cancer and death [139]. Aflatoxins, ochratoxins, trichothecenes, zearalenone, fumonisins, tremorgenic toxins, and ergot alkaloids are examples of toxic mycotoxins. Mycotoxins have been implicated in development of cancer by the WHO-International Agency for Research on Cancer in 1993. Naturally occurring aflatoxins are classified as carcinogenic to humans (Group 1) whereas ochratoxins and fumonisins are classified as possible carcinogens (Group 2B) [140]. Aflatoxins are the most studied group of mycotoxins [140]. Another important toxin, ochratoxin A (OTA) is a type of mycotoxin which is produced by several species of *Aspergillus* and *Penicillium* fungi and can be found in a wide variety of food matrices such as cereals, dried fruits, coffee, cocoa, spices, beer, wine and grape juice [141]. Hosseini *et al.* developed a AuNP-based aptasensor for detection of aflatoxin B1. The sensor measured AuNP aggregation due to desorption of the aflatoxin B1 aptamer from the surface of AuNPs after the aptamer-target interaction resulting in the color change of AuNPs from red to purple. A detection limit of 7 nM with a linear range from 80 to 270 nM was reported [142]. A similar approach was described by Luan *et al.* for the detection of aflatoxin B2 [143]. In the absence of aflatoxin B2, the random coil structure of the aptamer stabilizes the surface of AuNPs, which shows a red color under high NaCl conditions. In presence of aflatoxin B2, formation of aflatoxin B2-aptamer conjugate destabilizes the AuNPs from NaCl-induced aggregation, changing the color of the dispersion. The method was characterized by a linear dynamic range from 0.025 to 10 ng·mL^{-1}, and a detection limit of 25 pg·mL^{-1}. Xiao *et al.* demonstrated a colorimetric detection method based on disassembly of AuNP dimers for the detection of OTA (Figure 12). This system was characterized by a low detection of 0.05 nM, with a dynamic range from 0.2 to 250 nM [144]. The proposed sensor was applied for detection of OTA in red wine. The OTA concentration determined by this method was consistent with the result obtained with a commercially available ELISA kit. Soh *et al.* described a colorimetric method for detection of OTA by using aptamer controlled growth of AuNPs [145]. In this system the aptamer-target interactions control the amount of aptamer strands adsorbed on the surface of AuNPs. Depending on the surface coverage, AuNPs grow into morphologically varied nanostructures resulting in different colored solutions. AuNPs with low aptamer coverage produced red-colored solutions, whereas AuNPs with high aptamer coverage produced blue colored solutions. The detection limit for OTA using this method was 1 nM.

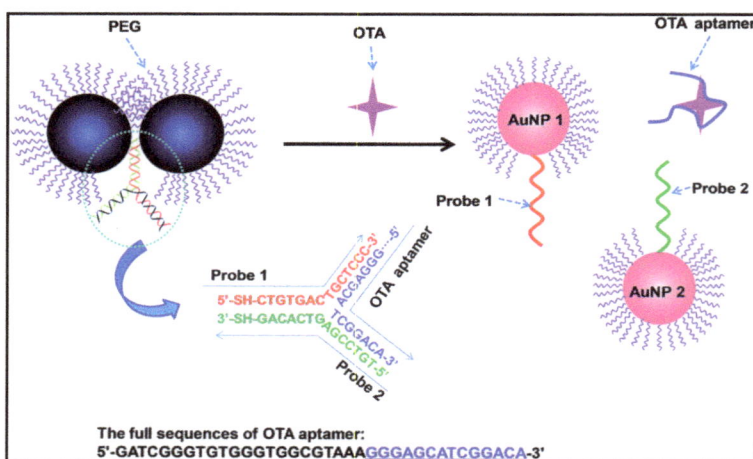

Figure 12. Sensing mechanism of the AuNP dimer-based colorimetric aptasensor for the detection of OTA (reproduced from [144] with permission of The Royal Society of Chemistry).

Other assays are using MNPs [146,147] to design immunomagnetic separation and detection platforms for mycotoxins. Electrochemical immunosensors and aptasensors employing magnetic NPs as immobilization support have been designed for the detection of mycotoxins such as OTA and aflatoxin [75,148]. Figure 13 shows an example of aptasensor platform for detection of OTA. A generic fluorescent aptasensing platform was designed by employing carboxy-modified fluorescent particles as a signal generating probe and magnetic particles as a solid separation support [149]. Table 1 provides the summary of analytical characteristics of portable NP-based sensors reported in literature for the detection of food contaminants.

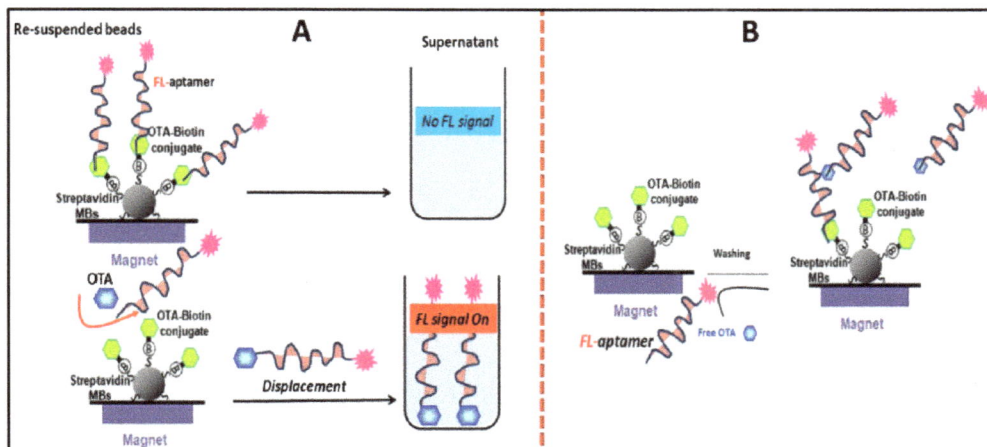

Figure 13. Schematic represents the principle of the assay for the fluorescence detection methodologies. (**A**) displacement assay; (**B**) competition assay. (With permission from Springer Science and Business Media: Analytical and Bioanalytical Chemistry. Development of an aptasensor based on a fluorescent particles-modified aptamer for ochratoxin A detection. 407, 2015, 7815–7822 Hayat, A.; Mishra, R.K.; Catanante, G.; Marty, J.L. Scheme 1.).

Table 1. Summary of analytical characteristics of reviewed studies on portable NP-based sensors for the detection of food contaminants.

	Analyte	Assay Format	NPs Used	Detection Principle	Detection Limit	Dynamic Range	Refs.
Microbial Contaminants	E. coli (XL1)	Colorimetric	AuNPs	Binding of bacteria to AuNPs leads the release of the enzyme attached on particles, restoring its activity	1×10^2 bacteria/mL	1×10^2–1×10^7 bacteria/mL	[40]
	E. coli O157:H7 / E. coli BL 21	Colorimetric	MNPs	The activity of intracellular enzymes measured	5 cfu/mL / 20 cfu/mL	-	[101]
	E. coli O157:H7	Colorimetric	AuNPs	Monoclonal antibody is conjugated to the colloidal AuNPs, and the sandwich format is utilized	1.8 cfu/mL	1.8 cfu–1.8×10^7 cfu/mL	[107]
	Salmonella	Colorimetric	MNPs and TiO$_2$ nanocrystals	Salmonella were captured by antibody-immobilized magnetic nanoparticles and separated from solution, subsequent binding of antibody-conjugated TNs to the MNP–Salmonella complexes increased absorption	>100 cfu/mL	10^8–10^2 cfu/mL	[114]
	E. coli / Sarcina lutea / Proteus vulgaris	Colorimetric	MNPs	Amine-functionalized (AF) MNPs used for rapid capture and removal of bacterial pathogens by using plate counting method	-	-	[115]
Pesticides	Paraoxon	Colorimetric	AuNPs	Bioassay based on AChE-catalysed enlargement of AuNPs co-entrapped with the enzyme on paper. Hydrolysis of the enzyme substrate generated thiocholine, which further reduced the Au(III) to AuNPs, inducing particle growth and resulting in an increase in color intensity	0.5 µM	500 nM–1 mM	[125]
	Sarin / Methyl-paraoxon / Acephate	Colorimetric	MNPs	ChO catalyzed the conversion of the product of the AChE reaction to hydrogen peroxide (H_2O_2). Produced H_2O_2 was then detected by the catalytic action of the MNPs on the oxidation of its substrate generating a color change.	1 nM / 10 nM / 5 µM	-	[126]
	Phosphorylated AChE	Electrochemical	ZrO$_2$ NPs	ZrO$_2$NPs were used as sorbents for enzyme capture while quantum dots (QDs) were used as tags to label anti-AChE antibody and form a sandwich-like immunoreaction. The immunocaptured QD were determined by electrochemical stripping analysis of Cd ions after an acid-dissolution step of the QDs	8.0 pM	10 pM to 4 nM	[127]
	Dimethoate / Dipterex / Carbaryl / Chlorpyrifos / Carbofuran	Chemiluminescent	AgNPs	The array is based on the triple-channel properties of luminol-functionalized AgNPs and hydrogen peroxide chemiluminescent (CL) system containing CL intensity	24 µg/mL	-	[128]

Table 1. *Cont.*

	Analyte	Assay Format	NPs Used	Detection Principle	Detection Limit	Dynamic Range	Refs.
Metals	Pb^{2+}	Colorimetric	AuNPs	Analyte induced aggregation/disaggregation phenomena of AuNPs	400 μM		[35]
	Hg^{2+}	Colorimetric	AuNPs	Analyte induced aggregation/disaggregation phenomena of AuNPs	0.8 nM	50–250 nM	[131]
	Hg^{2+}	Colorimetric	AuNPs	Analyte induced aggregation/disaggregation phenomena of AuNPs	8 nM	0.01–5 μM	[133]
	Hg^{2+}	Colorimetric	AuNPs	Analyte induced aggregation/disaggregation phenomena of AuNPs	50 nM	25–750 nM	[134]
	Hg^{2+}	Colorimetric	AuNPs	Analyte induced aggregation/disaggregation phenomena of AuNPs	53 nM	33–300 nM	[137]
	Pb^{2+}				1 6 nM	16×10^{-9} to 100×10^{-9} M	
	Hg^{2+}		AgNPs		16 nM	16–660 nM	
	Mn^{2+}					16×10^{-9}–50×10^{-8} M	
	Cd^{2+}	Colorimetric	AuNPs	Analyte induced aggregation/disaggregation phenomena of AuNPs	16.6 nM	0.5–16 μM	[138]
	Aflatoxin B1	Colorimetric and Chemiluminescence	AuNPs	Analyte induced aggregation/disaggregation phenomena of AuNPs	7 nM	80–270 nM	[142]
	Aflatoxin B2	Colorimetric	AuNPs	Analyte induced aggregation/disaggregation phenomena of AuNPs	25 pg/mL	0.025–10 ng/mL	[143]
	Ochratoxin A	Colorimetric	AuNPs	Analyte induced aggregation/disaggregation phenomena of AuNPs	0.05 nM	0.2 nM to 250 nM	[144]
Mycotoxins	Aflatoxin B1 (AFB)	Colorimetric	AuNPs	Competitive immunoassay between AFB modified magnetic beads and free AFB for AuNPs labelled antibodies.	12 ng/L	20–800 ng/L	[146]
	Ochratoxin A (OTA)	Fluorescent	Fluorescent particles	OTA-MBs (magnetic beads) were immobilized inside the wells and the analysis was performed by adding the fluorescent particles-modified aptamer which competed with the immobilized OTA and OTA in solution. The presence of OTA in solution prevented binding of the immobilized OTA to the aptamer, leading to a decrease of the fluorescence signal.	0.005 nM	0.1–150 nM	[149]

287

4. Conclusions and Future Directions

NPs and nanostructures have demonstrated ability to significantly enhance detection capabilities of analytical devices. A variety of platforms based on these materials have been reported, and many have shown high sensitivity and low detection limits. The design features indicate that these can be potentially used as portable instrumentation. However, most capabilities have been demonstrated with synthetic samples in laboratory conditions. Some detection protocols involve the use of sensitive reagents and multiple-step procedures which increase measurement time and cost and make field implementation difficult. Translation of this technology into the food safety and regulatory field requires rigorous validation with conventional methodologies, testing of real samples and careful evaluation of interferences. The effect of environmental parameters, storage and operational stability in field conditions should also be established. Moreover, concerns have been raised regarding the potential toxicity of nanomaterials. This aspect should be further considered before these platforms can be introduced into the marketplace.

Analysis of food is a difficult problem due to the inherent complexity of these samples. Challenges for implementation of emerging technologies as viable platforms for assessing food safety and quality are related to interferences and the need for sample preparation. Most developed sensors require sample preparation steps. Some advances have been made with the use of MNPs for immunoseparation. In the future, the integration of sample extraction and separation units with the sensing platforms would greatly improve portability for field use. The long term storage is another challenge especially for systems that include biological sensing components such as enzymes and antibodies. Achieving high specificity is also critical to minimize background signals and reduce the likelihood of false-positive results. Miniaturization, automation, multidetection capabilities and an effort to lower the cost per assay are some of the current trends in this field. The use of inexpensive materials such as paper to build these sensors has demonstrated potential as field-portable devices but their functionality for the analysis of complex food samples and the need for sample pretreatment are yet to be demonstrated. The performance of disposable SPC electrodes has been improved with the use of nanomaterials. Validation and testing of statistically-relevant sample numbers, comparability and inter-laboratory studies to demonstrate robustness of such platforms are the next critical steps for achieving industry acceptance and regulatory approvals. Future work to adapt these sensors so that they can be attached to food packaging to indicate contamination would be highly valuable for on-line control and safety of processed and stored food. Increased portability can also be achieved through connectivity and integration with largely used communication devices such as cell-phones and tablets. However, development of cybersensors for food monitoring is still in infancy and constitutes a fertile area for future investigation.

Acknowledgments: This material is based upon work supported by the National Science Foundation under Grant No. 0954919. Any opinions, findings, and conclusions or recommendations expressed in this material are those of the author(s) and do not necessarily reflect the views of the National Science Foundation.

Conflicts of Interest: The authors declare no conflict of interest.

References

1. Yáñez, L.; Ortiz, D.; Calderón, J.; Batres, L.; Carrizales, L.; Mejía, J.; Martínez, L.; García-Nieto, E.; Díaz-Barriga, F. Overview of human health and chemical mixtures: Problems facing developing countries. *Environ. Health Perspect.* **2002**, *110*, 901–909.

2. *Centers for Disease Control and Prevention (CDC)*; 2010. Available online: http://www.cdc.gov/foodborneburden/ 2011-foodborne-estimates.html (accessed on 5 November 2015).

3. Xiao, Y.; Patolsky, F.; Katz, E.; Hainfeld, J.F.; Willner, I. "Plugging into enzymes": Nanowiring of redox enzymes by a gold nanoparticle. *Science* **2003**, *299*, 1877–1881.

4. Cai, H.; Xu, Y.; Zhu, N.N.; He, P.G.; Fang, Y.Z. An electrochemical DNA hybridization detection assay based on a silver nanoparticle label. *Analyst* **2002**, *127*, 803–808.

5. Xu, J.J.; Zhao, W.; Luo, X.L.; Chen, H.Y. A sensitive biosensor for lactate based on layer-by-layer assembling MnO_2 nanoparticles and lactate oxidase on ion-sensitive field-effect transistors. *Chem. Commun.* **2005**, *6*, 792–794.

6. Fiorito, P.A.; Goncales, V.R.; Ponzio, E.A.; de Torresi, S.I.C. Synthesis, characterization and immobilization of Prussian blue nanoparticles. A potential tool for biosensing devices. *Chem. Commun.* **2005**, *3*, 366–368.

7. Zhuo, Y.; Yuan, R.; Chai, Y.Q.; Tang, D.P.; Zhang, Y.; Wang, N.; Li, X.; Zhu, Q. A reagentless amperometric immunosensor based on gold nanoparticles/thionine/ Nafion-membrane-modified gold electrode for determination of alpha-1-fetoprotein. *Electrochem. Commun.* **2005**, *7*, 355–360.

8. Wu, S.J.; Duan, N.; Ma, X.Y.; Xia, Y.; Wang, H.G.; Wang, Z.P.; Zhang, Q. Multiplexed Fluorescence Resonance Energy Transfer Aptasensor between Upconversion Nanoparticles and Graphene Oxide for the Simultaneous Determination of Mycotoxins. *Anal. Chem.* **2012**, *84*, 6263–6270.

9. Vidal, J.C.; Bonel, L.; Ezquerra, A.; Duato, P.; Castillo, J.R. An electrochemical immunosensor for ochratoxin A determination in wines based on a monoclonal antibody and paramagnetic microbeads. *Anal. Bioanal. Chem.* **2012**, *403*, 1585–1593.

10. Bonel, L.; Vidal, J.C.; Duato, P.; Castillo, J.R. An electrochemical competitive biosensor for ochratoxin A based on a DNA biotinylated aptamer. *Biosens. Bioelectron.* **2011**, *26*, 3254–3259.

11. Yang, C.; Wang, Y.; Marty, J.L.; Yang, X.R. Aptamer-based colorimetric biosensing of Ochratoxin A using unmodified gold nanoparticles indicator. *Biosens. Bioelectron.* **2011**, *26*, 2724–2727.

12. Rae, A. Real Life Applications of Nanotechnology in Electronics. *OnBoard Technol.* **2006**, *2006*, 28.

13. Duncan, T.V. Applications of nanotechnology in food packaging and food safety: Barrier materials, antimicrobials and sensors. *J. Coll. Interface Sci.* **2011**, *363*, 1–24.

14. Daniel, M.-C.; Astruc, D. Gold nanoparticles: Assembly, supramolecular chemistry, quantum-size-related properties, and applications toward biology, catalysis, and nanotechnology. *Chem. Reviews* **2004**, *104*, 293–346.

15. Sozer, N.; Kokini, J.L. Nanotechnology and its applications in the food sector. *Trends Biotechnol.* **2009**, *27*, 82–89.

16. Chellaram, C.; Murugaboopathi, G.; John, A.A.; Sivakumar, R.; Ganesan, S.; Krithika, S.; Priya, G. Significance of Nanotechnology in Food Industry. *APCBEE Proced.* **2014**, *8*, 109–113.

17. Martinez, A.W.; Phillips, S.T.; Butte, M.J.; Whitesides, G.M. Patterned paper as a platform for inexpensive, low-volume, portable bioassays. *Angew. Chem. Int. Ed.* **2007**, *46*, 1318–1320.

18. Reches, M.; Mirica, K.A.; Dasgupta, R.; Dickey, M.D.; Butte, M.J.; Whitesides, G.M. Thread as a Matrix for Biomedical Assays. *ACS Appl. Mater. Interfaces* **2010**, *2*, 1722–1728.

19. Martinez, A.W.; Phillips, S.T.; Whitesides, G.M.; Carrilho, E. Diagnostics for the developing world: Microfluidic paper-based analytical devices. *Anal. Chem.* **2009**, *82*, 3–10.

20. Zhao, W.; Ali, M.M.; Aguirre, S.D.; Brook, M.A.; Li, Y. Paper-Based Bioassays Using Gold Nanoparticle Colorimetric Probes. *Anal. Chem.* **2008**, *80*, 8431–8437.

21. Hayat, A.; Marty, J.L. Disposable screen printed electrochemical sensors: Tools for environmental monitoring. *Sensors* **2014**, *14*, 10432–10453.

22. Weng, Z.; Wang, H.; Vongsvivut, J.; Li, R.; Glushenkov, A.M.; He, J.; Chen, Y.; Barrow, C.J.; Yang, W. Self-assembly of core-satellite gold nanoparticles for colorimetric detection of copper ions. *Anal. Chim. Acta* **2013**, *803*, 128–134.

23. Zhou, Y.; Zhao, H.; He, Y.; Ding, N.; Cao, Q. Colorimetric detection of Cu^{2+} using 4-mercaptobenzoic acid modified silver nanoparticles. *Coll. Surf. A* **2011**, *391*, 179–183.

24. Sherry, L.J.; Jin, R.; Mirkin, C.A.; Schatz, G.C.; Van Duyne, R.P. Localized Surface Plasmon Resonance Spectroscopy of Single Silver Triangular Nanoprisms. *Nano Lett.* **2006**, *6*, 2060–2065.

25. Cao, Y.W.; Jin, A.R.; Mirkin, C.A. DNA-Modified Core-Shell Ag/Au Nanoparticles. *J. Am. Chem. Soc.* **2001**, *123*, 7961–7962.

26. Wei, H.; Chen, C.; Han, B.; Wang, E. Enzyme Colorimetric Assay Using Unmodified Silver Nanoparticles. *Anal. Chem.* **2008**, *80*, 7051–7055.

27. Rosi, N.L.; Mirkin, C.A. Nanostructures in Biodiagnostics. *Chem. Rev.* **2005**, *105*, 1547–1562.

28. Storhoff, J.J.; Lazarides, A.A.; Mucic, R.C.; Mirkin, C.A.; Letsinger, R.L.; Schatz, G.C. What Controls the Optical Properties of DNA-Linked Gold Nanoparticle Assemblies? *J. Am. Chem. Soc.* **2000**, *122*, 4640–4650.

29. Liu, M.; Jia, C.; Jin, Q.; Lou, X.; Yao, S.; Xiang, J.; Zhao, J. Novel colorimetric enzyme immunoassay for the detection of carcinoembryonic antigen. *Talanta* **2010**, *81*, 1625–1629.

30. Ambrosi, A.; Airò, F.; Merkoçi, A. Enhanced Gold Nanoparticle Based ELISA for a Breast Cancer Biomarker. *Anal. Chem.* **2010**, *82*, 1151–1156.

31. Saha, K.; Agasti, S.S.; Kim, C.; Li, X.; Rotello, V.M. Gold Nanoparticles in Chemical and Biological Sensing. *Chem. Rev.* **2012**, *112*, 2739–2779.

32. Lin, S.-Y.; Liu, S.-W.; Lin, C.-M.; Chen, C.-H. Recognition of Potassium Ion in Water by 15-Crown-5 Functionalized Gold Nanoparticles. *Anal. Chem.* **2002**, *74*, 330–335.

33. Lin, S.-Y.; Chen, C.-H.; Lin, M.-C.; Hsu, H.-F. A Cooperative Effect of Bifunctionalized Nanoparticles on Recognition: Sensing Alkali Ions by Crown and Carboxylate Moieties in Aqueous Media. *Anal. Chem.* **2005**, *77*, 4821–4828.

34. Obare, S.O.; Hollowell, R.E.; Murphy, C.J. Sensing Strategy for Lithium Ion Based on Gold Nanoparticles. *Langmuir* **2002**, *18*, 10407–10410.

35. Kim, Y.; Johnson, R.C.; Hupp, J.T. Gold Nanoparticle-Based Sensing of "Spectroscopically Silent" Heavy Metal Ions. *Nano Lett.* **2001**, *1*, 165–167.

36. Lin, S.Y.; Wu, S.H.; Chen, C. A simple strategy for prompt visual sensing by gold nanoparticles: General applications of interparticle hydrogen bonds. *Angew. Chem. Int. Ed.* **2006**, *45*, 4948–4951.

37. Lee, J.-S.; Han, M.S.; Mirkin, C.A. Colorimetric Detection of Mercuric Ion (Hg2+) in Aqueous Media using DNA-Functionalized Gold Nanoparticles. *Angew. Chem. Int. Ed.* **2007**, *46*, 4093–4096.

38. Wang, Z.; Lee, J.H.; Lu, Y. Label-Free Colorimetric Detection of Lead Ions with a Nanomolar Detection Limit and Tunable Dynamic Range by using Gold Nanoparticles and DNAzyme. *Adv. Mater.* **2008**, *20*, 3263–3267.

39. Phillips, R.L.; Miranda, O.R.; You, C.C.; Rotello, V.M.; Bunz, U.H. Rapid and Efficient Identification of Bacteria Using Gold-Nanoparticle–Poly (para-phenyleneethynylene) Constructs. *Angew. Chem. Int. Ed.* **2008**, *47*, 2590–2594.

40. Miranda, O.R.; Li, X.; Garcia-Gonzalez, L.; Zhu, Z.-J.; Yan, B.; Bunz, U.H.; Rotello, V.M. Colorimetric Bacteria Sensing Using a Supramolecular Enzyme-Nanoparticle Biosensor. *J. Am. Chem. Soc.* **2011**, *133*, 9650–9653.

41. Wang, S.G.; Singh, A.K.; Senapati, D.; Neely, A.; Yu, H.T.; Ray, P.C. Rapid Colorimetric Identification and Targeted Photothermal Lysis of Salmonella Bacteria by Using Bioconjugated Oval-Shaped Gold Nanoparticles. *Chem. Eur. J.* **2010**, *16*, 5600–5606.

42. Cao, Y.; Jin, R.; Mirkin, C.A. DNA-modified core-shell Ag/Au nanoparticles. *J. Am. Chem. Soc.* **2001**, *123*, 7961–7962.

43. Vilela, D.; González, M.C.; Escarpa, A. Sensing colorimetric approaches based on gold and silver nanoparticles aggregation: Chemical creativity behind the assay. A review. *Anal. Chim. Acta* **2012**, *751*, 24–43.

44. Yoosaf, K.; Ipe, B.I.; Suresh, C.H.; Thomas, K.G. *In Situ* Synthesis of Metal Nanoparticles and Selective Naked-Eye Detection of Lead Ions from Aqueous Media. *J. Phys. Chem. C* **2007**, *111*, 12839–12847.

45. Schofield, C.L.; Haines, A.H.; Field, R.A.; Russell, D.A. Silver and Gold Glyconanoparticles for Colorimetric Bioassays. *Langmuir* **2006**, *22*, 6707–6711.

46. Song, J.; Wu, F.; Wan, Y.; Ma, L. Colorimetric detection of melamine in pretreated milk using silver nanoparticles functionalized with sulfanilic acid. *Food Control* **2015**, *50*, 356–361.

47. Xiong, D.; Li, H. Colorimetric detection of pesticides based on calixarene modified silver nanoparticles in water. *Nanotechnology* **2008**, *19*, 465502.

48. Ma, Y.; Niu, H.; Zhang, X.; Cai, Y. One-step synthesis of silver/dopamine nanoparticles and visual detection of melamine in raw milk. *Analyst* **2011**, *136*, 4192–4196.

49. Lee, J.-S.; Lytton-Jean, A.K.; Hurst, S.J.; Mirkin, C.A. Silver nanoparticle-oligonucleotide conjugates based on DNA with triple cyclic disulfide moieties. *Nano Lett.* **2007**, *7*, 2112–2115.

50. Bulbul, G.; Hayat, A.; Andreescu, S. A generic amplification strategy for electrochemical aptasensors using a non-enzymatic nanoceria tag. *Nanoscale* **2015**, *7*, 13230–12338.

51. Ornatska, M.; Sharpe, E.; Andreescu, D.; Andreescu, S. Paper Bioassay Based on Ceria Nanoparticles as Colorimetric Probes. *Anal. Chem.* **2011**, *83*, 4273–4280.

52. Hayat, A.; Andreescu, S. Nanoceria Particles As Catalytic Amplifiers for Alkaline Phosphatase Assays. *Anal. Chem.* **2013**, *85*, 10028–10032.

53. Hayat, A.; Bulbul, G.; Andreescu, S. Probing phosphatase activity using redox active nanoparticles: A novel colorimetric approach for the detection of enzyme activity. *Biosens. Bioelectron.* **2014**, *56*, 334–339.

54. Asati, A.; Santra, S.; Kaittanis, C.; Nath, S.; Perez, J.M. Oxidase-Like Activity of Polymer-Coated Cerium Oxide Nanoparticles. *Angew. Chem. Int. Ed.* **2009**, *48*, 2308–2312.

55. Sharpe, E.; Frasco, T.; Andreescu, D.; Andreescu, S. Portable ceria nanoparticle-based assay for rapid detection of food antioxidants (NanoCerac). *Analyst* **2013**, *138*, 249–262.

56. Saha, B.; Evers, T.H.; Prins, M.W.J. How Antibody Surface Coverage on Nanoparticles Determines the Activity and Kinetics of Antigen Capturing for Biosensing. *Anal. Chem.* **2014**, *86*, 8158–8166.

57. Sharpe, E.; Bradley, R.; Frasco, T.; Jayathilaka, D.; Marsh, A.; Andreescu, S. Metal oxide based multisensor array and portable database for field analysis of antioxidants. *Sens. Actuat. B Chem.* **2014**, *193*, 552–562.

58. Novoselov, K.S.; Geim, A.K.; Morozov, S.; Jiang, D.; Zhang, Y.; Dubonos, S.V.; Grigorieva, I.V.; Firsov, A.A. Electric field effect in atomically thin carbon films. *Science* **2004**, *306*, 666–669.

59. Zhang, Y.; Small, J.P.; Amori, M.E.S.; Kim, P. Electric Field Modulation of Galvanomagnetic Properties of Mesoscopic Graphite. *Phys. Rev. Lett.* **2005**, *94*, 176803.

60. Novoselov, K.S.; Geim, A.K.; Morozov, S.V.; Jiang, D.; Katsnelson, M.I.; Grigorieva, I.V.; Dubonos, S.V.; Firsov, A.A. Two-dimensional gas of massless Dirac fermions in graphene. *Nature* **2005**, *438*, 197–200.

61. Schedin, F.; Geim, A.K.; Morozov, S.V.; Hill, E.W.; Blake, P.; Katsnelson, M.I.; Novoselov, K.S. Detection of individual gas molecules adsorbed on graphene. *Nat. Mater.* **2007**, *6*, 652–655.

62. Dan, Y.; Lu, Y.; Kybert, N.J.; Luo, Z.; Johnson, A.T.C. Intrinsic Response of Graphene Vapor Sensors. *Nano Lett.* **2009**, *9*, 1472–1475.

63. Robinson, J.T.; Perkins, F.K.; Snow, E.S.; Wei, Z.; Sheehan, P.E. Reduced Graphene Oxide Molecular Sensors. *Nano Lett.* **2008**, *8*, 3137–3140.

64. Chen, J.; Bo, Z.; Lu, G. Vertically-Oriented Graphene for Sensing and EnvironmentalApplications. In *Vertically-Oriented Graphene*; Springer International Publishing: Cham, Switzerland, 2015; pp. 67–77.

65. Huang, L.; Wu, J.; Zheng, L.; Qian, H.; Xue, F.; Wu, Y.; Pan, D.; Adeloju, S.B.; Chen, W. Rolling Chain Amplification Based Signal-Enhanced Electrochemical Aptasensor for Ultrasensitive Detection of Ochratoxin A. *Anal. Chem.* **2013**, *85*, 10842–10849.

66. Pan, D.; Gu, Y.; Lan, H.; Sun, Y.; Gao, H. Functional graphene-gold nano-composite fabricated electrochemical biosensor for direct and rapid detection of bisphenol, A. *Anal. Chim. Acta* **2015**, *853*, 297–302.

67. Zhou, L.; Wang, J.; Li, D.; Li, Y. An electrochemical aptasensor based on gold nanoparticles dotted graphene modified glassy carbon electrode for label-free detection of bisphenol A in milk samples. *Food Chem.* **2014**, *162*, 34–40.

68. Luo, X.; Pan, J.; Pan, K.; Yu, Y.; Zhong, A.; Wei, S.; Li, J.; Shi, J.; Li, X. An electrochemical sensor for hydrazine and nitrite based on graphene–cobalt hexacyanoferrate nanocomposite: Toward environment and food detection. *J. Electroanal. Chem.* **2015**, *745*, 80–87.

69. Yang, Y.; Asiri, A.M.; Du, D.; Lin, Y. Acetylcholinesterase biosensor based on a gold nanoparticle-polypyrrole- reduced graphene oxide nanocomposite modified electrode for the amperometric detection of organophosphorus pesticides. *Analyst* **2014**, *139*, 3055–3060.

70. Wan, Y.; Lin, Z.; Zhang, D.; Wang, Y.; Hou, B. Impedimetric immunosensor doped with reduced graphene sheets fabricated by controllable electrodeposition for the non-labelled detection of bacteria. *Biosens. Bioelectron.* **2011**, *26*, 1959–1964.

71. Kumar, S.V.; Huang, N.M.; Lim, H.N.; Zainy, M.; Harrison, I.; Chia, C.H. Preparation of highly water dispersible functional graphene/silver nanocomposite for the detection of melamine. *Sens. Actuat. B Chem.* **2013**, *181*, 885–893.

72. Sheng, L.; Ren, J.; Miao, Y.; Wang, J.; Wang, E. PVP-coated graphene oxide for selective determination of ochratoxin A via quenching fluorescence of free aptamer. *Biosens. Bioelectron.* **2011**, *26*, 3494–3499.

73. Zhang, Y.; Zuo, P.; Ye, B.-C. A low-cost and simple paper-based microfluidic device for simultaneous multiplex determination of different types of chemical contaminants in food. *Biosens. Bioelectron.* **2015**, *68*, 14–19.

74. Perreault, F.; de Faria, A.F.; Elimelech, M. Environmental applications of graphene-based nanomaterials. *Chem. Soc. Rev.* **2015**, *44*, 5861–5896.

75. Hayat, A.; Yang, C.; Rhouati, A.; Marty, J.L. Recent advances and achievements in nanomaterial-based, and structure switchable aptasensing platforms for ochratoxin A detection. *Sensors* **2013**, *13*, 15187–15208.

76. Hayat, A.; Catanante, G.; Marty, J.L. Current Trends in Nanomaterial-Based Amperometric Biosensors. *Sensors* **2014**, *14*, 23439–23461.

77. Van Dorst, B.; Mehta, J.; Bekaert, K.; Rouah-Martin, E.; de Coen, W.; Dubruel, P.; Blust, R.; Robbens, J. Recent advances in recognition elements of food and environmental biosensors: A review. *Biosens. Bioelectron.* **2010**, *26*, 1178–1194.

78. Kim, J.; Taitt, C.; Ligler, F.; Anderson, G. Multiplexed magnetic microsphere immunoassays for detection of pathogens in foods. *Sens. Instrum. Food Qual. Saf.* **2010**, *4*, 73–81.

79. Chen, J.; Alcaine, S.D.; Jiang, Z.; Rotello, V.M.; Nugen, S.R. Detection of *Escherichia coli* in Drinking Water Using T7 Bacteriophage-Conjugated Magnetic Probe. *Anal. Chem.* **2015**, *87*, 8977–8984.

80. Dominguez, R.B.; Alonso, G.A.; Muñoz, R.; Hayat, A.; Marty, J.-L. Design of a novel magnetic particles based electrochemical biosensor for organophosphate insecticide detection in flow injection analysis. *Sens. Actuat. B Chem.* **2015**, *208*, 491–496.

81. Jubete, E.; Loaiza, O.A.; Ochoteco, E.; Pomposo, J.A.; Grande, H.; Rodriguez, J. Nanotechnology: A Tool for Improved Performance on Electrochemical Screen-Printed (Bio)Sensors. *J. Sensors* **2009**, *2009*.

82. Shen, J.; Dudik, L.; Liu, C.-C. An iridium nanoparticles dispersed carbon based thick film electrochemical biosensor and its application for a single use, disposable glucose biosensor. *Sens. Actuat. B Chem.* **2007**, *125*, 106–113.

83. Rivas, G.A.; Rubianes, M.D.; Rodríguez, M.C.; Ferreyra, N.F.; Luque, G.L.; Pedano, M.L.; Miscoria, S.A.; Parrado, C. Carbon nanotubes for electrochemical biosensing. *Talanta* **2007**, *74*, 291–307.

84. Vasilescu, A.; Sharpe, E.; Andreescu, S. Nanoparticle-Based Technologies for the Detection of Food Antioxidants. *Curr. Anal. Chem.* **2012**, *8*, 495–505.

85. Metters, J.P.; Houssein, S.M.; Kampouris, D.K.; Banks, C.E. Paper-based electroanalytical sensing platforms. *Anal. Methods* **2013**, *5*, 103–110.

86. Abe, K.; Kotera, K.; Suzuki, K.; Citterio, D. Inkjet-printed paperfluidic immuno-chemical sensing device. *Anal. Bioanal. Chem.* **2010**, *398*, 885–893.

87. Liu, J.W.; Mazumdar, D.; Lu, Y. A simple and sensitive "dipstick" test in serum based on lateral flow separation of aptamer-linked nanostructures. *Angew. Chem. Int. Ed.* **2006**, *45*, 7955–7959.

88. Alkasir, R.S.; Rossner, A.; Andreescu, S. Portable Colorimetric Paper-Based Biosensing Device for the Assessment of Bisphenol A in Indoor Dust. *Environ. Sci. Technol.* **2015**, *49*, 9889–9897.

89. Alkasir, R.S.; Ganesana, M.; Won, Y.H.; Stanciu, L.; Andreescu, S. Enzyme functionalized nanoparticles for electrochemical biosensors: A comparative study with applications for the detection of bisphenol, A. *Biosens. Bioelectron.* **2010**, *26*, 43–49.

90. Dungchai, W.; Chailapakul, O.; Henry, C.S. Use of multiple colorimetric indicators for paper-based microfluidic devices. *Anal. Chim. Acta* **2010**, *674*, 227–233.

91. Dungchai, W.; Chailapakul, O.; Henry, C.S. Electrochemical Detection for Paper-Based Microfluidics. *Anal. Chem.* **2009**, *81*, 5821–5826.

92. Desmet, C.; Marquette, C.A.; Blum, L.J.; Doumèche, B. Paper electrodes for bioelectrochemistry: Biosensors and biofuel cells. *Biosens. Bioelectron.* **2016**, *76*, 145–163.

93. Liana, D.D.; Raguse, B.; Wieczorek, L.; Baxter, G.R.; Chuah, K.; Gooding, J.J.; Chow, E. Sintered gold nanoparticles as an electrode material for paper-based electrochemical sensors. *RSC Adv.* **2013**, *3*, 8683–8691.

94. Nie, Z.; Deiss, F.; Liu, X.; Akbulut, O.; Whitesides, G.M. Integration of paper-based microfluidic devices with commercial electrochemical readers. *Lab Chip* **2010**, *10*, 3163–3169.

95. Sharpe, E.; Hua, F.; Schuckers, S.; Andreescu, S.; Bradley, R. Effects of brewing conditions on the antioxidant capacity of twenty-four commercial green tea varieties. *Food Chem.* **2016**, *192*, 380–387.

96. Yaktine, A.; Pray, L. *Nanotechnology in Food Products: Workshop Summary*; National Academies Press: Washington, DC, USA, 2009.

97. Xihong, Z.; Chii-Wann, L.; Jun, W.; Deog Hwan, O. Advances in rapid detection methods for foodborne pathogens. *J. Microbiol. Biotechnol.* **2014**, *24*, 297–312.

98. Poltronieri, P.; Mezzolla, V.; Primiceri, E.; Maruccio, G. Biosensors for the Detection of Food Pathogens. *Foods* **2014**, *3*, 511–526.

99. Jokerst, J.C.; Adkins, J.A.; Bisha, B.; Mentele, M.M.; Goodridge, L.D.; Henry, C.S. Development of a paper-based analytical device for colorimetric detection of select. *Foodborne Pathog. Anal. Chem.* **2012**, *84*, 2900–2907.

100. Jodi Woan-Fei, L.; Ab Mutalib, N.-S.; Chan, K.-G.; Lee, L.-H. Rapid Methods for the Detection of Foodborne Bacterial Pathogens: Principles, Applications, Advantages and Limitations. *Front. Microbiol.* **2015**, *5*.

101. Hossain, S.Z.; Ozimok, C.; Sicard, C.; Aguirre, S.D.; Ali, M.M.; Li, Y.; Brennan, J.D. Multiplexed paper test strip for quantitative bacterial detection. *Anal. Bioanal. Chem.* **2012**, *403*, 1567–1576.

102. Sodha, S.; Heiman, K.; Gould, L.; Bishop, R.; Iwamoto, M.; Swerdlow, D.; Griffin, P.M. National patterns of *Escherichia coli* O157 infections, USA, 1996–2011. *Epidemiol. Infect.* **2015**, *143*, 267–273.

103. Wells, J.; Shipman, L.; Greene, K.; Sowers, E.; Green, J.; Cameron, D.; Downes, F.P.; Martin, M.L.; Griffin, P.M.; Ostroff, S.M.; *et al.* Isolation of *Escherichia coli* serotype O157: H7 and other Shiga-like-toxin-producing, *E. coli* from dairy cattle. *J. Clin. Microbiol.* **1991**, *29*, 985–959.

104. Feng, P. *Escherichia coli* serotype O157: H7: Novel vehicles of infection and emergence of phenotypic variants. *Emerg. Infect. Dis.* **1995**, *1*, 47–52.

105. Liu, Y.; Li, Y. An Antibody-Immobilized Capillary Column as a Bioseparator/Bioreactor for Detection of *Escherichia coli* O157:H7 with Absorbance Measurement. *Anal. Chem.* **2001**, *73*, 5180–5183.

106. Ngom, B.; Guo, Y.; Wang, X.; Bi, D. Development and application of lateral flow test strip technology for detection of infectious agents and chemical contaminants: A review. *Anal. Bioanal. Chem.* **2010**, *397*, 1113–1135.

107. Jung, B.Y.; Jung, S.C.; Kweon, C.H. Development of a rapid immunochromatographic strip for detection of *Escherichia coli* O157. *J. Food Prot.* **2005**, *68*, 2140–2143.

108. Wutor, V.; Togo, C.; Pletschke, B. Suitability of total coliform β-D-galactosidase activity and CFU counts in monitoring faecal contamination of environmental water samples. *Water SA* **2009**, *35*, 85–88.

109. Kilian, M.; Bülo, P. Rapid Diagnosis Of Enterobacteriaceae. *Acta Pathol. Microbiol. Scand. Sec. B Microbiol.* **1976**, *84*, 245–251.

110. Ratnam, S.; March, S.B.; Ahmed, R.; Bezanson, G.; Kasatiya, S. Characterization of *Escherichia coli* serotype O157:H7. *J. Clin. Microbiol.* **1988**, *26*, 2006–2012.

111. Stephen Inbaraj, B.; Chen, B.H. Nanomaterial-based sensors for detection of foodborne bacterial pathogens and toxins as well as pork adulteration in meat products. *J. Food Drug Anal.* **2015**. in press.

112. Hendriksen, R.S.; Vieira, A.R.; Karlsmose, S.; Lo Fo Wong, D.M.; Jensen, A.B.; Wegener, H.C.; Aarestrup, F.M. Global monitoring of Salmonella serovar distribution from the World Health Organization Global Foodborne Infections Network Country Data Bank: Results of quality assured laboratories from 2001 to 2007. *Foodborne Pathog. Dis.* **2011**, *8*, 887–900.

113. Amagliani, G.; Brandi, G.; Schiavano, G.F. Incidence and role of Salmonella in seafood safety. *Food Res. Int.* **2012**, *45*, 780–788.

114. Joo, J.; Yim, C.; Kwon, D.; Lee, J.; Shin, H.H.; Cha, H.J.; Jeon, S. A facile and sensitive detection of pathogenic bacteria using magnetic nanoparticles and optical nanocrystal probes. *Analyst* **2012**, *137*, 3609–3612.

115. Huang, Y.-F.; Wang, Y.-F.; Yan, X.-P. Amine-Functionalized Magnetic Nanoparticles for Rapid Capture and Removal of Bacterial Pathogens. *Environ. Sci. Technol.* **2010**, *44*, 7908–7913.

116. Singh, B.K.; Walker, A. Microbial degradation of organophosphorus compounds. *FEMS Microbiol. Rev.* **2006**, *30*, 428–471.

117. Kumar, S.V.; Fareedullah, M.; Sudhakar, Y.; Venkateswarlu, B.; Kumar, E.A. Current review on organophosphorus poisoning. *Arch. Appl. Sci. Res.* **2010**, *2*, 199–215.

118. Andreescu, S.; Marty, J.L. Twenty years research in cholinesterase biosensors: From basic research to practical applications. *Biomol. Eng.* **2006**, *23*, 1–15.

119. Pope, C.N. Organophosphorus pesticides: Do they all have the same mechanism of toxicity? *J. Toxicol. Environ. Health B Crit. Rev.* **1999**, *2*, 161–181.

120. Valdés, M.; Valdés González, A.; García Calzón, J.; Díaz-García, M. Analytical nanotechnology for food analysis. *Microchim. Acta* **2009**, *166*, 1–19.

121. Pérez-López, B.; Merkoçi, A. Nanomaterials based biosensors for food analysis applications. *Trends Food Sci. Technol.* **2011**, *22*, 625–639.

122. Ellman, G.L.; Courtney, K.D.; Andres, V., Jr.; Featherstone, R.M. A new and rapid colorimetric determination of acetylcholinesterase activity. *Biochem. Pharmacol.* **1961**, *7*, 88–95.

123. Hossain, S.M.Z.; Luckham, R.E.; Smith, A.M.; Lebert, J.M.; Davies, L.M.; Pelton, R.H.; Filipe, C.D.; Brennan, J.D. Development of a Bioactive Paper Sensor for Detection of Neurotoxins Using Piezoelectric Inkjet Printing of Sol-Gel-Derived Bioinks. *Anal. Chem.* **2009**, *81*, 5474–5483.

124. Hossain, S.M.Z.; Luckham, R.E.; McFadden, M.J.; Brennan, J.D. Reagentless Bidirectional Lateral Flow Bioactive Paper Sensors for Detection of Pesticides in Beverage and Food Samples. *Anal. Chem.* **2009**, *81*, 9055–9064.

125. Luckham, R.E.; Brennan, J.D. Bioactive paper dipstick sensors for acetylcholinesterase inhibitors based on sol-gel/enzyme/gold nanoparticle composites. *Analyst* **2010**, *135*, 2028–2035.

126. Liang, M.; Fan, K.; Pan, Y.; Jiang, H.; Wang, F.; Yang, D.; Lu, D.; Feng, J.; Zhao, J.; Yang, L.; *et al.* Fe_3O_4 Magnetic Nanoparticle Peroxidase Mimetic-Based Colorimetric Assay for the Rapid Detection of Organophosphorus Pesticide and Nerve Agent. *Anal. Chem.* **2013**, *85*, 308–312.

127. Liu, G.; Wang, J.; Barry, R.; Petersen, C.; Timchalk, C.; Gassman, P.L.; Lin, Y. Nanoparticle-based electrochemical immunosensor for the detection of phosphorylated acetylcholinesterase: An exposure biomarker of organophosphate pesticides and nerve agents. *Chemistry* **2008**, *14*, 9951–9959.

128. He, Y.; Xu, B.; Li, W.; Yu, H. Silver nanoparticle-based chemiluminescent sensor array for pesticide discrimination. *J. Agric. Food Chem.* **2015**, *63*, 2930–2934.

129. Kim, H.N.; Ren, W.X.; Kim, J.S.; Yoon, J. Fluorescent and colorimetric sensors for detection of lead, cadmium, and mercury ions. *Chem. Soc. Rev.* **2012**, *41*, 3210–3244.

130. Zhao, C.; Zhong, G.; Kim, D.-E.; Liu, J.; Liu, X. A portable lab-on-a-chip system for gold-nanoparticle-based colorimetric detection of metal ions in water. *Biomicrofluidics* **2014**, *8*, 052107.

131. Du, J.; Wang, Z.; Fan, J.; Peng, X. Gold nanoparticle-based colorimetric detection of mercury ion via coordination chemistry. *Sens. Actuat. B Chem.* **2015**, *212*, 481–486.

132. Chansuvarn, W.; Tuntulani, T.; Imyim, A. Colorimetric detection of mercury(II) based on gold nanoparticles, fluorescent gold nanoclusters and other gold-based nanomaterials. *TrAC Trends Anal. Chem.* **2015**, *65*, 83–96.

133. Zhou, Y.; Dong, H.; Liu, L.; Li, M.; Xiao, K.; Xu, M. Selective and sensitive colorimetric sensor of mercury (II) based on gold nanoparticles and 4-mercaptophenylboronic acid. *Sens. Actuat. B Chem.* **2014**, *196*, 106–111.

134. Chen, G.-H.; Chen, W.-Y.; Yen, Y.-C.; Wang, C.-W.; Chang, H.-T.; Chen, C.-F. Detection of Mercury(II) Ions Using Colorimetric Gold Nanoparticles on Paper-Based Analytical Devices. *Anal. Chem.* **2014**, *86*, 6843–6849.

135. Sener, G.; Uzun, L.; Denizli, A. Colorimetric Sensor Array Based on Gold Nanoparticles and Amino Acids for Identification of Toxic Metal Ions in Water. *ACS Appl. Mater. Interfaces* **2014**, *6*, 18395–18400.

136. Alizadeh, A.; Khodaei, M.M.; Hamidi, Z.; Shamsuddin, M.B. Naked-eye colorimetric detection of Cu^{2+} and Ag^+ ions based on close-packed aggregation of pyridines-functionalized gold nanoparticles. *Sens. Actuat. B Chem.* **2014**, *190*, 782–791.

137. Annadhasan, M.; Muthukumarasamyvel, T.; Sankar Babu, V.R.; Rajendiran, N. Green Synthesized Silver and Gold Nanoparticles for Colorimetric Detection of Hg^{2+}, Pb^{2+}, and Mn^{2+} in Aqueous Medium. *ACS Sustain. Chem. Eng.* **2014**, *2*, 887–896.

138. Sung, Y.-M.; Wu, S.-P. Colorimetric detection of Cd(II) ions based on di-(1H-pyrrol-2-yl)methanethione functionalized gold nanoparticles. *Sens. Actuat. B Chem.* **2014**, *201*, 86–91.

139. Cheli, F.; Pinotti, L.; Campagnoli, A.; Fusi, E.; Rebucci, R.; Baldi, A. Mycotoxin Analysis, Mycotoxin-Producing Fungi Assays and Mycotoxin Toxicity Bioassays in Food Mycotoxin Monitoring and Surveillance. *Ital. J. Food Sci.* **2008**, *20*, 447–462.

140. Hussein, H.S.; Brasel, J.M. Toxicity, metabolism, and impact of mycotoxins on humans and animals. *Toxicology* **2001**, *167*, 101–134.

141. Fernández-Cruz, M.L.; Mansilla, M.L.; Tadeo, J.L. Mycotoxins in fruits and their processed products: Analysis, occurrence and health implications. *J. Adv. Res.* **2010**, *1*, 113–122.

142. Hosseini, M.; Khabbaz, H.; Dadmehr, M.; Ganjali, M.R.; Mohamadnejad, J. Aptamer-Based Colorimetric and Chemiluminescence Detection of Aflatoxin B1 in Foods Samples. *Acta Chim. Slov.* **2015**, *62*, 721–728.

143. Luan, Y.; Chen, J.; Xie, G.; Li, C.; Ping, H.; Ma, Z.; Lu, A. Visual and microplate detection of aflatoxin B2 based on NaCl-induced aggregation of aptamer-modified gold nanoparticles. *Microchim. Acta* **2015**, *182*, 995–1001.

144. Xiao, R.; Wang, D.; Lin, Z.; Qiu, B.; Liu, M.; Guo, L.; Chen, G. Disassembly of gold nanoparticle dimers for colorimetric detection of ochratoxin, A. *Anal. Methods* **2015**, *7*, 842–825.

145. Soh, J.H.; Lin, Y.; Rana, S.; Ying, J.Y.; Stevens, M.M. Colorimetric Detection of Small Molecules in Complex Matrixes via Target-Mediated Growth of Aptamer-Functionalized Gold Nanoparticles. *Anal. Chem.* **2015**, *87*, 7644–7652.

146. Wang, X.; Niessner, R.; Knopp, D. Magnetic Bead-Based Colorimetric Immunoassay for Aflatoxin B1 Using Gold Nanoparticles. *Sensors* **2014**, *14*, 21535–21548.

147. Urusov, A.E.; Petrakova, A.V.; Vozniak, M.V.; Zherdev, A.V.; Dzantiev, B.B. Rapid Immunoenzyme Assay of Aflatoxin B1 Using Magnetic Nanoparticles. *Sensors* **2014**, *14*, 21843–21857.

148. Ricci, F.; Volpe, G.; Micheli, L.; Palleschi, G. A review on novel developments and applications of immunosensors in food analysis. *Anal. Chim. Acta* **2007**, *605*, 111–129.

149. Hayat, A.; Mishra, R.; Catanante, G.; Marty, J. Development of an aptasensor based on a fluorescent particles-modified aptamer for ochratoxin A detection. *Anal. Bioanal. Chem.* **2015**, *407*, 7815–7822.

Theoretical Basis and Application for Measuring Pork Loin Drip Loss Using Microwave Spectroscopy

Alex Mason, Badr Abdullah, Magomed Muradov, Olga Korostynska, Ahmed Al-Shamma'a, Stefania Gudrun Bjarnadottir, Kathrine Lunde and Ole Alvseike

Abstract: During cutting and processing of meat, the loss of water is critical in determining both product quality and value. From the point of slaughter until packaging, water is lost due to the hanging, movement, handling, and cutting of the carcass, with every 1% of lost water having the potential to cost a large meat processing plant somewhere in the region of €50,000 per day. Currently the options for monitoring the loss of water from meat, or determining its drip loss, are limited to destructive tests which take 24–72 h to complete. This paper presents results from work which has led to the development of a novel microwave cavity sensor capable of providing an indication of drip loss within 6 min, while demonstrating good correlation with the well-known EZ-Driploss method ($R^2 = 0.896$).

Reprinted from *Sensors*. Cite as: Mason, A.; Abdullah, B.; Muradov, M.; Korostynska, O.; Al-Shamma'a, A.; Bjarnadottir, S.G.; Lunde, K.; Alvseike, O. Theoretical Basis and Application for Measuring Pork Loin Drip Loss Using Microwave Spectroscopy. *Sensors* **2016**, *16*, 182.

1. Introduction

From the point at which an animal is slaughtered during the meat production process, it is inevitable that water will be lost from the carcass. This is a key concern for meat producers as this water content is said to contribute to the juiciness and tenderness of meat products [1,2], which impacts on consumer opinion, thus affecting demand and saleable value. While the loss of product quality and appeal is often difficult to measure due to its subjective nature, Table 1 demonstrates the average basic constituents of common meat products, with many containing greater than 75% water [3].

Since most meat products are sold on the basis of their weight, it stands to reason that loss of water is directly proportional to a loss in revenue. It is estimated that for large production facilities (*i.e.*, those processing in the order of thousands of animals per day), for every 1% of water lost, this could equate to €50,000 (this estimate is based up on a large processing plant, but of course is dependent on the volume of meat trimmings produced in addition to the market value of meat at the time of processing) per day in lost revenue. For this reason, meat producers are keen to employ new tools which enable them to monitor the production process more quickly and effectively than current methods allow, therefore permitting minimization of water loss. This paper builds significantly upon previous proof of concept work [4,5] by the authors in the use of a novel microwave spectroscopy technique to measure drip loss in pork. In particular, this paper demonstrates how the technique compares with the most widely used industry method for drip loss measurement (*i.e.*, the EZ-Driploss method).

Table 1. Average constituents of common meat cuts.

Meat	Cut	Water %	Protein %	Fat %	Ash %
Pork	Boston butt	74.9	19.5	4.7	1.1
	Loin	75.3	21.1	2.4	1.2
	Cutlets/chops [1]	54.5	15.5	29.4	0.8
	Ham	75.0	20.2	3.6	1.1
	Side cuts	60.3	17.8	21.1	0.85
Beef	Shank	76.4	21.8	0.7	1.2
	Sirloin steak [1]	76.4	21.8	0.7	1.2
Chicken	Hind leg	73.3	20.0	5.5	1.2
	Breast	74.4	23.3	1.2	1.1

[1] With adhering adipose tissue.

2. Water Holding Capacity

The term drip loss, or its reciprocal parameter Water Holding Capacity (WHC), is used by the food industry to refer to the ability of meat products to retain water, with much of our current knowledge on the topic being based on fundamental research by Hamm [6] in 1960, followed by Offer and Trinick [7] in 1983. There is a general agreement [8,9] that water in meat can exist in three forms: (1) bound; (2) immobilized; and (3) free, with these representing as much as 5%, 15%, and 85% of total water content, respectively. Bound water is tightly bound to proteins and is not free to move around, cannot be frozen, and is not affected by chemical changes (e.g., pH), while immobilized water shares similar properties, albeit with weaker protein bonding. Free water, on the other hand, is held loosely in the capillary space between and within proteins and, unlike bound and immobilized water, is easily lost. Therefore, anything which alters the protein structure or spacing will affect the ability of the meat to retain free water. Examples of factors which might impact protein structure, and hence drip loss, include *post mortem* rigor (the steric effect) [10,11]; pre-slaughter stress [8,12]; pH [1,13]; and common processing techniques [14], such as heating, grinding, cutting, pressing, and freezing.

Measuring the drip loss of meat at various stages during the production process could enable impact assessments of the factors causing water loss. In principle this would allow optimization of processes in addition to the sorting of carcasses or cuts so that resources are allocated to optimal products. Such action would reduce lost revenue and ensure consistent meat quality and tenderness. In practice, however, measurement of drip loss is challenging, as the current commercially available methods are destructive, manual, and time consuming. Furthermore, despite efforts described by Honikel [15], there is no international standard method, which makes comparison of results derived from the various techniques difficult; Table 2 presents an overview of these techniques.

A number of attempts to provide a sensor technique to standardize and automate the measurement of drip loss or WHC have been made. X-ray diffraction has been used extensively within the food industry for foreign object detection (e.g., metal, glass, and plastic shards) in addition to recent systems such as the MeatMaster (FOSS, Denmark) which can give online prediction of fat content in meat trimmings. It has also proven to give excellent resolution in relation to the spacing between muscle filaments [16–18]; since it is said that most of the water in meat is held between the muscle filaments (or myofibrils) this can give an indication of WHC of samples. Despite this, X-ray systems have been unable to demonstrate a method for online prediction of WHC, most likely because the technique, itself, is exacting, requiring careful dissection of muscle slips and long exposure times. Notably, however, recent work from O'Farrell et al. [19] has demonstrated an energy-dispersive diffraction system with a correlation of $R^2 = 0.72$ when compared with the industry standard EZ-Driploss method which is described in Table 2. While this is promising, particularly because the technique offers measurement speeds of minutes rather than hours as described in some works, X-ray systems are usually deployed at only one or two locations across a production line (e.g., at the point of packaging if checking for foreign objects). For effective drip loss measurement, many points of measurement are required which precludes the use of X-ray largely due to the costly nature of the equipment in addition to concerns regarding worker exposure to radiation.

Near-Infrared Spectroscopy (NIRS) has also been considered in relation to the issue of drip loss. As with X-ray systems, NIRS devices are becoming popular in the meat industry for online compositional analysis; three examples are the QVision 500 Analyzer (TOMRA, Norway), ProFoss (Foss, Norway), and Spektron (Prediktor, Norway). However, work by Kapper [20] and O'Farrell et al. [19] have demonstrated poor correlation between NIRS measurements and drip loss; Kapper noted R^2 between 0.36 and 0.73 depending on meat color and O'Farrell's work demonstrated $R^2 = 0.47$, despite a number of outlier data points (10%) being removed in the latter work.

A further method has also been demonstrated by Lee et al. [21] using an electrical conductivity measurement to attempt correlation of drip loss. This method showed some promise, with three categories of drip loss used to determine the effectiveness of the technique: <2%, 2%–6%, and >6%. It was demonstrated that 80% of the time the technique could correctly categorize meat samples from a production plant, albeit under laboratory conditions. It is notable however that in the vast majority of production facilities, the invasive nature of the electrodes used in this method would preclude it from online testing due to the potential for product contamination and spoilage.

Table 2. Manual methods of measuring water holding capacity, or its reciprocal parameter, drip loss, used in the commercial and research environments.

Method	Description
EZ-Driploss	The day after slaughter the muscles to be analyzed are taken from the carcass. Within one hour, a 25 mm slice is cut at a right angle to the muscle fiber direction, with samples being taken from this slice using a cork borer, again cutting in the fiber direction. The cylindrical sample, 25 mm in diameter and 25 mm in height, is weighed and then placed in a special container equipped with a lid to avoid evaporation and loss of water. The container is stored for 24 h at 4 °C–6 °C before being weighed again; the WHC is determined by ratio of the two weight measurements [22].
Filter Paper Press	This method involves the pressing of a meat sample into a filter paper; typically a defined pressure is recommended and the amount of released water is determined by weighing the meat sample or the filter paper before and after pressing. Hamm suggested a more rigorous protocol in 1972, which involves small meat samples (0.3 g) being pressed onto a filter paper at a pressure of 35 kg/cm^2 between two plates. Five minutes later, meat samples are removed. The areas covered by the flattened meat sample and the stain from the meat sample are marked and measured [23,24].
Centrifuge	A weighed meat sample (3–4 g) is centrifuged at 100,000 xg for 1 h in a stainless steel tube. The water released from the meat is decanted off as quickly as possible (in order to avoid re-absorption). The meat sample is removed from the tube with forceps, dried with tissue paper, and then reweighed to determine liquid loss. If the residue is dried in the tube at 105 °C, the total water content of the sample can be determined, and WHC can be expressed as released or bound water as a percentage of total water. The need for a high-speed centrifuge makes it almost impossible to use this type of method in a slaughterhouse [25].
Bag	Meat samples (weighing approximate 100 g) are cut from a carcass and immediately weighed. The samples are then placed in a bag and hung in an airtight container using a hook under the lid. After the required storage time at the temperature under investigation (usually 24–48 h at 1 °C–4 °C) samples are weighed again [15].
Absorption	Cotton-rayon material is inserted into a "+" shaped incision in the longissimus muscle through the subcutaneous fat layer. The incision is approx. 2.4 inches deep at a well-defined place (e.g., 12th rib) and is left for either 15 min at 15 min post-mortem or 15 min at 24 h post-mortem. Absorption is calculated as the difference between the final weight plus exudates and the initial dry weight of the material. Notably this technique is the quickest of all those listed here, however it also requires a skilled operator to enable repeatable incisions and measurement [26].

3. Microwave Spectroscopy

Sensors which operate at microwave frequencies are widely used in a variety of industrial sectors in addition to having been demonstrated in the research domain. Examples include structural analysis [27,28], water quality monitoring [29–31], and medical applications [32–37]. Aside from research considering quality classification of fresh [38,39] and cured meats [40,41], there is little evidence of microwave sensors making a significant impact in the food industry. This point is supported by a recent comprehensive review of electromagnetic wave sensors (from radio frequencies to X-ray) conducted by Damez and Clerjon [42].

Microwave sensors provide the opportunity for a rapid non-invasive and robust method of materials analysis. The authors have also demonstrated that the technique can take on many physical forms, including resonant cavities [43], planar structures [30,37], and fluidic devices [36], which makes it highly adaptable to a range of situations and applications. Furthermore, the technology to generate and detect microwave signals is inexpensive; it is featured in the many millions of smart phone devices, tablets, and portable computers, for example, that make use of wireless communications. This is a particularly attractive feature for the food industry, given the desire for high-resolution drip loss information from across production facilities, which is unlikely to be cost effective with technologies such as X-ray. Added to this, the technique is non-ionizing, utilizing less than 10 mW of power, significantly less than modern wireless communication devices and is, therefore, thought to be safe to use within food production without fear of harming the product or nearby workers.

The principle of monitoring using microwave sensors, in the context of this work, is based on the interaction of electromagnetic (EM) waves with a sample under test. When this sample is exposed to EM irradiation it alters the velocity of the signal, attenuates, or reflects it. If one considers a hollow structure with conducting walls (*i.e.*, a cavity), it will resonate when it is excited at an appropriate EM frequency provided some means for this to occur is introduced, for example, via a small antenna placed inside it. Resonant modes occur inside the cavity when the electric or magnetic components of the EM signal form standing waves, which are dependent on the dimensions of the cavity and the dielectric properties of the test sample. The resonant frequency for TE_{nml} and TM_{nml} modes in a rectangular waveguide [44] can be calculated using Equation (1), where c is the speed of light, μ_r is of the relative permeability, ε_r is the relative permittivity, p_{nm} is the value of the Bessel function for the TE or TM modes of a rectangular waveguide, a is the width of the cavity, b is the height of the cavity and d is the depth of the cavity.

$$f_{mnl} = \frac{c}{2\pi\sqrt{\mu_r\varepsilon_r}}\sqrt{\left(\frac{m\lambda}{a}\right)^2 + \left(\frac{n\lambda}{b}\right)^2 + \left(\frac{l\pi}{d}\right)^2} \tag{1}$$

Any number of such antennae may be placed within the cavity for the purposes of transmission and reception of EM energy, however the most typical configurations involve one and two port (thus, one or two antennae) cavities since often the materials placed within them are assumed to be relatively homogeneous and therefore further ports serve little purpose.

In a one port configuration it is possible, using a Vector Network Analyzer (VNA), to measure the power which is reflected from the cavity; this is often referred to simply as an S_{11} measurement. In a two port configuration, one can also measure power transmitted through the cavity; this is referred to as an S_{21} measurement. This is illustrated in Figure 1a, which shows a 3D model of the cavity designed specifically for this work using Ansys High Frequency Structure Simulator (HFSS) finite element modelling package; Figure 1b shows the sample model which is an EZ-Driploss sample container. With reference to the full description of this drip loss measurement method given in Table 2, the EZ-Driploss container holds a cylindrical meat sample in the larger top section, with water lost over the 24 h measurement period being collected at the bottom of the thin tube. The cavity is designed such that only the larger top section resides within the cavity as it serves no purpose to measure any fluids lost from the meat samples. EZ-Driploss containers were used to hold the sample during the course of this work to allow direct correlation of the standard EZ-Driploss measurement against the data acquired from the microwave cavity.

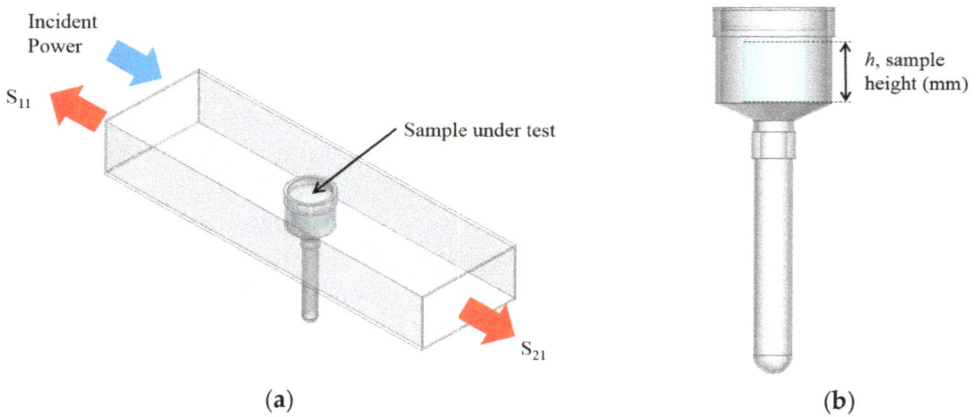

Figure 1. (**a**) A 3D model of the cavity designed for this work; and (**b**) the modelled EZ-Driploss sample container which is used to hold meat samples during the measurement procedure.

From Equation (1) it is shown that all EM modes have the same dependence upon $\sqrt{\varepsilon_r}$, so when the cavity is excited over an appropriate range of frequencies and the resultant spectra is captured, the resonant peaks corresponding to these modes will shift, typically in frequency and amplitude, as the permittivity is varied. This can be demonstrated by taking the model illustrated in Figure 1a and varying the sample height, h, shown in Figure 1b such that the sample in this case is water when the cavity it resonating in the TE_{010} mode, as represented in Figure 2. It is assumed that this approximates the composition of most fresh meat products immediately *post mortem* since, as noted in Table 1, water is the major constituent. Figure 3 shows the relationship for both the signal amplitude and resonant frequency shift as h is varied in the range 2–16 mm. A high correlation, using this modelling approach, is demonstrated for both signal amplitude ($R^2 = 0.874$) and resonant frequency ($R^2 = 0.978$) and shows the responsiveness of the technique to variations in water, which provides the basis for its use in monitoring the drip loss of meat samples.

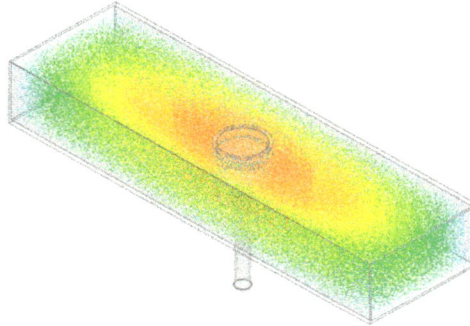

Figure 2. The resonant cavity model with electric field distribution overlay when resonant at approx. 1.5 GHz, which is where the TE_{010} mode is present within the cavity according to Equation (1). Notably the electric field is concentrated, as noted by the red/orange coloration, around the EZ-Driploss container which ensures maximum interaction of the EM signal with the target sample.

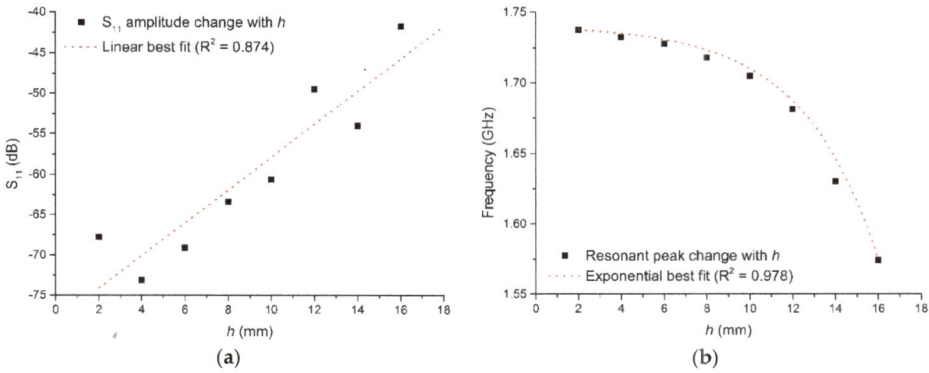

Figure 3. Cavity response when varying sample height, h, for (**a**) the S_{11} signal amplitude, and (**b**) the frequency at which the resonant TE_{010} mode occurs.

4. Experimental Methodology

The purpose of the experimental work outlined in this paper was to compare and correlate drip loss measurements from the microwave cavity illustrated in Section 3 with the current dominant method used in industry, *i.e.*, the EZ-Driploss method [22].

To this end, a total of 24 pork carcasses were selected after pre-sorting based on a wide range of pH and breeds in an attempt to obtain a large variation of drip loss across the samples. While it was impossible to guarantee with absolute certainty a wide range of drip loss from the selected samples (*i.e.*, there is no current online method available for this purpose), the link between pH and ability of meat to retain water is widely reported [8]. Furthermore, the carcasses were selected from Noroc and Landrace breeds since there is typically a pH difference between them; namely, Landrace often have a lower pH than Noroc breeds.

To prepare the samples, a loin was taken from each carcass, with each loin having a slice of approximate 20 mm thickness taken from the middle. Each loin was split into two portions; one portion was retained by researchers at Animalia to establish a baseline or control EZ-Driploss measurement, and the other was provided to the researchers from Liverpool John Moores University who had developed the microwave cavity sensor. All work measurements took place simultaneously at Animalia's pilot plant facility, located in Oslo, Norway.

From these loin portions, two 25 mm diameter core samples were taken with a borer. Each core sample was then placed into a separate EZ-Driploss polypropylene container and lid was closed prior to measurement commencing. Therefore, four core samples were taken from each loin sample and measured using the EZ-Driploss method, giving 96 core samples in total. The procedure for sample preparation is illustrated in Figure 4. Care was taken when preparing samples to avoid deposits of fat and other visible inconsistences in the product, which is standard practice when employing the EZ-Driploss method.

All core samples were weighed prior to being stored for a period of 24 h at between 4 °C and 6 °C. After this period, all core samples were weighed once more and drip loss was calculated using Equation (2), where W_c is the weight of an empty EZ-Driploss container, W_t is the weight of the container with meat and exudate and W_l is the weight of the container with liquid only. While only one of the core samples taken from each loin could be measured using the microwave method (therefore, 24 measurements in total), an average value was determined for the EZ-Driploss method from across the four core samples taken from each loin. All measurements (*i.e.*, 96 EZ-Driploss and 24 using the microwave cavity) took place over three days, which allowed time for system configuration between measurements and sample preparations.

$$Drip\ Loss_{EZ}\ (\%) = \frac{W_l - W_c}{W_t - W_c} \times 100 \qquad (2)$$

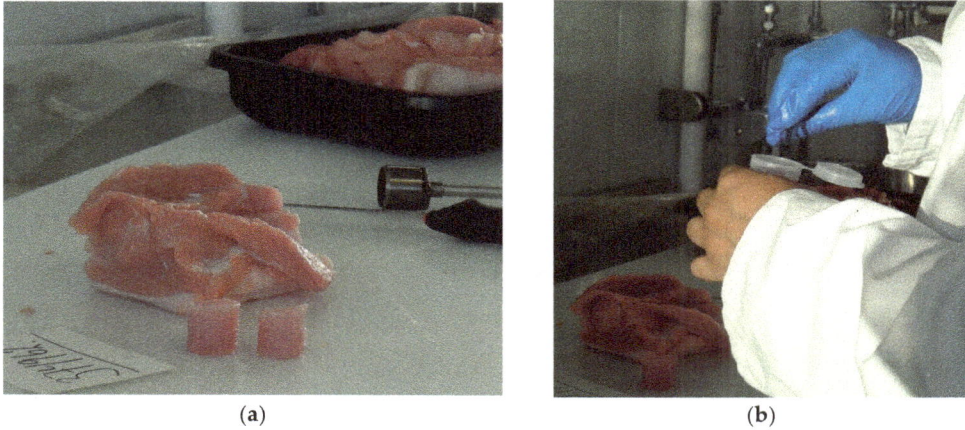

Figure 4. (a) Preparation of the 25 mm diameter core samples from the 20 mm thick slice of loin; and (b) placement of a sample in an EZ-Driploss container.

The microwave cavity sensor was used immediately after sample preparation, with the sample being placed inside the cavity as shown in Figure 5a. The measurement equipment was configured inside a refrigerated chamber in order to ensure that measurements were undertaken at similar temperatures to which samples were stored. Both reflected (S_{11}) and transmitted (S_{21}) power measurements were taken for 30 min per sample; the microwave spectrum was captured between 1 and 6 GHz every minute. This timing was established as a result of a preliminary study [4] which measured similar samples over a 24 h period. A Rohde and Schwarz ZVL6 VNA was used for capturing spectral data, which was automated via a bespoke National Instruments LabVIEW® interface as shown in Figure 5b.

(a)   (b)

Figure 5. (**a**) Experimental setup showing the microwave cavity connected to a Rohde and Schwarz Vector Network Analyzer; and (**b**) a bespoke LabVIEW® interface capturing spectral data.

5. Results and Discussion

5.1. Comparison of Drip Loss Measurements

As noted in Section 4, drip loss measurements following the EZ-Driploss method were conducted on two portions of the same meat, with one portion being used only for drip loss measurement and the other being used for measurements with the microwave cavity sensor. The comparison of EZ-Driploss measurements obtained is shown in Figure 6, and demonstrates that there is a reasonable agreement ($R^2 = 0.76$) between the two portions of the same pork loin which gives confidence that the sample preparation methods are similar. It does however also serve to highlight the potential for error or variation when using the EZ-Driploss method since the meat itself is heterogeneous, and despite the best efforts of the operator, this is likely to be a factor. It is also notable that results from EZ-Driploss testing can vary from operator to operator, which is noted by a number of authors, including Christensen [45] for example. Therefore, when considering measurement data from the microwave sensor system, or any other for that matter, it is important to remember that the EZ-Driploss test, despite being a widely used and accepted method for drip loss measurement, harbors considerable inherent variability.

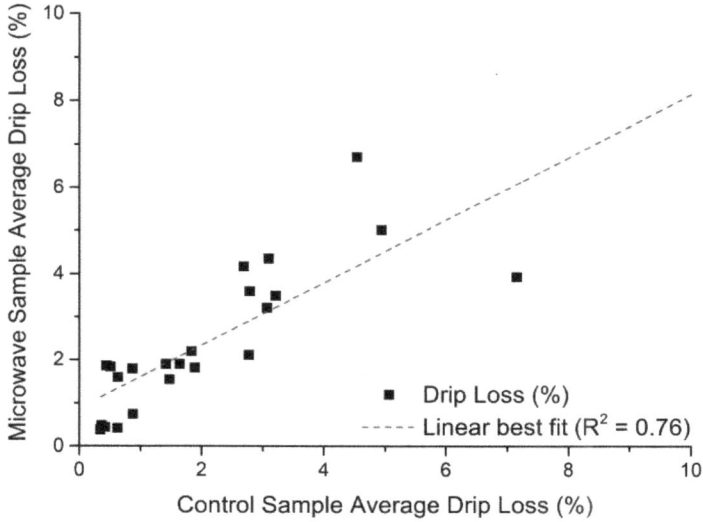

Figure 6. Correlation of drip loss in control sample and the portion of pork loin used for microwave measurement in this work, where $R^2 = 0.76$.

5.2. Microwave Cavity Measurements

When using microwave sensors for measuring properties or changes in materials, it is often possible to derive a correlation directly from single measurements of the sample(s) under test and, therefore, one would be able to rapidly develop a calibration curve to define the sensor performance. However, due to the heterogeneous nature of meat and the small physical size of the sample, this typical approach yielded only a weak correlation at 4.23 GHz (S_{11}), where $R^2 = 0.62$. This is illustrated in Figure 7.

Fortunately, data was collected for each sample over a time period of 30 min (where timing begins at the moment when the sample is inserted into the cavity sensor), with measurements taken at 1 min intervals. This gave the opportunity to consider whether the sensor responded to any immediate change in the sample after it was placed within the EZ-Driploss container. This yielded some rather interesting findings, as illustrated in Figure 8, whereby change in the S_{21} spectra gave a relationship to the end drip loss measurement for each sample. In particular, Figure 8 shows the microwave spectra between 5.4 and 6.0 GHz for two samples; one with a low drip loss (0.42%) and one with a high drip loss (7.15%) as determined by subsequent EZ-Driploss measurements. Over the 30 min measurement period both samples exhibited a reduction in resonant frequency, most notable in the range 5.47 to 5.50 GHz, as well as at 5.636 to 5.656 GHz. This reduction in resonant frequency is indicative of a change in the bulk relative permittivity of the meat sample, possibly due to diffusion or redistribution of water, post-preparation, while the sample is housed in the EZ-Driploss container. It is thought that such diffusion or redistribution of water would occur more rapidly in samples with a high drip loss owing to the availability of free water, thus enabling the sensor to be utilized for the purposes of determining the drip loss of the sample.

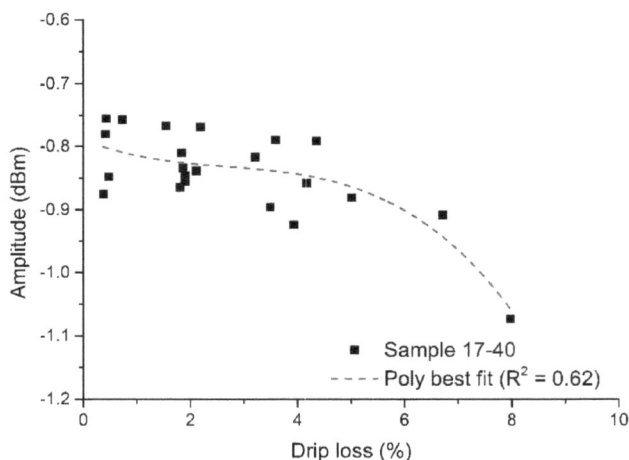

Figure 7. Correlation of the microwave cavity sensor at 4.23 GHz, with S_{11} amplitude changing as a function of drip loss ($R^2 = 0.62$).

This principle was applied to all 24 samples measured using the microwave sensor technique, using the frequency shift gradient between the first and subsequent S_{21} measurements between 5.47 and 5.50 GHz as a reference against which to correlate with the EZ-Driploss results.

An aim of this process was to establish the minimum time in which a result with acceptable agreement to the EZ-Driploss measurement could be obtained. While a measurement time of 30 min is clearly favorable when compared to the current 24 h required of the EZ-Driploss test, it is of industrial value to reduce this time as much as possible. To this end, the captured data was analyzed with a view to understanding the point at which the correlation between the microwave cavity and EZ-Driploss measurements fell significantly. Referring to Figure 9, the maximum R^2 value is obtained over the full 30 min measurement ($R^2 = 0.967$). It is also noted that at 6 min, the R^2 value (0.896) is still acceptable; below this measurement time the correlation drops significantly to a minimum of 0.401. A comparison of the data produced for 30 and 6 min measurements is evidenced in Figure 10.

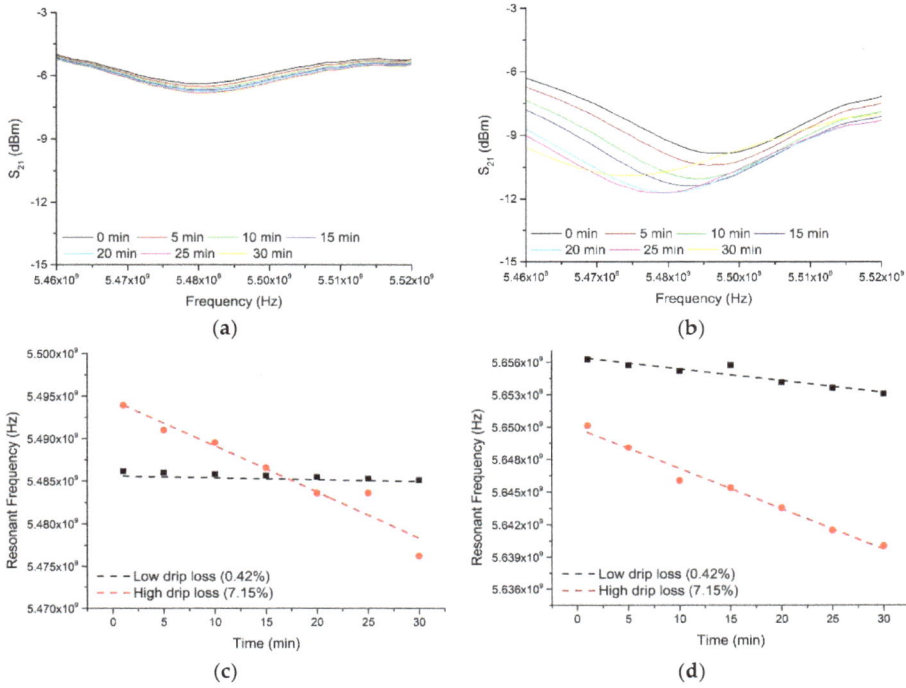

Figure 8. The microwave spectra between 5.46 and 5.52 GHz for two samples; one with (**a**) low drip loss (0.42%) and one with (**b**) high drip loss (7.15%) as determined by EZ-Driploss measurements. The change in resonant frequency in the range (**c**) 5.47 to 5.50 GHz and (**d**) 5.636 to 6.656 GHz are demonstrated and compared for the high and low drip loss samples.

While there is still some way to go to produce a commercially viable rapid non-invasive drip loss measurement system, this work demonstrates that it is possible to provide a measurement with good correlation to the existing industry standard EZ-Driploss test within 6 min. The correlation between the results evidenced in this paper exceeds that of other reported non-invasive systems such as X-ray or NIRS. Furthermore, microwave sensor systems are much cheaper by comparison, since the components to produce them are often based upon wireless electronics available in consumer devices (e.g., Wi-Fi, mobile phones, *etc.*). Concerns regarding safety are also alleviated since the system used in this work utilizes low power (<10 dBm) non-ionizing radiation.

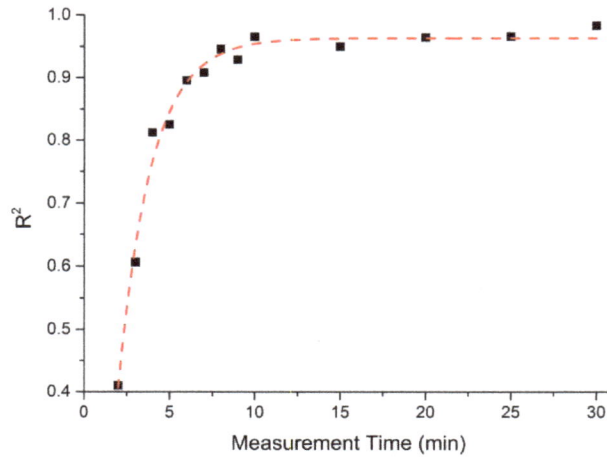

Figure 9. The relationship between R^2 value and measurement time of the sample inside the microwave cavity sensor.

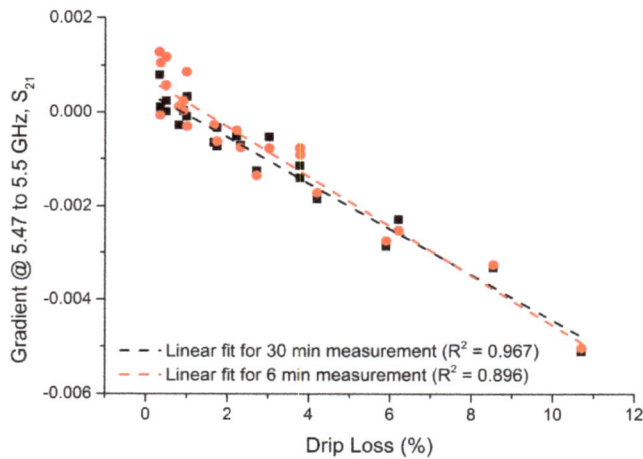

Figure 10. Correlation of the microwave cavity sensor when sample is measured for 30 min ($R^2 = 0.967$) and 6 min ($R^2 = 0.8955$).

6. Conclusions/Outlook

This paper presents a novel microwave cavity sensor for the measurement of drip loss, correlating results obtained from 24 pork loin samples against the widely used EZ-Driploss test, which typically takes 24 h to yield a result. It was shown that the sensor can provide a maximum correlation of $R^2 = 0.967$ with a 30 min measurement time, or a weaker correlation of $R^2 = 0.896$ with a 6 min measurement time. Not only does the sensor provide results in good agreement with the EZ-Driploss method, it demonstrates comparable, if not better, performance than reported alternative automated techniques, such as NIRS, X-ray and electrode based methods.

Future work in developing this technique could consider a number of directions, including the application of the technique to a broader spectrum of meat types (e.g., beef and lamb) in addition to considering the translation of the method for online use. Owing to the highly flexible nature of microwave spectroscopy, particularly in terms of the format of the sensor, the team foresees that it may be possible to identify and sort carcasses with high drip loss at an early stage post-slaughter. This would have significant implications for the industry in relation to meat production costs.

Acknowledgments: Acknowledgments: This work has been conducted as part of the EUREKA project entitled "Increased Efficiency: Moving from Assumed Quality to Online Measurement and Process Control" (INFORMED, Project Number 210516/O10) financed by Research Council Norway (Sustainable Innovation in Food and Bio-based Industries, BIONAER, programme).

Author Contributions: Author Contributions: The initiation of this project was made by Ole Alvseike and Professor Ahmed Al-Shamma'a, with the project ultimately being led by the team from Animalia. Researchers at LJMU, namely Alex Mason and Badr Abdullah were responsible for development of the microwave sensor technology which represents the core theme of this paper. Development of experimental methodology and selection of samples was performed by Kathrine Lunde, and preparation of samples supported by Stefania Bjarnadottir. Microwave measurements were performed in Norway by Alex Mason and Badr Abdullah. Subsequent data analysis was performed by Alex Mason and supported by Magomed Muradov as part of his PhD studies. Finally, writing of the paper was led by Alex Mason and Olga Korostynska, and amended as a result of input from across the team.

Conflicts of Interest: Conflicts of Interest: The authors declare no conflict of interest.

References

1. McClain, P.E.; Mullins, A.M. Relationship of water binding and pH to tenderness of bovine muscles. *J. Anim. Sci.* **1969**, *29*, 268–271.
2. Maltin, C.; Balcerzak, D.; Tilley, R.; Delday, M. Determinants of meat quality: Tenderness. *Proc. Nutr. Soc.* **2003**, *62*, 337–347.
3. Belitz, H.D.; Grosch, W. Food chemistry. In *Technology and Engineering*; Media, S.S.B., Ed.; Springer-Verlag Berlin Heidelberg: Berlin, Germany, 2013; p. 532.
4. Abdullah, B.M.; Cullen, J.D.; Korostynska, O.; Mason, A.; Al-Shamma'a, A.I. Assessing water-holding capacity (WHC) of meat using microwave spectroscopy. In *Sensing Technology: Current Status and Future Trends I*; Mason, A., Mukhopadhyay, S.C., Jayasundera, K.P., Bhattacharyya, N., Eds.; Springer International Publishing: Gewerbestrasse, Switzerland, 2014; Volume 7, pp. 117–140.

5. Abdullah, B.M.; Mason, A.; Cullen, J.D.; Al-Shamma'a, A.I. Water-holding capacity assessment of meat using an electromagnetic sensing method. In Proceedings of the RF and Microwave Conference (RFM), 2013 IEEE International, Penang, Malaysia, 9–11 December 2013; pp. 11–16.

6. Hamm, R. Biochemistry of meat hydration. *Adv. Food Res.* **1960**, *10*, 355–463.

7. Offer, G.; Trinick, J. On the mechanism of water holding in meat: The swelling and shrinking of myofibrils. *Meat Sci.* **1983**, *8*, 245–281.

8. Huff-Lonergan, E.; Lonergan, S.M. Mechanisms of water-holding capacity of meat: The role of postmortem biochemical and structural changes. *Meat Sci.* **2005**, *71*, 194–204.

9. Puolanne, E.; Halonen, M. Theoretical aspects of water-holding in meat. *Meat Sci.* **2010**, *86*, 151–165.

10. Pearce, K.L.; Rosenvold, K.; Andersen, H.J.; Hopkins, D.L. Water distribution and mobility in meat during the conversion of muscle to meat and ageing and the impacts on fresh meat quality attributes—A review. *Meat Sci.* **2011**, *89*, 111–124.

11. Ke, S. *Effect of pH and Salts on Tenderness and Water-Holding Capacity of muscle Foods*; University of Massachusetts—Amherst: Amherst, MA, USA, 2006.

12. Kristensen, L.; Purslow, P.P. The effect of ageing on the water-holding capacity of pork: Role of cytoskeletal proteins. *Meat Sci.* **2001**, *58*, 17–23.

13. Melody, J.L.; Lonergan, S.M.; Rowe, L.J.; Huiatt, T.W.; Mayes, M.S.; Huff-Lonergan, E. Early postmortem biochemical factors influence tenderness and water-holding capacity of three porcine muscles. *J. Anim. Sci.* **2004**, *82*, 1195–1205.

14. Den Hertog-Meischke, M.J.; van Laack, R.J.; Smulders, F.J. The water-holding capacity of fresh meat. *Vet. Q.* **1997**, *19*, 175–181.

15. Honikel, K.-O. Reference methods supported by oecd and their use in mediterranean meat products. *Food Chem.* **1997**, *59*, 573–582.

16. Swatland, A.J.; Belfry, S. Post-mortem changes in the shape and size of myofibrils from skeletal-muscle of pigs. *Mikroskopie* **1985**, *42*, 26–34.

17. Diesbourg, L.; Swatland, H.J.; Millman, B.M. X-ray diffraction measurements of postmortem changes in the myofilament lattice of pork. *J. Anim. Sci.* **1988**, *66*, 1048–1054.

18. Irving, T.C.; Swatland, H.J.; Millman, B.M. Effect of ph on myofilament spacing in pork measured by x-ray diffraction. *Can. Inst. Food Sci. Technol. J.* **1990**, *23*, 79–81.

19. O'Farrell, M.; Bouquet, G.; Tshcudi, J.; Bakke, K.A.H.; Egelandsal, B.; Lunde, K. Measuring water holding capacity—A comparison between a miniature near infrared system and an energy dispersive x-ray scattering system. *NIR News* **2014**, *25*, 11–14.

20. Kapper, C.; Klont, R.E.; Verdonk, J.M.A.J.; Urlings, H.A.P. Prediction of pork quality with near infrared spectroscopy (NIRS): 1. Feasibility and robustness of NIRS measurements at laboratory scale. *Meat Sci.* **2012**, *91*, 294–299.

21. Lee, S.; Norman, J.M.; Gunasekaran, S.; van Laack, R.L.J.M.; Kim, B.C.; Kauffman, R.G. Use of electrical conductivity to predict water-holding capacity in post-rigor pork. *Meat Sci.* **2000**, *55*, 385–389.

22. Rassmussen, A.; Andersson, M. New method for determination of drip loss in pork muscles. In Proceedings of the 42nd International Congress of Meat Science and Technology, Lillehammer, Norway, 1–6 September 1996.

23. Hamm, R. Kolloidchemie des fleisches—Das wasserbindungsvermoegen des muskeleiweisses. In *Theorie und Praxis*; Paul Parey: Berlin, Germany, 1972.

24. Grau, R.; Hamm, R. Eine einfache methode zur bestimmung der wasserbindung im muskel. *Naturwissenschaften* **1953**, *40*, 29–30.

25. Pearson, A.M.; Dutson, T.R. *Quality Attributes and Their Measurement in Meat, Poulty and Fish Products*; Springer: Berlin, Germany, 1995.

26. Walukonis, C.; Morgan, T.; Gerrard, D.; Forrest, J. *A Technique for Predicting Water-Holding Capacity in Early Post-Mortem Muscle*; Purdue University: West Lafayette, IN, USA, 2002.

27. Adous, M.; Quéffélec, P.; Laguerre, L. Coaxial/cylindrical transition line for broadband permittivity measurement of civil engineering materials. *Meas. Sci. Technol.* **2006**, *17*.

28. Ali, A.; Hu, B.; Ramahi, O. Intelligent detection of cracks in metallic surfaces using a waveguide sensor loaded with metamaterial elements. *Sensors* **2015**, *15*, 11402–11416.

29. Korostynska, O.; Ortoneda-Pedrola, M.; Mason, A.; Al-Shamma'a, A.I. Flexible electromagnetic wave sensor operating at GHz frequencies for instantaneous concentration measurements of NaCl, KCl, $MnCl_2$ and CuCl solutions. *Meas. Sci. Technol.* **2014**, *25*.

30. Korostynska, O.; Mason, A.; Al-Shamma'a, A.I. Flexible microwave sensors for real-time analysis of water contaminants. *J. Electromagn. Waves Appl.* **2013**, *27*, 2075–2089.

31. Korostynska, O.; Mason, A.; Ortoneda-Pedrola, M.; Al-Shamma'a, A. Electromagnetic wave sensing of NO_3 and COD concentrations for real-time environmental and industrial monitoring. *Sens. Actuators B Chem.* **2014**, *198*, 49–54.

32. Lu, G.; Yang, F.; Tian, Y.; Jing, X.; Wang, J. Contact-free measurement of heart rate variability via a microwave sensor. *Sensors* **2009**, *9*, 9572–9581.

33. Dei, D.; Grazzini, G.; Luzi, G.; Pieraccini, M.; Atzeni, C.; Boncinelli, S.; Camiciottoli, G.; Castellani, W.; Marsili, M.; Lo Dico, J. Non-contact detection of breathing using a microwave sensor. *Sensors* **2009**, *9*, 2574–2585.

34. Fok, M.; Bashir, M.; Fraser, H.; Strouther, N.; Mason, A. A novel microwave sensor to detect specific biomarkers in human cerebrospinal fluid and their relationship to cellular ischemia during thoracoabdominal aortic aneurysm repair. *J. Med. Syst.* **2015**, *39*, 1–5.

35. Korostynska, O.; Mason, A.; Al-Shamma'a, A. Microwave sensors for the non-invasive monitoring of industrial and medical applications. *Sens. Rev.* **2014**, *34*, 182–191.

36. Blakey, R.; Nakouti, I.; Korostynska, O.; Mason, A.; Al-Shamma'a, A. Real-time monitoring of pseudomonas aeruginosa concentration using a novel electromagnetic sensors microfluidic cell structure. *IEEE Trans. Biomed. Eng.* **2013**, *60*, 3291–3297.

37. Mason, A.; Korostynska, O.; Ortoneda-Pedrola, M.; Shaw, A.; Al-Shamma'a, A. A resonant co-planar sensor at microwave frequencies for biomedical applications. *Sensors Actuators A Phys.* **2013**, *202*, 170–175.

38. Castro-Giráldez, M.; Aristoy, M.C.; Toldrá, F.; Fito, P. Microwave dielectric spectroscopy for the determination of pork meat quality. *Food Res. Int.* **2010**, *43*, 2369–2377.

39. Castro-Giráldez, M.; Botella, P.; Toldrá, F.; Fito, P. Low-frequency dielectric spectrum to determine pork meat quality. *Innov. Food Sc. Emerg. Technol.* **2010**, *11*, 376–386.

40. Fulladosa, E.; Duran-Montge, P.; Serra, X.; Picouet, P.; Schimmer, O.; Gou, P. Estimation of dry-cured ham composition using dielectric time domain reflectometry. *Meat Sci.* **2013**, *93*, 873–879.

41. Bjarnadottir, S.G.; Lunde, K.; Alvseike, O.; Mason, A.; Al-Shamma'a, A.I. Assessing quality parameters in dry-cured ham using microwave spectroscopy. *Meat Sci.* **2015**, *108*, 109–114.

42. Damez, J.-L.; Clerjon, S. Quantifying and predicting meat and meat products quality attributes using electromagnetic waves: An overview. *Meat Sci.* **2013**, *95*, 879–896.

43. Mason, A.; Korostynska, O.; Wylie, S.; Al-Shamma'a, A.I. Non-destructive evaluation of an activated carbon using microwaves to determine residual life. *Carbon* **2014**, *67*, 1–9.

44. Pozar, D.M. Rectangular cavity modes. In *Microwave Engineering*, 3rd ed.; John Wiley and Sons: New York, NY, USA, 2005; p. 120.

45. Christensen, L.B. Drip loss sampling in porcine m. Longissimus dorsi. *Meat Sci.* **2003**, *63*, 469–477.

MDPI AG

Klybeckstrasse 64

4057 Basel, Switzerland

Tel. +41 61 683 77 34

Fax +41 61 302 89 18

http://www.mdpi.com/

Sensors Editorial Office

E-mail: Sensors@mdpi.com

http://www.mdpi.com/journal/sensors

www.ingramcontent.com/pod-product-compliance
Lightning Source LLC
Chambersburg PA
CBHW080133240326
41458CB00128B/6398